T0138024

Advances in Pattern Recognition

Advances in Pattern Recognition is a series of books which brings together current developments in all areas of this multi-disciplinary topic. It covers both theoretical and applied aspects of pattern recognition, and provides texts for students and senior researchers.

Springer also publishers a related journal, **Pattern Analysis and Applications.** For more details see: springeronline.com

The book series and journal are both edited by Professor Sameer Singh of Exeter University, UK.

Also in this series:

Principles of Visual Information Retrieval
Michael S. Lew (Ed.)
1-85233-381-2

Statistical and Neural Classifiers: An Integrated Approach to Design
Šarūnas Raudys
1-85233-297-2

Advanced Algorithmic Approaches to Medical Image Segmentation
Jasjit Suri, Kamaledin Setarehdan and Sameer Singh (Eds)
1-85233-389-8

NETLAB: Algorithms for Pattern Recognition
Ian T. Nabney
1-85233-440-1

Object Recognition: Fundamentals and Case Studies
M. Bennamoun and G.J. Mamic
1-85233-398-7

Computer Vision Beyond the Visible Spectrum
Bir Bhanu and Ioannis Pavlidis (Eds)
1-85233-604-8

Lee Middleton and Jayanthi Sivaswamy

Hexagonal Image Processing

A Practical Approach

With 116 Figures

Lee Middleton, PhD
ISIS, School of Electronics and Computer Science,
University of Southampton, UK

Jayanthi Sivaswamy, PhD
IIIT-Hyderabad, India

Series editor
Professor Sameer Singh, PhD
Department of Computer Science, University of Exeter, Exeter, EX4 4PT, UK

British Library Cataloguing in Publication Data
A catalogue record for this book is available from the British Library

Advances in Pattern Recognition ISSN 1617-7916

ISBN 978-1-4471-5884-4
Springer Science+Business Media
springeronline.com

Additional material to this book can be downloaded from http://extras.springer.com

Softcover re-print of the Hardcover 1st edition 2005

34/3830-543210 Printed on acid-free paper SPIN 10984727

To my parents,
Lee

To Munni and to the loving memory of Appa,
Jayanthi

Foreword

The sampling lattice used to digitize continuous image data is a significant determinant of the quality of the resulting digital image, and therefore, of the efficacy of its processing. The nature of sampling lattices is intimately tied to the tessellations of the underlying continuous image plane. To allow uniform sampling of arbitrary size images, the lattice needs to correspond to a regular - spatially repeatable - tessellation. Although drawings and paintings from many ancient civilisations made ample use of regular triangular, square and hexagonal tessellations, and Euler later proved that these three are indeed the only three regular planar tessellations possible, sampling along only the square lattice has found use in forming digital images. The reasons for these are varied, including extensibility to higher dimensions, but the literature on the ramifications of this commitment to the square lattice for the dominant case of planar data is relatively limited. There seems to be neither a book nor a survey paper on the subject of alternatives. This book on hexagonal image processing is therefore quite appropriate.

Lee Middleton and Jayanthi Sivaswamy well motivate the need for a concerted study of hexagonal lattice and image processing in terms of their known uses in biological systems, as well as computational and other theoretical and practical advantages that accrue from this approach. They present the state of the art of hexagonal image processing and a comparative study of processing images sampled using hexagonal and square grids. They address the hexagonal counterparts of a wide range of issues normally encountered in square lattice-based digital image processing - data structures for image representation, efficient pixel access, geometric and topological computations, frequency domain processing, morphological operations, multiscale processing, feature detection, and shape representation. The discussions of transformations between square and hexagonal lattice-based images and of hybrid systems involving both types of sampling are useful for taking advantage of both in real-life applications. The book presents a framework that makes it easy to implement hexagonal processing systems using the square grid as the base, e.g., to accommodate existing hardware for image acquisition and display, and gives sample computer code for some commonly encountered computations.

This book will serve as a good reference for hexagonal imaging and hexagonal image processing and will help in their further development. I congratulate the authors on this timely contribution.

Professor Narendra Ahuja
August, 2004

Preface

The field of image processing has seen many developments in many fronts since its inception. However, there is a dearth of knowledge when it comes to one area namely the area of using alternate sampling grids. Almost every textbook on Digital Image Processing mentions the possibility of using hexagonal sampling grids as an alternative to the conventional square grid. The mention, however, is usually cursory, leading one to wonder if considering an alternative sampling grid is just a worthless exercise. Nevertheless, the cursory mention also often includes a positive point about a hexagonal grid being advantageous for certain types of functions. While it was curiosity that got us interested in using hexagonal grids, it was the positive point that spurred us to study the possibility of using such a grid further and deeper. In this process we discovered that while many researchers have considered the use of hexagonal grids for image processing, most material on this topic is available only in the form of research papers in journals or conference proceedings. In fact it is not possible to find even a comprehensive survey on this topic in any journal. Hence the motivation for this monograph.

In writing this book, we were mindful of the above point as well as the fact that there are no hardware resources that currently produce or display hexagonal images. Hence, we have tried to cover not only theoretical aspects of using this alternative grid but also the practical aspects of how one could actually perform hexagonal image processing. For the latter, we have drawn from our own experience as well that of other researchers who have tried to solve the problem of inadequate hardware resources.

A large part of the work that is reported in the book was carried out when the authors were at the Department of Electrical and Electronic Engineering, The University of Auckland, New Zealand. The book took its current shape and form when the authors had moved on to the University of Southampton (LM) and IIIT-Hyderabad (JS). Special thanks to Prof. Narendra Ahuja for readily agreeing to write the foreword. Thanks are due to the anonymous reviewers whose feedback helped towards making some key improvements to the book.

Lee: Thanks are first due to Prof. Mark Nixon and Dr John Carter who were understanding and provided me time to work on the book. Secondly thanks go to my, then, supervisor Jayanthi for believing in the idea I came to her office with. Thirdly, I would like to thank the *crew* at Auckland University for making my time there interesting: adrian, anthony, bev, bill, brian, brad, bruce, colin, david, dominic, evans, evan, geoff ($\times 2$), jamie, joseph, nigel, russell m, and woei. Finally, thanks go to Sylvia for being herself the whole time I was writing the manuscript.

Jayanthi: Thanks to Richard Staunton for many helpful comments and discussions, to Prof. Mark Nixon for the hospitality. I am also grateful to Bikash for clarifications on some of the finer points and to Professors K Naidu, V U Reddy, R Sangal and other colleagues for the enthusiastic encouragement and support. The leave from IIIT Hyderabad which allowed me to spend concentrated time on writing the book is very much appreciated. The financial support provided by the DST, Government of India, the Royal Society and the British Council partly for the purpose of completing the writing is also gratefully acknowledged. Finally, I am indebted to Prajit for always being there and cheering me on.

Contents

1 Introduction ... 1
- 1.1 Scope of the book ... 2
- 1.2 Book organisation ... 3

2 Current approaches to vision ... 5
- 2.1 Biological vision ... 5
  - 2.1.1 The human sensor array ... 6
  - 2.1.2 Hierarchy of visual processes ... 9
- 2.2 Hexagonal image processing in computer vision ... 10
  - 2.2.1 Acquisition of hexagonally sampled images ... 11
  - 2.2.2 Addressing on hexagonal lattices ... 15
  - 2.2.3 Processing of hexagonally sampled images ... 18
  - 2.2.4 Visualisation of hexagonally sampled images ... 21
- 2.3 Concluding Remarks ... 24

3 The Proposed HIP Framework ... 27
- 3.1 Sampling as a tiling ... 27
- 3.2 Addressing on hexagonal lattices ... 35
  - 3.2.1 Hexagonal addressing scheme ... 35
  - 3.2.2 Arithmetic ... 43
  - 3.2.3 Closed arithmetic ... 49
- 3.3 Conversion to other coordinate systems ... 52
- 3.4 Processing ... 54
  - 3.4.1 Boundary and external points ... 54
  - 3.4.2 Distance measures ... 56
  - 3.4.3 HIP neighbourhood definitions ... 59
  - 3.4.4 Convolution ... 61
  - 3.4.5 Frequency Domain processing ... 62
- 3.5 Concluding remarks ... 68

4 Image processing within the HIP framework ... 71
4.1 Spatial domain processing ... 71
4.1.1 Edge detection ... 71
4.1.2 Skeletonisation ... 79
4.2 Frequency Domain Processing ... 82
4.2.1 Fast algorithm for the discrete Fourier transform ... 83
4.2.2 Linear Filtering ... 91
4.3 Image pyramids ... 96
4.3.1 Subsampling ... 97
4.3.2 Averaging ... 98
4.4 Morphological processing ... 100
4.5 Concluding remarks ... 103

5 Applications of the HIP framework ... 105
5.1 Saccadic search ... 105
5.1.1 Saccadic exploration ... 106
5.1.2 Discussion ... 109
5.2 Shape extraction ... 111
5.2.1 Shape extraction system ... 111
5.2.2 Critical point extraction ... 111
5.2.3 Attention window ... 112
5.2.4 Feature extraction ... 113
5.2.5 Integration ... 113
5.2.6 Discussion ... 115
5.3 Logo shape discrimination ... 117
5.3.1 Shape extraction system ... 117
5.3.2 Image conversion ... 120
5.3.3 Local energy computation ... 120
5.3.4 Feature vectors ... 123
5.3.5 LVQ classifier ... 123
5.3.6 Discussion ... 124
5.4 Concluding remarks ... 125

6 Practical aspects of hexagonal image processing ... 127
6.1 Resampling ... 128
6.1.1 True hexagonal lattice ... 128
6.1.2 Irregular hexagonal lattice ... 136
6.1.3 Hexagonal to square resampling ... 138
6.2 Display of hexagonal images ... 141
6.2.1 Approximation with rectangular hyperpixels ... 142
6.2.2 Approximation with hexagonal hyperpixels ... 143
6.2.3 Approximation via polygon generation ... 144
6.2.4 Displaying HIP images ... 145
6.3 Concluding remarks ... 148

7 Processing images on square and hexagonal grids - a comparison . . . 151
7.1 Sampling density . . . 151
7.2 Comparison of line and curve representation . . . 155
7.2.1 Algorithmic comparison . . . 156
7.2.2 Down-sampling comparison . . . 165
7.2.3 Overview of line and curve experiments . . . 167
7.3 General computational requirement analysis . . . 168
7.4 Performance of image processing algorithms . . . 175
7.4.1 Edge detection . . . 175
7.4.2 Skeletonisation . . . 180
7.4.3 Fast Fourier transform . . . 182
7.4.4 Linear Filtering . . . 186
7.4.5 Image pyramids . . . 192
7.5 Concluding Remarks . . . 195

8 Conclusion . . . 197

A Mathematical derivations . . . 201
A.1 Cardinality of A_λ . . . 201
A.2 Locus of aggregate centres . . . 202
A.3 Conversion from HIP address to Her's 3-tuple . . . 206
A.4 Properties of the HDFT . . . 208
A.4.1 Linearity . . . 208
A.4.2 Shift/translation . . . 209
A.4.3 Convolution theorem . . . 210

B Derivation of HIP arithmetic tables . . . 211
B.1 HIP addition . . . 211
B.2 HIP multiplication . . . 213

C Bresenham algorithms on hexagonal lattices . . . 215
C.1 HIP Bresenham line algorithm . . . 215
C.2 Hexagonal Bresenham circle algorithm . . . 215

D Source code . . . 219
D.1 HIP addressing . . . 219
D.2 HIP data structure . . . 227
D.3 HIP resampling . . . 230
D.4 HIP visualisation . . . 236

References . . . 243

Index . . . 251

1 Introduction

The perceptual mechanisms used by different biological organisms to negotiate the visual world are fascinatingly diverse. Even if we consider only the sensory organs of vertebrates, such as the eye, there is much variety. From the placement of the eyes (lateral as in humans or dorsal as in fish and many birds) to the shape of the pupil, and the distribution of photoreceptors. The striking aspect about nature is the multiplicity in the designs and solutions devised for gathering visual information. This diversity also continues in the way the visual information is processed. The result of this multiplicity is that the visual world perceived by different organisms is different. For instance, a frog's visual world consists only of darting objects (which are all, hopefully, juicy flies), whereas a monkey's and a human's visual world is richer and more colourful affording sight of flies, regardless of whether they are immobile *or* airborne.

In contrast, computer vision systems are all very similar, be it in gathering visual information or in their processing. The sensing systems are designed similarly, typically based on square or rectangular arrays of sensors which are individually addressed to access the visual information. Just about the only thing that differs among sensors is the spectrum of light information that can be captured. Furthermore, the information is processed using algorithms evaluated and tested over a considerable amount of time since, like most sciences, computer vision requires the repeatability of the performance of algorithms as a fundamental tenet. Hence, from both an algorithmic and physical view all computer vision systems can be said to *see* the world with the *same eyes*.

This monograph endeavours to study the effect of changing one aspect of the sensing methodology used in computer vision, namely the sampling lattice. The change considered is from a square to a hexagonal lattice. Why hexagons? Two simple reasons: geometry and nature. Hexagonal lattices have been of interest to humans for over two millennia. Geometers from Pythagorean times have studied hexagons and found them to have special properties, including membership in the exclusive set of three regular polygons with which one can tile the plane, the other two being a square and a triangle. A honeycomb is the best 2-D example of a hexagonal lattice in nature and has fascinated people, including scientists, and been studied for a long time. This has led to the well known honeycomb conjecture. This conjecture, put simply, states that the best way to partition a plane into regions of equal area is with a region that is a regular hexagon. This conjecture has existed at least since Pappus of Alexandria but has eluded a formal proof until very recently, when Prof. Thomas Hales [1,2] proved it elegantly in 1999. In gathering information about the visual world, we believe the task at hand is similar to the problem underlying the honeycomb conjecture: capture the visual information with a set of identical sensors arranged in a regular grid structure on a planar surface. Taking cues from science and nature it is then interesting to ask what happens when you use a hexagonal (instead of a square) lattice to gather visual information. To use the previous analogy, this is viewing the world with *different eyes*. This alternative view of the visual world may present researchers with some advantages in representation and processing of the visual information. Furthermore, such a study may illuminate the importance of the role which the sensors play in computer vision.

1.1 Scope of the book

As stated, the aim of this monograph is to study the effect of changing the sampling lattice from a square to a hexagonal one. Based on lattice geometry, the hexagonal lattice has some advantages over the square lattice which can have implications for processing images defined on it. These advantages are as follows:

- *Isoperimetry.* As per the isoperimetric theorem, a hexagon encloses more area than any other closed planar curve of equal perimeter, except a circle. This implies that the sampling density of a hexagonal lattice is higher than that of a square lattice.
- *Additional equidistant neighbours.* Every hexagon in the lattice and hence a hexagonal pixel in an image has six equidistant neighbours with a shared edge. In contrast, a square pixel has only four equidistant neighbours with a shared edge or a corner. This implies that curves can be represented in a better fashion on the hexagonal lattice and following an edge will be easier.
- *Uniform connectivity.* There is only one type of neighbourhood, namely N_6, possible in the hexagonal lattice unlike N_4 and N_8 in the square lattice. This implies that there will be less ambiguity in defining boundaries and regions.

The goal of this monograph is then to verify the above and to understand the overall impact of changing the sampling lattice underlying a digital image, from both a theoretical and a practical perspective. Towards this goal, we will first seek out answers from what has been done in this field by other researchers in a period that spans roughly 40 years. We will also seek to further our understanding by developing a framework for hexagonal image processing and studying specific issues using the framework. For the sake of brevity, the terms square image and hexagonal image will be used throughout to refer to images sampled on a square lattice and hexagonal lattice, respectively.

In general, we will examine the entire gamut of issues pertaining to processing hexagonally sampled images. These start from fundamental ones such as appropriate data structures, definitions of neighbourhoods and distance functions which are essential for examining and developing processing methodologies in both the spatial and frequency domains. Applications using some of these methodologies are also of interest as they are the end goal of any image processing system. The coverage is intended to be comprehensive enough to help develop more extensive studies as well as applications. However, an exhaustive coverage is neither intended nor possible, given the current state of development of this field.

1.2 Book organisation

This monograph is divided into eight chapters and four appendices.

Chapter 2 provides an overview of the relevant background in both biological and computer vision. The latter focusses exclusively on hexagonal image processing and summarises the work that has been reported in the literature up till now.

Chapter 3 is concerned with developing a comprehensive framework for hexagonal image processing. The approach to the development concentrates on aspects required for an efficient framework: addressing and processing. Fundamental aspects of processing in both spatial and frequency domains using the developed framework, are examined and discussed. Overall this chapter is quite theoretical in nature. For non-mathematically inclined readers, the key point to examine is how the addressing scheme works, as this is central to the remaining chapters in the book.

Chapter 4 provides many examples of processing hexagonally sampled images. The proposed framework is employed for this and the examples cover most traditional problems in the field of image processing. In the spatial domain, this includes edge detection and skeletonisation. In the frequency domain, this includes the development of an algorithm for the fast computation of the discrete Fourier transform and linear filtering. Further examples included in the chapter are operations using image pyramids and mathematical morphology.

Several applications of the proposed hexagonal framework are illustrated in Chapter 5. A biologically-inspired application involves finding interesting points in an image. The rest of the applications presented are applicable to problems in content-based image retrieval. This includes one which uses a search methodology to find the shape of an object and discriminates shapes of objects based on the local energy. The applications discussed in this chapter employ the fundamental approaches outlined in Chapter 3.

The practical aspects of processing hexagonal images is investigated in Chapter 6. Two alternatives to the conventional systems which process square sampled images, are considered. These are, namely, a complete system for hexagonal image processing and a mixed system where some of the processing uses square sampled images while others use hexagonally sampled images. These alternative systems require solutions for image acquisition and visualisation which are developed and presented along with accompanying code (in Python).

To help understand the effect of changing the sampling lattice from a square to a hexagonal one, a comprehensive comparison between processing images sampled using these lattices is provided in Chapter 7. The comparison is performed from a computational perspective as well as based on visual quality analysis. Several of the examples illustrated in Chapter 3 are also compared in both square and hexagonal image processing frameworks. Conclusions are drawn about the relative merits and demerits of processing within the two frameworks.

The final chapter provides a discussion of the future directions for hexagonal image processing that merit immediate attention.

Appendix A provides derivations and proofs of various results which are mentioned throughout the book. Appendix B provides a derivation of the arithmetic tables required in the proposed framework. The Bresenham algorithms for drawing lines and circles on the hexagonal lattice are given in Appendix C. Finally, Appendix D provides some useful code for resampling and visualisation of hexagonal images and for performing all the arithmetic called for in the proposed framework.

2

Current approaches to vision

Many advances in the physical sciences have come from examination of the world around us. By performing experiments our understanding of the governing laws of the universe expands and thus we are able to build systems to take advantage of our newly acquired knowledge. The early understanding of the nature of light came from understanding and trying to mimic our eyes. The models were incorrect but they provided the foundation for all future research which truly was performed standing upon giants' shoulders. So it was in the early days of the science of image processing. The primary motivation was to recreate the abilities of the visual system in modern computing devices.

In the years since its inception the science of image processing has forked many times, each time with a resulting name change. Many of these forks disregarded the influence of the visual system when devising image processing algorithms. However, due to the rapid rise in computational power in recent times, it is possible to accurately model portions of the brain. This has led to a resurgence in research into image processing using the results from biological visual system.

It is with these ideas in mind that this chapter provides an overview of both biological and computer vision. The study of the visual system will lead to an architecture in which the brain performs its visual processing. This generic architecture will be then be applied to the study of conventional vision. Central to this thrust is the specific hexagonal arrangement implicit in the visual system's sensor array. Logically, this arrangement affects all other aspects of the visual system. In line with the historical perspective, the biological system will be discussed first followed by current computer vision approaches.

2.1 Biological vision

The complexities of the brain and all its subsystems have fascinated mankind for an extremely long time. The first recorded references to brain dissection and study date to Galen in the 2nd century AD though there is evidence, in papyri, that the Egyptians were also interested in brain function [3]. The awareness of a distinct subsystem associated with vision dates from the 18th century [4].

This section will provide a brief overview of some key aspects of the human visual system (HVS) which occupies two thirds of the human brain's volume. The first of part of the HVS is the eye. The eye performs a similar function to a smart sensor array in a modern camera. The key feature of the visual system is that it performs the processing of the information using a hierarchy of cooperative processes.

2.1.1 The human sensor array

The visual system, according to Kepler who founded the modern study of the eye in 1604 [5], begins when"the image of the external world is projected onto the pink superficial layer of the retina". Later, Descartes [6] studied the optics of the eye and Helmholtz [7] studied the retina. This early work promoted the view that the eye performed in the same way as a pinhole camera (or camera obscura). Advances made since then however, have led to the view held today that the eye is more sophisticated and functions like a mini-brain. We will now briefly explain the reasons for this view.

The eye is roughly spherical with a slightly protruding part that is exposed while the remaining part sits in the eye socket. The light enters the eye through the pupil behind the cornea and is projected by a lens onto the inner spherical surface at the rear part of the eye. Here, the light is converted into electrical signals in an array of interconnected nerve cells known as the retina. An interesting aspect of the retina is its structure. The superficial layers, which are transparent, consist of neurons while the photoreceptors are found at the deepest layer. In the thinnest part of the retina, called the fovea, the neurons are moved aside to let light pass through directly to the photoreceptors. There is also a region of the retina known as the optic disc where there is an absence of photoreceptors to permit neural wiring to carry information out to the brain. This gives rise to a blind spot.

There are two distinct sorts of photoreceptors, namely, rods and cones, with their nomenclature stemming from their shapes. Their functions are mutually complementary as summarised in Table 2.1.

A remarkable feature of the photoreceptive layer of the retina is that the rods and cones are distributed non-uniformly as illustrated in Figure 2.1. There is a radial distribution of these receptors: cones are concentrated in the central foveal region and, as one moves away from the centre, rods are found in abundance but gradually diminish in number. The foveal region, rich in cones, specialises in high resolution, colour vision under bright illumination such as during the day. This region however, is very small in extent. The region outside the fovea is rod-rich and hence contributes towards vision under low levels of illumination such as during night time. The field of view afforded by

Table 2.1. Differences between rods and cones.

Rods	Cones
high sensitivity	low sensitivity
more photopigment	less photopigment
high amplification	lower amplification
slow response	fast response
low resolution	high resolution
achromatic	chromatic (red, green, blue)
night vision	day vision

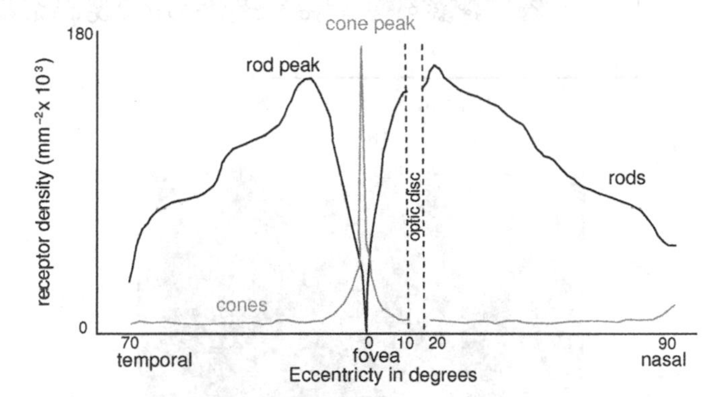

Fig. 2.1. Distribution of rods and cones in the retina (redrawn from Osterberg [8]).

high resolution and colour vision sensors is complemented by a combination of eye and head movements. The arrangement of the photoreceptors along the spherical retinal surface, is illustrated in Figure 2.2(a). Here, the larger circles correspond to the rods and the smaller circles to the cones. A significant fact to notice is that the general topology in this diagram is roughly hexagonal. This is because, as we shall see later, all naturally deformable circular structures pack best in two dimensions within a hexagonal layout such as found in honeycombs. An example of an enlarged portion of the foveal region of the retina, showing this behaviour, is given in Figure 2.2(b).

The signals from the photoreceptors are preprocessed by a neuronal assembly made of four major types of neurons: bipolar, horizontal, amacrine, and ganglion. Of these cells, the horizontal and amacrine are purely used as lateral connections joining remote regions. The lateral connections enable receptors to influence each other and help in contrast correction and adaptation to sudden changes in ambient illumination. The ganglion cells are specialised for processing different aspects of the visual image such as movement, fine spatial detail, and colour. Two of the widely studied types of ganglion cell are the magno and parvo cells. Functionally speaking, these two types of cells give rise to the formation of two distinct pathways (called the M and P pathways)

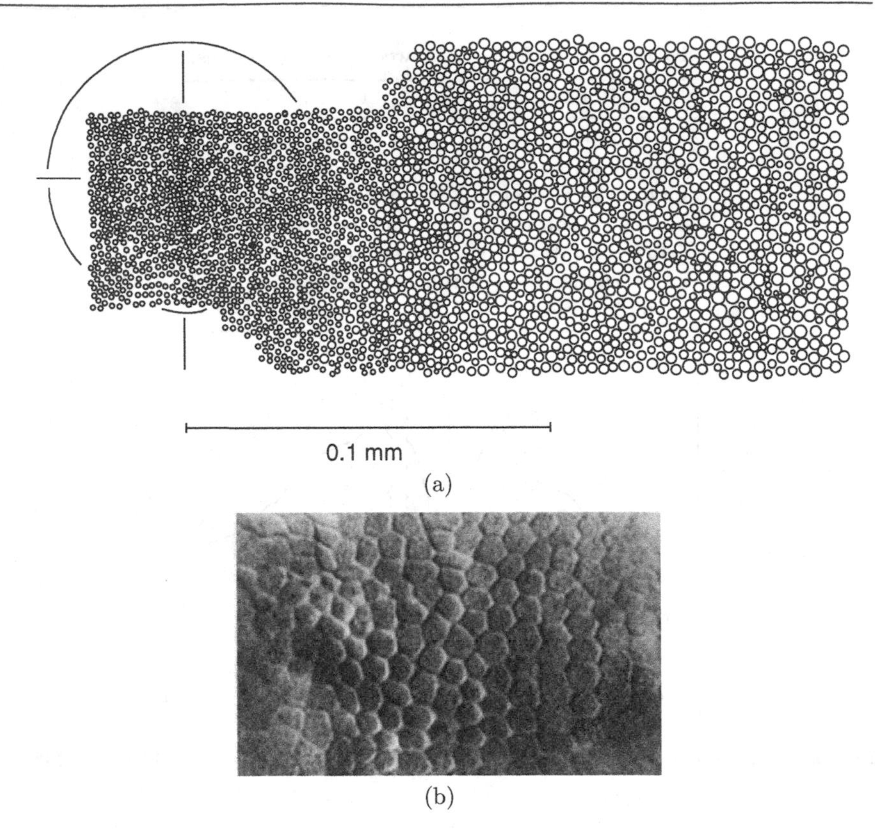

Fig. 2.2. (a) Arrangement of rods and cones in eye adapted from Pirenne 1967 [9] (b) A close up of the foveal region (reprinted from Curcio et al. [10]. Copyright 1987 AAAS).

through which visual information is passed to the brain and processed. The magno cells have large receptive fields due to their large dendritic arbours, and respond relatively transiently to sustained illumination. Thus, they respond to large objects and follow rapid changes in stimulus. For this reason it is believed that magno cells are concerned with the gross features of the stimulus and its movement. On the other hand, the more numerous parvo ganglion cells have smaller receptive fields and selectively respond to specific wavelengths. They are involved with the perception of form and colour and are considered responsible for the analysis of fine detail in an image. The ganglion cells are collected together in a mylenated sheath at the optic disk to pass the visual information to the next stage in the visual system in the brain.

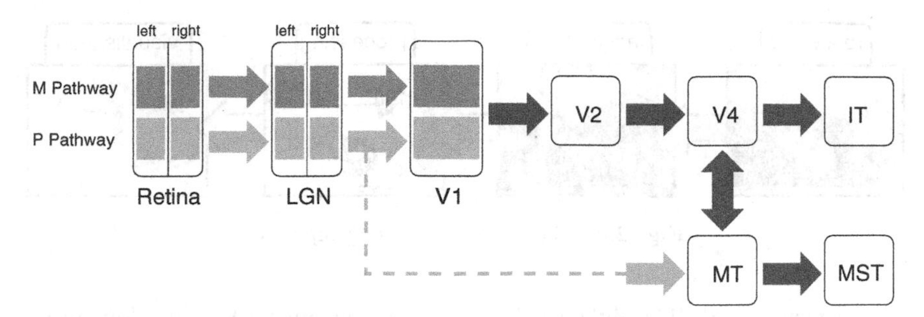

Fig. 2.3. Information flow from the retina to different cortical areas.

2.1.2 Hierarchy of visual processes

The visual system is the most complex of all the various sensory systems in the human brain. For instance, the visual system contains over 30 times the number of neurons associated with auditory processing. The sampled and pre-processed visual input is passed on to a mid-brain structure called the lateral geniculate nucleus (LGN) and then to the visual cortex. A diagram illustrating the information flow in the human visual system is shown in Figure 2.3.

The LGN is a six-layered structure that receives inputs from both the left and right visual fields via a crossover mechanism. The mapping of information from different eyes to the LGN is retinotopic which means cells in adjacent retinal regions project to cells in adjacent regions in the LGN. Additionally, inputs received by the two eyes from adjacent regions of the visual field are mapped to LGN to preserve the spatial relationships. Hence, the information mapping to LGN is spatiotopic as well. This nature of mapping continues to the first stage of the visual cortex (V1) which is about 2mm thick and also consists of a six-layer structure. The M and P pathways project to distinct sub-layers within this layer. The main function of V1 is to decompose the results from the LGN into distinct features which can used by other parts of the visual system.

Studies of the receptive fields of cells in V1 have found the cells to be considerably more complex than the cells in the retina and the LGN [11]. For instance, the LGN cells respond to spots (circles) of light whereas the simple cells in V1 respond to bars of light at specific orientations. The complex cells in V1 appear to pool outputs of several simple cells as they respond to orientation but not to the position of the stimulus. The cells in V1 are further organised into large functional structures. These include orientation-specific columns, ocular dominance columns, and colour-sensitive blobs. Neurons with similar responses but in different vertically oriented systems are linked by long range horizontal connections. Information thus flows both between the layers and between the columns, while remaining within the layer. This pattern of interconnections seems to link columnar systems together. For instance, such linkage might produce an effect of grouping all orientations in a specific region

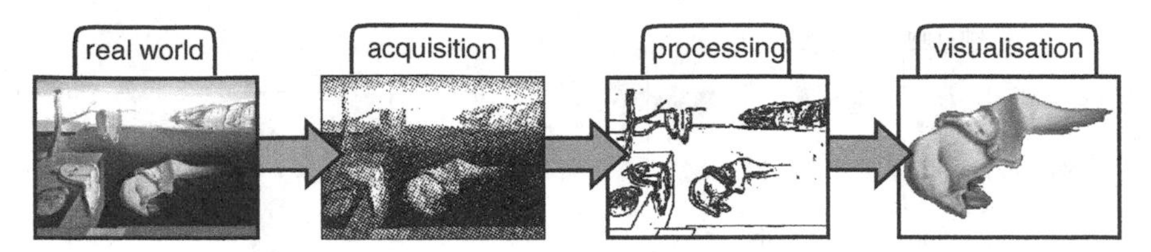

Fig. 2.4. The image processing pipeline.

of the visual field. The linked columns serve the purpose of an elementary computational module. They receive varied inputs, transform them, and send the results to a number of different regions in the HVS.

As we move further up the visual pathway, there is an increased pooling of information from lower levels and more complex forms of specialised processing carried out in different regions.

2.2 Hexagonal image processing in computer vision

Image processing in computer vision systems consists of three important components: acquisition, processing and visualisation. A simple image processing system with the information flowing from one end to another is depicted in Figure 2.4.

The acquisition stage deals with generating image data from a real world source. This can be performed via a camera, a scanner, or some more complex input device. The data may require additional processing before the acquisition is complete. In terms of the discussion of the visual system in Section 2.1.2 the acquisition stage is equivalent to the eye(s). Processing involves manipulation of the image data to yield meaningful information. This could be the application of a simple linear filtering algorithm using a convolution operator or something more complicated such as extracting a detailed description of the structures contained in the image. The visualisation stage is very useful and often essential for human observers to make sense of the processed information. Underpinning these three components is the use of a lattice or a grid on which the visual information is defined. The acquisition stage uses this to capture the image of the real world while the processing stage uses it to define appropriate data structures to represent and manipulate the image. The lattice of interest here is the hexagonal lattice, hence we will restrict the literature survey to hexagonal image processing.

The beginning of digital image processing is generally traced to the early 1960s, at which time it was spurred by the need to enhance images transmitted by the Ranger 7 [12]. These images were sampled on a square lattice. Interest in images defined on hexagonal lattices can also be traced to the 1960s. McCormick, reporting in 1963, considered a *rhombic* array, which is a hexagonal lattice, in addition to a rectangular array for a thinning algorithm to process digital images of bubble chamber negatives [13]. The work was part of the design of the ILLIAC III, a parallel computer developed exclusively for pattern recognition. Another theoretical work on 2-D signals from this period is that of Petersen [14] who found the hexagonal lattice to be the optimal arrangement for sampling of 2-D bandlimited signals. However, as we shall show next, the work on hexagonal image processing has not been sustained or intense compared to square image processing, the reasons for which are open to speculation.

The overview of the work that has been carried out by researchers in the last 40 years on hexagonal image processing is organised for convenience, along the lines of the information flow shown in Figure 2.4.

2.2.1 Acquisition of hexagonally sampled images

There are two main approaches to acquiring hexagonally sampled images. Since conventional acquisition devices acquire square sampled images, the first approach is to manipulate the square sampled image, via software, to produce a hexagonally sampled image. The second approach is to use dedicated hardware to acquire the image. We will discuss each of these in turn.

Software-based acquisition

Manipulating data sampled on one lattice to produce data sampled on a different lattice is termed resampling. In the current context, the original data is sampled on a square lattice while the desired image is to be sampled on a hexagonal lattice.

The approach of Hartman [15] used a hexagonal lattice and triangular pixels. The construction process is illustrated in Figure 2.5(a), where black squares indicate square pixels. Two square pixels that are vertically adjacent, are averaged to generate each individual triangular pixel. Hartman observed that the resulting pixel does not produce perfect equilateral triangles but instead a triangle with base angles of 63.4° and a top angle of 53.2°.

With the goal of deriving an image transform that was mathematically consistent with the primary visual cortex, Watson [16] proposed the hexagonal orthogonal-oriented pyramid. The square to hexagonal conversion process used the affine relationship between the square and hexagonal lattice points. The idea is illustrated in Figure 2.5(b). This means that rectangular images are skewed to form hexagonal images. After the skewing process, a hexagon becomes elongated in an oblique direction. A consequence of this stretching is that the hexagonal shape no longer exhibits as many degrees of rotational symmetry. For the target application of image compression, under consideration in the work, this distortion was deemed unimportant.

An approximation to the hexagonal lattice that is simple and easy to generate is a brick wall. Here, the pixels in alternate rows are shifted by half a pixel to simulate the hexagonal lattice. In Fitz and Green [17] this approach is taken. First they perform weighted averaging and subsampling on an image to halve the resolution in both directions and then they perform the pixel shift in alternate rows. The resulting image is like a brick wall made of square pixels. A different implementation of the brick wall is found in Overington's work [18]. It was noted that a hexagonal lattice is achievable with an array of square pixels where the horizontal separation is 8 pixels and the vertical separation is $5\sqrt{2}$, with alternate rows being staggered by 4 pixels. This can be approximated by a brick wall of rectangles having an 8×7 aspect ratio. Overington observes that there are no measurable errors in this approach, even though the shapes are incorrect. The process to generate hexagonal images in this case then involves first generating 8 rows of data from each 7 rows in the original image. Alternate rows are computed using the mean of adjacent pairs of data. These two steps are illustrated in Figure 2.6(a) and 2.6(b). The first step is a simple interpolation with a linear kernel and the second step is a nearest neighbour interpolation.

Another approach to generating hexagonally sampled images is via the use of quincunx sampling. These samples are arranged as in a chessboard as illustrated in Figure 2.7. Laine [19] followed this approach and used linear interpolation to double the image size in the horizontal direction and triple it in the vertical direction. The main purpose of the interpolation step is to scale the image in a way that emphasises the hexagonal arrangement. The interpolated image was masked to produce the quincunx pattern, following which the remaining data was mapped onto a hexagonal grid.

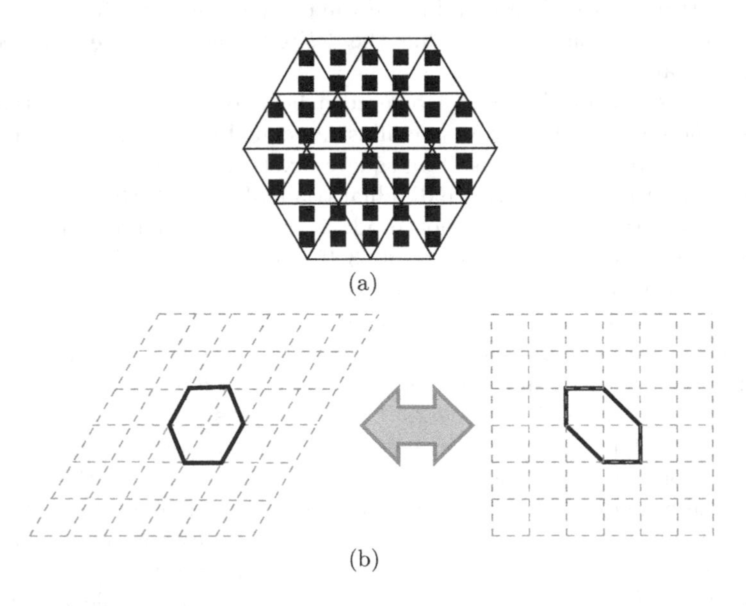

Fig. 2.5. Hexagonal resampling schemes of (a) Hartman and Tanimoto (1984) (b)Watson and Ahumada (1989).

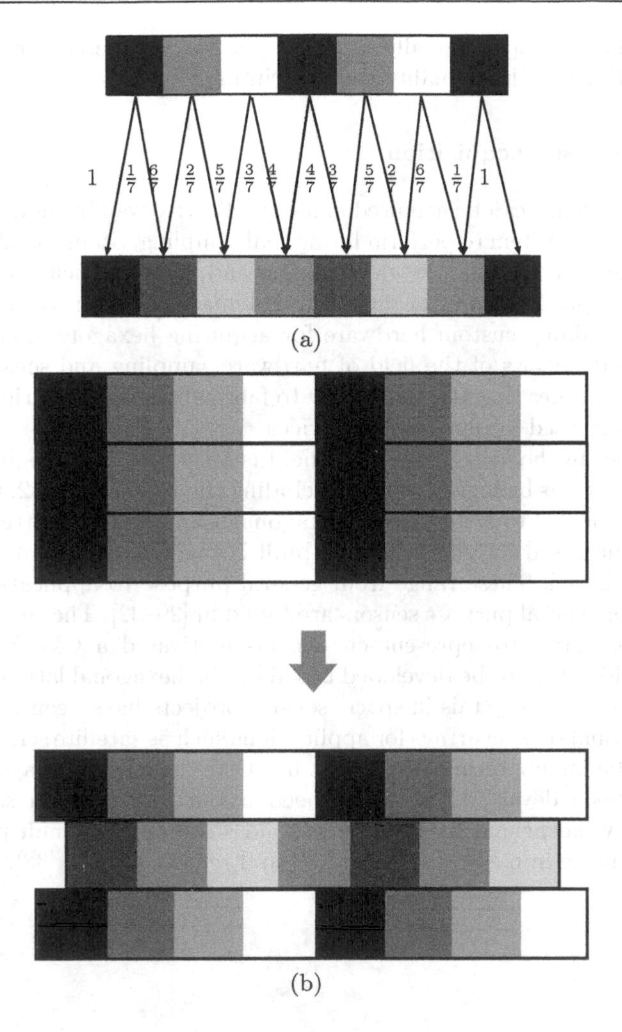

Fig. 2.6. Hexagonal resampling scheme of Overington (1992) (a) making 8 rows from 7 (b) combining alternate rows.

The resampling approach of Her [20] is also an approximate one. It is similar in idea to the work of Hartman. Here, interpolation is performed on the original image to halve the vertical resolution. After comparing a variety of different interpolation functions, Her found that bi-cubic interpolation performed the best, although the computational speed was low. Hence, bi-linear sampling was advocated as a viable alternative.

More recently, least squares approximation of splines has been proposed for resampling square images onto a hexagonal lattice [21–23]. This is an exact method and is computationally intensive. It has been used for suppressing aliasing artifacts in high quality colour printing.

Hardware based acquisition

A hexagonal image can be acquired in a cost-effective way by modifying an existing hardware system to perform hexagonal sampling. Staunton [24] designed a pipeline architecture to take video images and, by introducing a delay to alternate lines, produced a hexagonal sampled image. There has also been much interest in building custom hardware for acquiring hexagonal images. There are two good reviews of the field of hardware sampling and sensors [25, 26]. Staunton [25] notes that the technology to fabricate hexagonal grids does exist as it is widely used for large RAM devices.

A pioneer in the hexagonal sensor field is Mead [27] who has built sensors mimicking various biological sensors including the retina. In 1982, Gibson and Lucas [28] referred to a specialised hexagonal scanner. The last ten years has witnessed increased activity in custom-built hexagonal sensors, many of which are CMOS based. These range from general purpose to application specific. Examples of general purpose sensors are found in [29–32]. The superior ability of hexagonal grids to represent curves has motivated a CMOS fingerprint sensing architecture to be developed based on the hexagonal lattice [33]. With the ability to grow crystals in space, several projects have been performed to grow hexagonal sensing arrays for applications such as satellite sensors [34] and replacing the human retina [35] after it has been damaged. Hexagonal sensors have also been developed for high speed colour and position sensing [36]. Interestingly, hexagonal sensors (be it solid state or photomultiplier based) also find a place in medical imaging [37] and remote sensing [38].

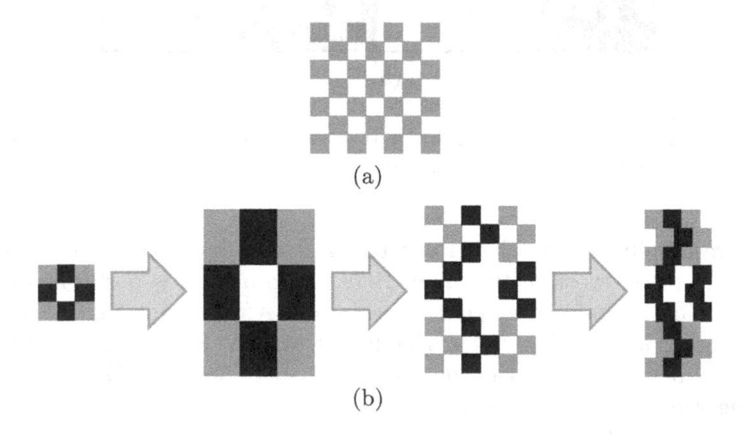

Fig. 2.7. Quincunx sampling (a) arrangement (b) hexagonal image construction.

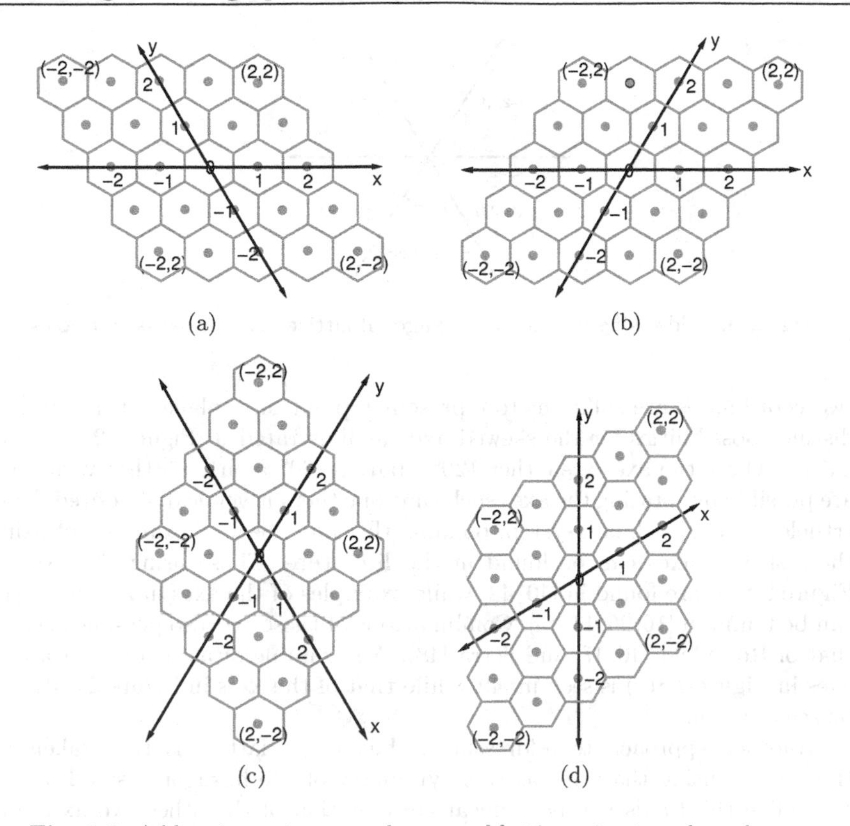

Fig. 2.8. Addressing points on a hexagonal lattice using two skewed axes.

Non-square grids, such as quincunx grids have also been found to be advantageous for high-speed imaging due to the reduction in aliasing and improved resolution after interpolation [39].

2.2.2 Addressing on hexagonal lattices

Unlike the square lattice, the points in a hexagonal lattice do not easily lend themselves to be addressed by integer Cartesian coordinates. This is because the points are not aligned in two orthogonal directions. Due to the nature of the hexagonal lattice, an alternative choice for the coordinate axes would be the axes of symmetry of the hexagon. This is convenient as it will provide purely integer coordinates for every point in the lattice. Since there are more than two axes of symmetry, many schemes have been developed for addressing points on a hexagonal lattice.

The simplest way in which to address points on a hexagonal lattice is to use a pair of skewed axes which are aligned along axes of rotational symmetry of the hexagon. This will yield integer coordinates and is efficient as

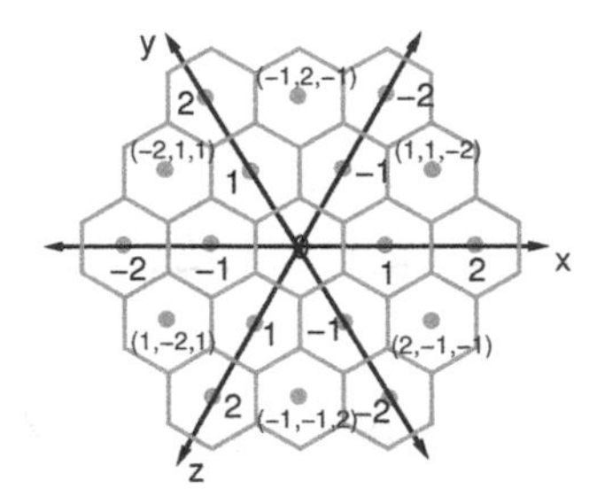

Fig. 2.9. Addressing points on a hexagonal lattice using three skewed axes.

two coordinates are sufficient to represent a point on a plane. There are two distinct possibilities for the skewed axes as illustrated in Figures 2.8(a) and 2.8(b), where the axes are either 120° apart or 60° apart. Further variations are possible by rotating the axes such that one them is vertical, if desired. Nevertheless, the coordinate system remains the same. Several examples of using these skewed axes can be found in the literature. Those using the axes in Figure 2.8(a) are found in [40–42] while examples of the axes in Figure 2.8(b) can be found in [16,26,43–45]. Combinations of these are also possible such as that of Rosenfeld [46,47] and Serra [48]. An example of usage of the rotated axes in Figure 2.8(c) is seen in [49] while that of the axes in Figure 2.8(d) can be seen in [50].

Another approach to addressing a hexagonal lattice is that taken by Her [20, 51] using the three axes of symmetry of the hexagon instead of two axes. The third axis will be a linear combination of the other two axes and hence this coordinate system suffers from redundancy. However, this symmetric hexagonal coordinate frame has advantages when it comes to operations which involve a large degree of symmetry such as rotation and distance measures. The corresponding coordinate scheme uses a tuple of coordinates (l, m, n) which correspond to the distance from the lines $x = 0$, $y = 0$, and $z = 0$ respectively, and they obey the following rule:

$$l + m + n = 0$$

It is clear that the distance between any two neighbouring points in this scheme is 1. Additionally, as this scheme uses all three axes of symmetry, it is possible to reduce the coordinates to any of the skewed axis coordinate systems. Thus, any theories or equations derived for the two skewed axes schemes can then be applied to Her's tuple. An illustration of the coordinate system is given in Figure 2.9. The coordinate scheme can also be considered to be the projection of a 3-dimensional Cartesian coordinate scheme, $\mathbb{R}^3$, onto an oblique plane through the origin, with equation $x+y+z = 0$. Hence, many of the geometrical properties of $\mathbb{R}^3$ can be readily exploited in this coordinate scheme. However, according to Her [20], this coordinate scheme leads to more

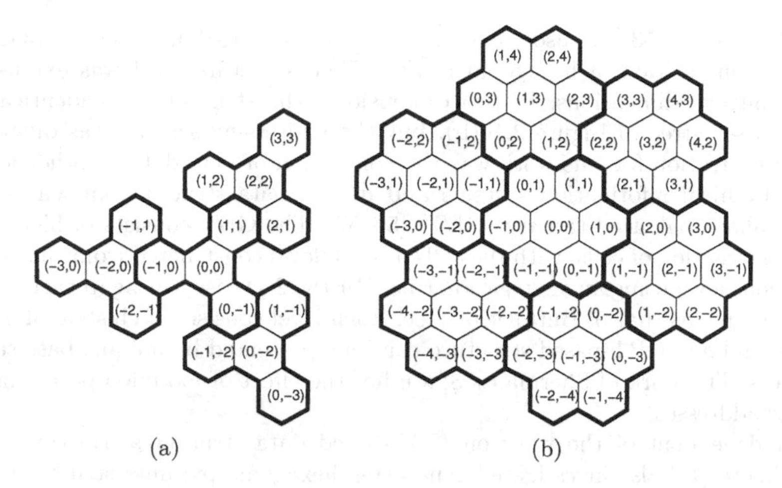

Fig. 2.10. The hierarchical addressing scheme proposed by Burt [52].

cumbersome data structures than using skewed axes and can lead to increased processing time, especially for non-symmetric operations.

The approach taken by Overington [18] is a little different from the approaches outlined previously. The entire hexagonal array is treated as if it is a rectangular array and Cartesian coordinates are directly employed to address all points. However, for practical use, it must be remembered that either odd or even rows should correspond to a half pixel shift. He notes that it is possible to approximately predetermine local operators with this shift taken into account.

There are other methods for addressing a hexagonal lattice and the motivation behind these lie in the exploitation of hierarchical hexagonal structures and their properties. The earliest of these was the work of Burt [52] who examined the construction of trees and pyramids for data sampled on hexagonal lattices. He distinguished four distinct constructions as being compact and symmetric. The coordinate scheme used a tuple, (i, j, k), and is called a pyramidal address. The last coordinate, k, is the level of the pyramid and the remaining two coordinates (i, j) were found using a skewed axis such as the one illustrated in Figure 2.8(b). The coordinate scheme is relative to each layer of the pyramid. Figure 2.10 illustrates two examples of this scheme for hexagonal lattices employing skewed axes (120°). The first two layers of the pyramid are shown highlighted with thick lines. The pyramid address of these pixels has a third coordinate of 0 or 1 indicating the first or second layer. The second of these examples which is hexagonal, Figure 2.10(b), was called a sept-tree. It was considered to be particularly promising as it was the largest compact (there are no nodes outside the pattern which are closer to the centroid than those inside the pattern) tile in either square or hexagonal lattices.

Gibson [28, 53–55] also looked at an addressing regime based on pyramidal decomposition of hexagonal lattices. The structure used was extensible to arbitrary dimensions. In two dimensions, the structure was identical to Burt's sept-tree in Figure 2.10(b), but the addressing scheme was different. The motivation for this work was the conversion of map data to other forms for graphical information systems and used a data structure known as the generalised balanced ternary (GBT) [56, 57]. The GBT consists of hierarchical aggregation of cells, with the cells at each level constructed from those at a previous level using an aggregation rule. For the 2-D case, the aggregation rule is the same as that of Burt's sept-tree. Each level consists of clusters of seven cells, and any GBT-based structure can be represented by unique, base-seven indices. The work of Sheridan [58] studies the effect of modulo operations on these addresses.

Independent of the work on GBT-based data structures, Hartman and Tanimoto [15] also investigated generating hexagonal pyramid structures, for the purpose of modelling the behaviour of the orientation- and location-specific cells in the primary visual cortex. The structure they devised was based on aggregations of triangular pixels which resulted in an image which was hexagonal in shape. Each pixel in the image was labelled uniquely by a three-tuple of coordinates: the first being the level of the pyramid and the other two being the position within the level. The position was measured on a coordinate system with the x-axis at $0°$ and the y-axis at $-60°$.

2.2.3 Processing of hexagonally sampled images

Contributions to the body of knowledge about processing aspects of hexagonally sampled images can be found at the level of theory as well as applications.

At the theoretical level, the earliest work is by Petersen [14], who considered it as a possible alternative sampling regime for a 2-D Euclidean space. Petersen concluded that the most efficient sampling schemes were not based on square lattices. The contribution of Mersereau [41, 59] for hexagonally processed signals is extensive. Based on earlier studies of general sampling [14], he proposed a hexagonal sampling theorem and used it to study properties of linear shift invariant systems for processing hexagonally sampled signals. The hexagonal discrete Fourier transform (HDFT) was also formulated by Mersereau along with a fast algorithm for its computation. The Fourier transform kernel is non-separable on a hexagonal lattice, hence an entirely different approach based on the work of Rivard [60] was employed to speed up the computations. Despite the non-separability problem, the fast algorithm was found to be 25% faster than the equivalent fast Fourier transforms for the square grid. Other contributions by Mersereau include recursive systems and of the design of FIR filters for hexagonal sampled signals.

Overington's [18] work is focused upon modelling the human visual system. Motivated by a keen interest in the limits of the visual system and its implications, hexagonal lattice was chosen as it models the retinal arrangement.

The problems that he examined are varied. He studied such areas as image preprocessing, optical flow analysis, stereo vision, colour processing, texture analysis, and multi-resolution image analysis. A good review of his work can be found in [61].

Work on morphological processing of hexagonally sampled images dates back to Golay [62] who proposed a parallel computer based on hexagonal modules which could be connected to perform different morphological operations. It was also shown that it required fewer interconnections compared to a similar square based architecture [63]. A good review of this work is found in Preston [64]. Serra has also contributed in a major way to the knowledge of mathematical morphology on a hexagonal lattice. He showed a distinct preference for hexagonal images by applying all the algorithms to hexagonal lattices prior to square lattices in his book [65,66]. The uniform connectivity and other topological properties of the hexagonal lattice are cited as reasons to make it an ideal candidate for morphological operations. Staunton [24, 25, 67–70] is another researcher who has worked extensively in both the hardware and software aspects of hexagonal image processing. As mentioned earlier, he has designed specialised hardware to generate hexagonal images and worked on a variety of applications, such as edge operators and thinning, that can help make a hexagonal system a replacement for automated inspection systems. For a good summary of the wide-ranging work see [25].

Distance transforms have been studied by several researchers starting from 1968. Rosenfeld [46] studied distance functions on digital pictures using a variety of coordinate systems. Simple morphological operators were also evaluated in these coordinate systems via some simple applications. Distance transforms on hexagonal grids were studied extensively later as well, with integer and non-integer arithmetic [71–73] and extended for 3-D images [74]. A parallel algorithm for deriving a convex covering for non-convex shapes, defined on a hexagonal grid, proposes to use the distance transform for shape analysis [75].

Thinning algorithms for hexagonal and square grids were first considered by McCormick [13] in the context of parallel implementations of thinning. The hexagonal grid was found to offer a specific edge for parallelisation over a square grid since there are only six neighbours compared to eight. Deutsch [76] also implemented a thinning algorithm on hexagonal, square and triangular lattices. The results on the hexagonal lattice were found to be better in terms of the number of edge points and resistance to noise. These results were confirmed later by Staunton [67,68] and shown to be quite efficient under parallel implementation.

Frequency domain processing of hexagonally sampled images has motivated several researchers to investigate various transforms. The work on HDFT by Mersereau was revisited in 1989 by Nel [45] who also introduced a fast Walsh transform. The fast Walsh transform was derived using a similar formulation to that for the fast Fourier transform. An analytical derivation of the DCT for hexagonal grids is available in [77]. An unusual work in the area of fast algorithms for DFT takes hexagonal images as input but computes the frequency domain samples on a square grid [78]. The formulation allows the 2-D DFT computation by two 1-D DFTs which is, as noted before, not possible on a hexagonal grid. Fitz and Green [17] have used this algorithm for fingerprint classification and report that the algorithm is not as efficient as the square FFT, but is more efficient in memory usage since fewer data points are required for the same resolution. Another work which has a square lattice formulation to compute the hexagonal DFT is given in [79]. Hexagonal aggregates identical to those defined by Gibson [28] have been used in [80] to develop a fast algorithm for computing the DFT. The radix-7 algorithm is very efficient with computational complexity of $N log_7 N$ for an image containing a total of N pixels.

Multiresolution image processing on hexagonal images is an area which is seeing a slow pace of activity although the theoretical foundations have been laid for over a decade via filter bank design [81,82] and wavelet bases [83,84]. It has been pointed out that the advantage of using a hexagonal grid for multiresolution decomposition is the possibility of having *oriented* subbands which is not possible in a square lattice. Simoncelli's work [82] on perfect reconstruction filter banks has been used as a basis in [40] for sub-band coding and in [19,85] for tumour detection in mammograms. Both these works use a pyramidal decomposition of the image with three sub-band orientations namely at 30°, 90° and 120°.

The hexagonal, orthogonal, oriented pyramid of Watson [16] was used for an image coding scheme. The main motivation was to be comparable to the receptive fields of the human visual system. It was built upon an earlier work called the cortex transform. The levels of compression achieved in this scheme were greater than the equivalent schemes for square images.

The orientation selectivity afforded by hexagonal lattice and multiscale representations have been used a bit differently to define a set of ranklets for feature detection [86]. Such ranklets on square lattices resemble the Haar wavelets. Results of applying them to face detection are said to be good and consistent with those using square ranklets.

The work of Gibson and Lucas [28,53–56] was primarily in the addressing of hexagonal lattices. However, several applications were performed upon their addressing scheme. The first was the rasterisation of digitised maps for a geographical information system. Another application was automated target recognition where the aim was to recognise tanks from a cluttered background for the United States military. Snyder [26] revisited addressing of hexagonal images and defined his own addressing scheme. Within the context of this addressing scheme, neighbourhoods were defined and associated operators such as convolution and gradient were developed.

The contribution of Her [20,51], like Gibson and Lucas, is primarily in the addressing scheme. His 3-coordinate scheme exploits the implicit symmetry of the hexagonal lattice. Her also introduced a series of geometric transformations such as scaling, rotation and shearing on the hexagonal 3-coordinate system. Operators were also devised for efficient rounding operations to find nearest integer grid points in a hexagonal lattice, which are useful in geometric transformations.

Kimuro [87] used a spherical hexagonal pyramid to analyse the output from an omni-directional camera. The results were promising and were better than such systems designed using square images. Surface area estimates of scanned images, have been computed using hexagonal and other tilings by Miller [44]. Hexagonal tiles were found to perform with less error than the other tiles. Sheridan [58] revisited the work of Gibson and Lucas and labelled his version of the generalised balanced ternary as the Spiral Honeycomb Mosaic and studied modulo arithmetic upon this structure. Texture characterisation using co-occurrence matrices have also been studied on the hexagonal lattice [88]. It was found that there were benefits for textures which have information at multiples of 60°. Additionally, texture synthesis using Markov random fields has been proposed which permit classes of texture which are not simple to generate using square images [89]. A family of compactly supported hexagonal splines are proposed in [23]. It has an increased number of symmetry compared to the B-splines used on square lattices. Among the suggested applications are zooming using a hexagonal spline transform and resampling.

Quantitative studies comparing image quality on square and hexagonal lattices have also been carried out in [90–93]. Quantisation error measures and profiles have been developed and used to compare the square and hexagonal sampling in [92, 93]. The results reported for line representation show the hexagonal grid somewhat favourably. However, it is cautioned that the specific algorithm to be used should be taken into consideration for an accurate evaluation. [91] presents an interesting comparison of the square and hexagonal lattices with respect to image quality measurements. They propose a metric in terms of the area enclosed within the Fourier transform of the Wigner-Seitz cell. Effective resolution is a term that has been often overused in digital camera literature. Almansa [90] presents a quantitative means to measure the effective resolution of image acquisition systems which can be used as a basis of comparison of square and hexagonal sampling as well as for improving image resolution. He also gives a good overview of hexagonal image processing in his doctoral dissertation [94].

2.2.4 Visualisation of hexagonally sampled images

The final part of the review we will present is on the visualisation aspect of hexagonal image processing. Visualisation is strongly tied to advances in display technology. There have been many approaches over the years to this problem and some of these will be now presented. The presentation will be roughly chronological.

The earliest approach to displaying hexagonal images dates back to the 1960s and Rosenfeld [46]. His intuitive approach, illustrated in Figure 2.11, was to use an evenly spaced grid with alternate rows offset by one point. The images were purely monochrome, so an *on* pixel was illustrated with a '.'

Fig. 2.11. Rosenfeld's method for displaying a hexagonally sampled image [46].

and an *off* pixel was illustrated with a ' '. Intensities and other information, however, could be conveyed by employing different characters. The results were generated using a dot matrix printer. As the horizontal spacing is fixed, the sole manipulation that could be performed was the vertical spacing. This approach generated hexagonal lattices that appeared very regular.

In the 1980s there were several significant advances in technology which directly impacted on the visualisation of hexagonal images. The first of these was in the printing industry and the second was in the visual display industry. In the print industry, devices were developed which afforded a much higher resolution and more control of the printed output than previously available in dot matrix devices. These had far reaching effects on print media, specifically newspapers, and led to research into how to display continuous tone photographs on the printed page. This area was known as half-toning. Stevenson [42] proposed the use of a hexagonal layout of dots. The dots, being squirts of ink, are circular and hence a hexagonal lattice was more robust to errors and compensated the characteristics of the non-ideal printing device. His method was analogous to that of Rosenfeld except that individual dots were used to indicate the hexagonal pixels instead of entire characters. These dots had a controllable radius and were slightly bigger than an individual print element. In this way, alternate rows could be offset to produce the required hexagonal lattice arrangement. This allowed hexagonal images to be printed with much denser arrangements of hexagonal pixels. A variation of this method of display is used in newspapers even today. Recently, resampling square to hexagonal images using splines has also been used to reduce alias artifacts in colour printing [22].

The visual display industry also experienced significant advances in the 1980s with the arrival of high resolution visual display units which were capable of displaying millions of colours. This had an effect on the display of hexagonal images. For instance, the work of Hartman and Tanimoto [15] made up a hexagonal image using accumulations of screen pixels which approximated an equilateral triangle. These were coloured with the desired intensity of the hexagonal pixel. Wüthrich [50] used an accumulation of screen pixels which, when taken together, looked hexagonal in shape. This is illustrated in Figures 2.12(a) and 2.12(b). The logical consequence of this approach is that the screen resolution is significantly reduced when compared with square

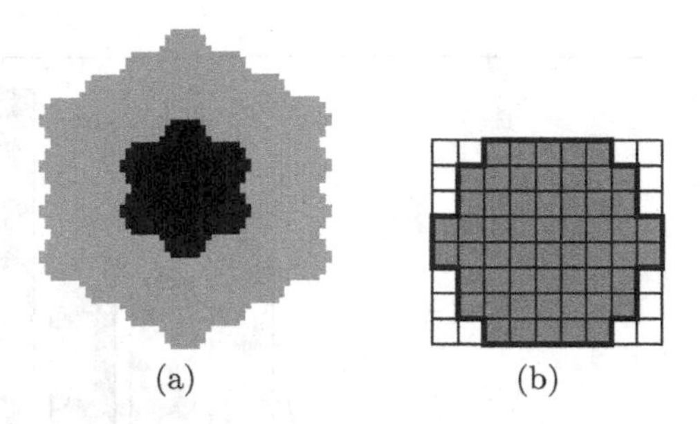

Fig. 2.12. Wüthrich's method for displaying a hexagonally sampled image [50] (a) an image (b) an individual picture element.

Fig. 2.13. Her's method for displaying a hexagonally sampled image [20].

images. The design of these hexagonally shaped aggregations of pixels is also problematical.

Her [20] used two pixels to approximate the hexagonal pixels in a display. Using the assumption that the individual screen pixels are oblate (slightly elongated in the horizontal direction), an approximation to a hexagonal pixel could be made with just two pixels. This gave the individual hexagonal pixels an aspect ratio of 1:1.64. For an equilateral triangle drawn through the centres of three adjacent hexagons, this gave base angles of 51° and a top angle of 78°. A one-pixel shift in alternate rows is all that is required here to generate the image. A hexagonal image using these pseudo-hexagonal pixels is illustrated in Figure 2.13. This method has the advantage that it is easy to compare results with square sampled images as the only modification required is to remove the offset on alternate rows.

The work of Staunton [49] uses a consistent display scheme throughout opting for the brick wall approximation. Unlike Her however, larger pixels which have a width and height tuned to give the same aspect ratio as a real hexagonal pixel are employed. For ideal pixels which have a vertical separation of 1 the corresponding horizontal separation is $\frac{2}{\sqrt{3}}$. Thus, it is possible to plot a rectangular accumulation of pixels that has an aspect ratio of 1.15:1. Overington [18] uses different approaches to display hexagonal images depending on the application. For coarse images, a methodology similar to Rosenfeld is utilised. However, in the case of high resolution images, a methodology similar to Staunton is preferred. Overington remarks that the brick wall display process has little error and can be used rather than going to the trouble of using a real hexagonal grid.

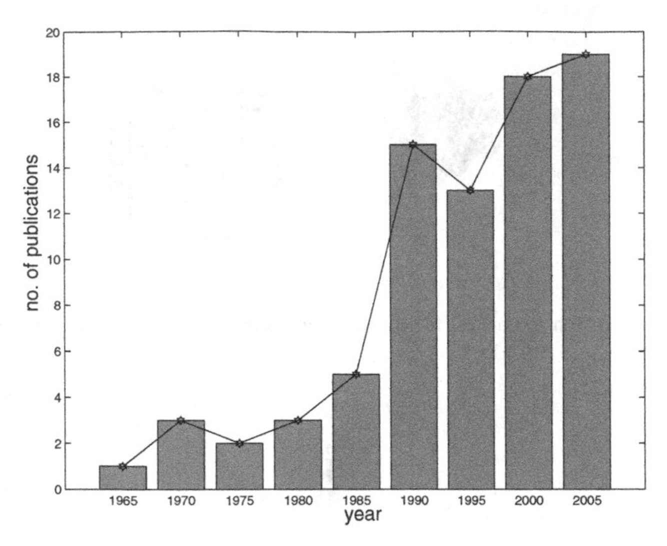

Fig. 2.14. Publications about hexagonal image processing over the last 40 years.

The work of Gray et al. [95] was in the perception of differently shaped pixels. They experimented with a variety of different pixel shapes and arrangements for displaying images. Experiments included a conventional square image and a hexagonal image, both displayed using a brick wall approach. For pixel shapes, real hexagons, using the method of Wüthrich, and many more complex shapes were tested. The results demonstrated that the human visual system has a preference for a pixel shape that has dominant diagonals in it. This includes diamonds and hexagons.

2.3 Concluding Remarks

The aim of this chapter has been to provide an overview of the visual systems. The two systems that were discussed were the human visual system (HVS) and computer vision systems.

The brief discussion of the HVS revealed it to be a complex one. In the language of modern computer vision, the HVS appears to employ *smart* sensing in two ways: (i) it is adaptable to ambient illumination due to the use of two types of receptors and long range connections and (ii) it is capable of a wide field of high resolution imaging, achieved by optimising the sensing methodology via a combination of movable sensory system and non-uniform sampling. Other remarkable aspects of the HVS are the hexagonal arrangement of photoreceptors (especially in the fovea) and the hierarchical nature of representation of visual information and its processing.

The review of computer vision was restricted to hexagonal image processing. A chronological picture of the research activity in this field is shown in Figure 2.14. From the review and this figure we can see that the field of hexagonal image processing is as old as image processing itself. Researchers in this field have had disparate motivations for considering hexagonal sampling, ranging from modelling HVS to parallel implementation of thinning and fast algorithm for Fourier Transform to efficient map data handling for geographical information systems. Almost all traditional areas of image processing have been investigated by researchers in this field.

The length of time for which hexagonal sampling has held interest among researchers and the wide range of areas covered, we believe, points to the fact that there is not only significant interest but also merit in processing hexagonally sampled images. However, the second point to note from the figure is that the level of activity is low in general, notwithstanding the upward trend that is apparent in the last 15 years. This is true both in terms of the number of researchers and in the number of publications that has come out in this area over 40 odd years. The reason for this can only be described as the weight of *tradition*. Hardware for image acquisition and visualisation have traditionally been based on a square lattice. The reason for the former is ascribed to the complications in designing a hexagonal arrangement of sensors, notwithstanding the availability of technology for hexagonal layouts in large RAMs.

The scenario in computer vision as a whole is witnessing a change. There appears to be renewed interest (both at research and commercial levels) in fovea-type image sensors and non-uniform sampling, features which we have noted to be hallmarks of HVS. These developments are motivated by the search for new solutions to many difficult problems in vision as well as the need to innovate. In keeping with this trend, we take a fresh look at hexagonal image processing as a whole in the remaining parts of this monograph and propose a complete framework. This idea was initially introduced in [96–101] and is being expounded further in this monograph. The framework is intended to serve as a test bed for studying the impact of changing the sampling lattice on image processing.

3

The Proposed HIP Framework

This chapter introduces the proposed HIP framework for processing hexagonally sampled images. The fundamental difference between square and hexagonally sampled images is in the geometry of the underlying sampling grids. While the points in the square grid lie along directions which are mutually orthogonal, the points in the hexagonal grid lie along directions which are not. A direct implication of this difference is that Cartesian coordinates are a natural choice to represent points in square images but unsuitable in hexagonal images. This lack of orthogonality leads to several distinct possibilities based on different sets of skewed axes, and thus representation. However, such representations are cumbersome and do not fully exploit the hexagonal nature of the underlying lattice. Devising an efficient representation scheme for hexagonal images is necessary as it can drastically affect the overall performance of a system. Towards this purpose, we take a fresh look at the sampling process from the point of view of tiling theory.

3.1 Sampling as a tiling

The notion put forward in this section is that trying to sample a 2-D spatial signal is similar to trying to tile the Euclidean plane. In both cases, the central motivation is to completely cover the space in an easily reproducible fashion. However, the rationale behind the covering process is different. In a tiling, the plane is covered in order to study the properties of the individual tile. By contrast, sampling aims to efficiently capture as much information as possible about the underlying signal extending over the plane. The sampling process is performed by dividing the Euclidean plane into regular and reproducible regions and analysing the signal in each region. It is possible to study sampling within the context of tiling, as sampling is just a form of specialised tiling. With this in mind, sampling can be studied and perhaps improved by first understanding some of the general principles of tiling.

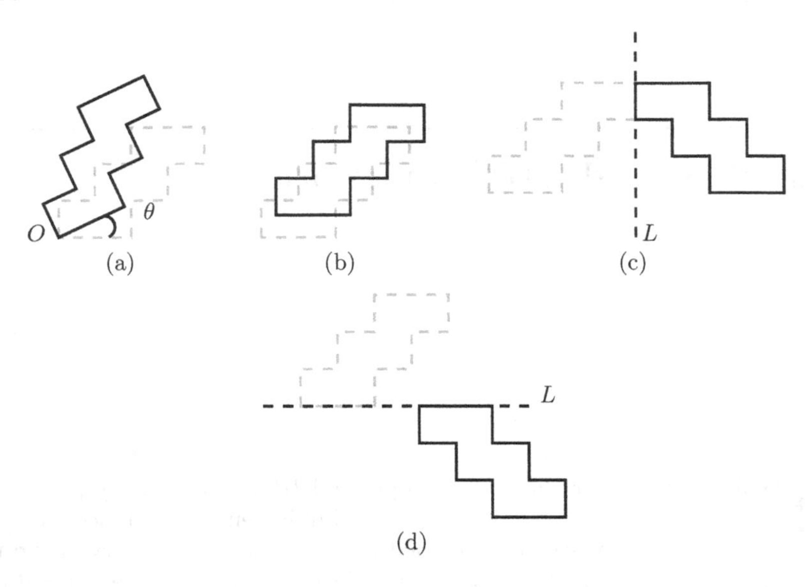

Fig. 3.1. The four valid isometries: (a) rotation (b) translation (c) reflection (d) glide reflection.

Formally, a tiling, $\mathcal{T}$, is a collection of closed sets, $\mathcal{T} = \{T_1, T_2, \cdots\}$, which cover the Euclidean plane without gaps or overlaps, $T_i \cup T_j \neq \emptyset$ where $i \neq j$ [102]. Each T_i is a series of points that completely define the tile and all such occurrences of the tile. The set of points within T_i which completely describe a single tile is known as the prototile of the tiling $\mathcal{T}$. The concatenation of the tiles is the entire Euclidean plane. A packing is a weaker version of a tiling where gaps are permitted while covering the space.

A tiling in which all the individual tiles are of the same size and shape is called a monohedral tiling. When the tiles are regular polyhedral shapes, we have a regular tiling. It has been shown that only three such monohedral regular tilings are possible [102]. These use triangles, squares, or hexagons as the prototile. The conventional images with square pixels and the images treated in this book which use hexagonal pixels are thus examples of regular monohedral tilings. By extension, a tiling with two prototiles is called a dihedral tiling.

Regular tilings were the first kinds of tilings to be studied in any detail. The first systematic study of such tilings dates to Kepler [103] in 1619. Since by definition there can be no gaps in a tiling, the corners of individual tiles must meet at a point with the sum of all the interior angles at that point being 2π radians. The interior angle at each corner of a regular polygon is $\frac{(n-2)\pi}{n}$ radians, where n is the number of sides. Consequently, for a tiling made up of r such polygons the following relationship holds:

$$\frac{n_1 - 2}{n_1} + \cdots + \frac{n_r - 2}{n_r} = 2 \tag{3.1}$$

Here n_i is the number of sides of one of the unique polygons in the tiling. There are only 17 unique tilings which satisfy this equation and these are often called the Archimedean tilings in reference to the first recorded discoverer of these tilings. If all the individual tiles are identical, there can be only three possible scenarios consisting of a meeting point of 6 triangles, 4 squares, or 3 hexagons. It can be seen that a regular pentagon cannot be used to create a regular tiling because its interior angle of $\frac{3\pi}{5}$ will not divide into 2π thereby failing the criteria in equation (3.1).

Many of the important properties of tilings depend upon the notion of symmetry which can be explained in terms of an isometry. An isometry is a mapping of the Euclidean plane onto itself which preserves all distances [104]. The mapping is denoted by $\sigma : \mathbb{R}^2 \rightarrow \mathbb{R}^2$. There are only four types of isometries [105]:

1. rotation about a point O through an angle θ (Figure 3.1(a))
2. translation in a given direction by a given distance (Figure 3.1(b))
3. reflection in a line L (Figure 3.1(c))
4. glide reflection in which reflection in a line L is combined with a translation by a given distance (Figure 3.1(d))

Isometries of types 1 and 2 are called direct because they preserve the sense of the points. This means that if the points are labelled in an anti-clockwise sense then they will remain so after the isometry has been performed. Isometries of types 3 and 4 are called indirect. For a particular isometry, σ, being performed on a set S the symmetries of the set are the isometries which map onto the original set, or $\sigma(S) = S$. The reflectional symmetries for a square and a hexagon are illustrated in Figure 3.2. Apart from the reflectional symmetries the hexagon has rotational symmetry at multiples of $\frac{\pi}{3}$ radians and the square has rotational symmetry at multiples of $\frac{\pi}{2}$ radians. Thus, the hexagon exhibits 6-fold rotational symmetry and the square exhibits 4-fold rotational symmetry. Furthermore, each shape has an identity isometry which maps the original shape onto itself. All told there are 12 (six reflections, five rotations, and identity) symmetries for the hexagon and 8 (four reflections, three rotations, and identity) for the square [106].

If a tiling admits any of the four isometries along with the identity then it is called symmetric. If it contains at least two translations in non-parallel directions then the tiling is called periodic. Regular periodic tilings of the plane are also called tessellations. Periodic tilings are easy to describe. If the two non-parallel translations are represented as vectors $\mathbf{v}_1$ and $\mathbf{v}_2$, the set of all translations, $S(\mathcal{T})$, is $n\mathbf{v}_1 + m\mathbf{v}_2$ where $n, m \in \mathbb{Z}$. All these translations naturally arise by combining n of translation $\mathbf{v}_1$ and m of translation $\mathbf{v}_2$. Starting from a fixed origin the set of translations $n\mathbf{v}_1 + m\mathbf{v}_2$ forms a lattice. The most commonly occurring lattice is the set of points in the Euclidean plane with integer coordinates. This is known as the unit square lattice and

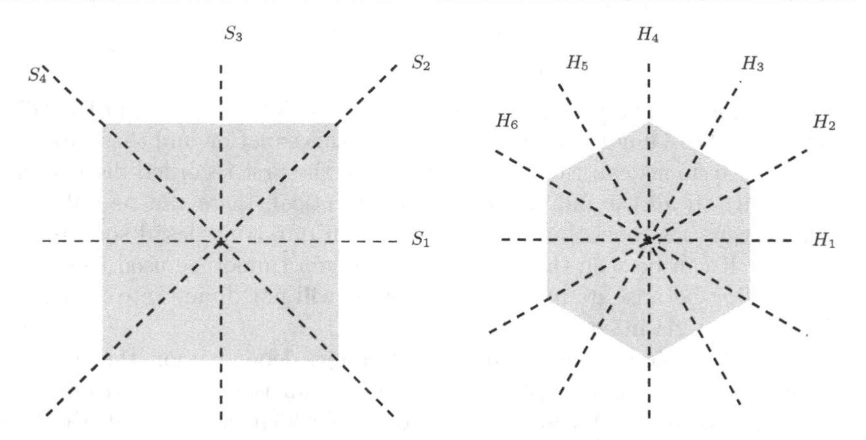

Fig. 3.2. The reflectional symmetries of a square ($S_1 \rightarrow S_4$) and a hexagon ($H_1 \rightarrow H_6$).

is defined by $\mathbf{v}_1 = (1, 0)$ and $\mathbf{v}_2 = (0, 1)$. More generally however, a lattice can be considered to consist of the vertices of a parallelogram. An illustration of this is given in Figure 3.3. In the figure, the entire tiling can be easily extended by repeated copying of the central parallelogram. Indeed, this was the method used to draw this seemingly complex looking tiling.

Uniform sampling has its beginning in the definition of the sampling theorem by Nyquist [107] and the later consolidation at Bell Labs by Shannon [108]. The extension of this work to n-dimensional spaces was carried out by Petersen and Middleton [14]. To remain consistent with the description of tilings the remainder of this discussion will be limited to 2-D samplings. Also

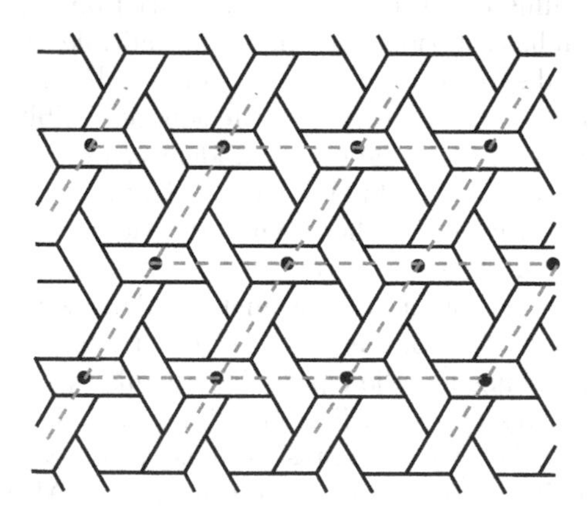

Fig. 3.3. A periodic dihedral tiling with its corresponding lattice and one possible period parallelogram.

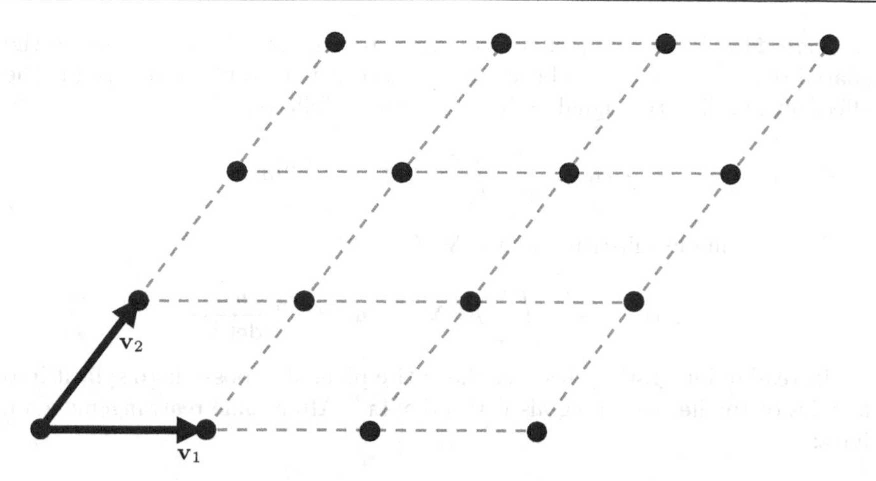

Fig. 3.4. A possible uniform sampling of the Euclidean plane.

note that the description of sampling henceforth follows that of Petersen [14] closely.

Consider a continuous function, $f(x_1, x_2)$, that is defined in $\mathbb{R}^2$. In order to uniformly sample this function, the domain of the function must be divided up in a regular fashion and samples taken. An intuitive way to perform this is to use rectangular sampling and derive a sampled version of the function $f_s(n_1, n_2) = f(n_1T_1, n_2T_2) = f(\mathbf{Vn})$. T_1 and T_2 are real numbered constants that provide the horizontal and vertical sampling intervals respectively. There is no reason why rectangular sampling must be used and in fact, any valid basis of $\mathbb{R}^2$ can be used. One such basis is $V = \{\mathbf{v}_1, \mathbf{v}_2\}$. The individual sample points are integer multiples of these basis vectors, $n_1\mathbf{v}_1 + n_2\mathbf{v}_2$, in a fashion similar to that of rectangular sampling, just described. An example is illustrated in Figure 3.4. In the diagram, the sampling points are generators for a grid made up of parallelograms. Note the similarity between the definition of this grid and the lattice previously described for tilings.

Sampling theory [14] can be used to determine necessary conditions for reliably sampling a continuous signal $f(\mathbf{x})$. The conditions are based on the relationship between the sampled spectra and the original signal's spectra. Given a continuous signal $f(x)$ its Fourier transform $F(\boldsymbol{\Omega})$, the relationships between the two are defined as follows:

$$F(\boldsymbol{\Omega}) = \int_{-\infty}^{\infty} f(\mathbf{x}) e^{-\mathrm{j}\boldsymbol{\Omega}^T\mathbf{x}} d\mathbf{x}$$

$$f(\mathbf{x}) = \frac{1}{4\pi^2} \int_{-\infty}^{\infty} F(\boldsymbol{\Omega}) e^{\mathrm{j}\boldsymbol{\Omega}^T\mathbf{x}} d\boldsymbol{\Omega}$$

Here, $\mathbf{\Omega} \in \mathbb{R}^2$ is a frequency domain vector and $\mathbf{x} \in \mathbb{R}^2$ is a vector in the spatial domain. Using the above relationship, it is possible to determine the effect of sampling the signal on its spectrum as follows:

$$f_s(\mathbf{n}) = \frac{1}{4\pi^2} \int_{-\infty}^{\infty} F(\mathbf{\Omega}) e^{\mathrm{j}\mathbf{\Omega}^T(\mathbf{Vn})} d\mathbf{\Omega}$$

Using a simple substitution $\boldsymbol{\omega} = \mathbf{V}^T\mathbf{\Omega}$ yields:

$$f_s(\mathbf{n}) = \frac{1}{4\pi^2} \int_{-\infty}^{\infty} F((\mathbf{V}^T)^{-1}\boldsymbol{\omega}) e^{\mathrm{j}\boldsymbol{\omega}^T\mathbf{n}} \frac{d\boldsymbol{\omega}}{|\det \mathbf{V}|}$$

Instead of integrating this over the entire plane it is possible to split it into a series of smaller sub integrals with area $4\pi^2$. After some rearrangement we have:

$$f_s(\mathbf{n}) = \frac{1}{4\pi^2} \int_{-\pi}^{\pi} \left[\frac{1}{|\det \mathbf{V}|} \sum_{\mathbf{k}} F((\mathbf{V}^T)^{-1}(\boldsymbol{\omega} - 2\pi\mathbf{k})) e^{\mathrm{j}\boldsymbol{\omega}^T\mathbf{n}} e^{-2\pi\mathrm{j}\mathbf{k}^T\mathbf{n}} \right] d\boldsymbol{\omega}$$

Now, $e^{-2\pi\mathrm{j}\mathbf{k}^T\mathbf{n}}$ always has unit value for all possible values of $\mathbf{n}$ and $\mathbf{k}$. Hence, by defining a new variable $F_s(\boldsymbol{\omega})$, the sampled signal can be computed analogously to an inverse Fourier transform as follows:

$$F_s(\boldsymbol{\omega}) = \frac{1}{|\det \mathbf{V}|} \sum_{\mathbf{k}} F((\mathbf{V}^T)^{-1}(\boldsymbol{\omega} - 2\pi\mathbf{k}))$$

or, more simply:

$$F_s(\mathbf{V}^T\mathbf{\Omega}) = \frac{1}{|\det \mathbf{V}|} \sum_{\mathbf{k}} F(\mathbf{\Omega} - \mathbf{Uk})$$

where $\mathbf{U}$ is the reciprocal lattice, or frequency domain lattice . We have thus derived the familiar result in sampling theory, namely, that the spectra of the sampled signal $F_s(\mathbf{V}^T\mathbf{\Omega})$ is a periodic extension of the original continuous signal $F(\mathbf{\Omega})$. Note that the sampling lattice $\mathbf{V}$ and the reciprocal lattice are related as $\mathbf{U}^T\mathbf{V} = 2\pi\mathbf{I}$. For the sampling lattice given in Figure 3.4 the reciprocal lattice in the frequency domain is illustrated in Figure 3.5. The basis vectors in the two domains are seen to be mutually orthogonal, which is due to the fact that the Fourier transform is an orthogonal projection. The gray circles illustrate the individual spectra of the the sampled function.

It is evident from the figure that, to avoid aliasing, there should be no overlap between adjacent copies of the spectra. Hence, it is important that the original signal is band, or wave-number (using the terminology of Petersen [14]), limited. This means that the spectrum of the original function $F(\mathbf{\Omega})$ is equal to zero outside some region of finite extent, B, which is known as the baseband or region of support. There is no constraint on the shape of

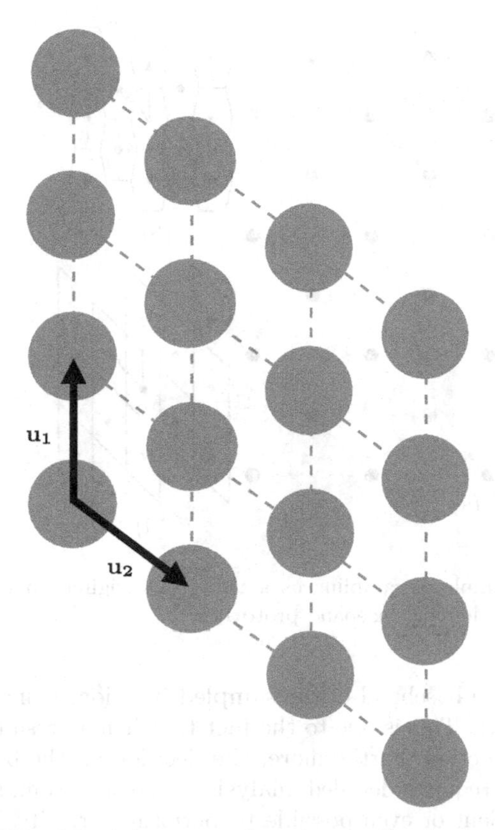

Fig. 3.5. A frequency domain sampling lattice.

B, though certain shapes make the reconstruction simpler. It is possible to vary the spatial sampling matrix $\mathbf{V}$ so that there is no overlap among the periodically repeated versions of $F(\boldsymbol{\Omega})$. Consequently, there is no aliasing.

The main point to draw from this discussion is that in terms of sampling, it is desirable that no aliasing is present for perfect reconstruction. In the language of tiling, this is equivalent to requiring the sampled signal's spectrum to be a monohedral packing. Furthermore, the matrices which define the spatial and frequency domain sampling lattices are required to be isometries.

As an example, Figure 3.6(a) illustrates a simple sampling of the spatial domain. The chosen sampling lattice is roughly hexagonal with the horizontal spacing being twice the vertical. The original lattice is illustrated in gray dashed lines. The frequency domain lattice is illustrated in the diagrams on the right. The lattice however, says nothing about the nature of the baseband in the frequency domain. Figures 3.6(b) to 3.6(e) show four possible monohedral tilings using different prototiles. However, all of these are not equally desirable. Whilst a prototile of a particular, though unusual, shape may provide

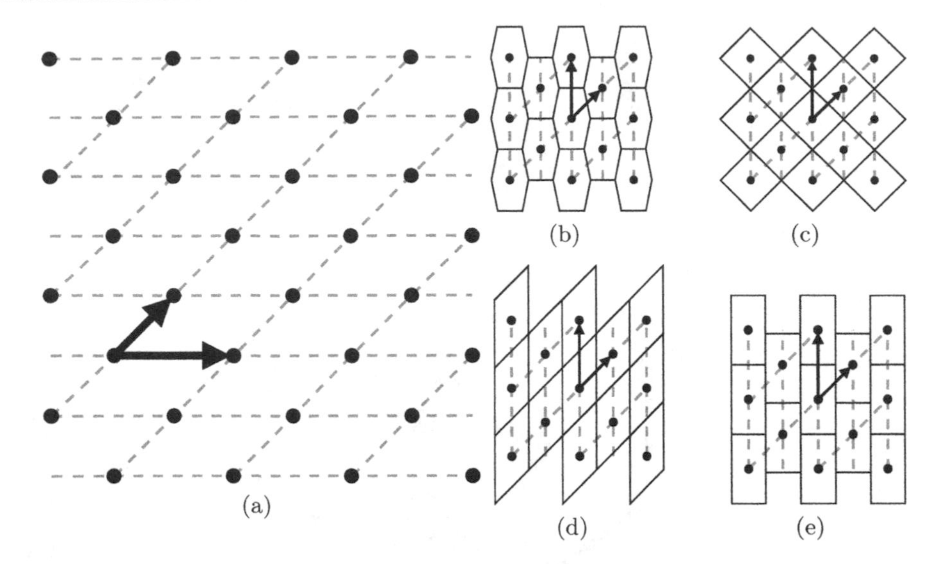

Fig. 3.6. An example of sampling as a tiling (a) original spatial sampling (b-e) possible frequency domain baseband prototiles.

a better fit for the baseband of the sampled function, it may be impractical to use such shapes. This is due to the fact that it may result in a more complicated reconstruction. Furthermore, the decision on the best packing for a particular signal requires detailed analysis of the spectrum, something which may not be prudent or even possible to perform in reality. It suffices to say then that the packing that is appropriate should be general purpose, simple to reconstruct, and provide high density of samples. There have been several studies of these problems [14, 109]. The conclusion from those studies is that a regular hexagonal tiling is the best solution as it simultaneously provides greater density of information and easy reconstruction.

Practical limitations in reconstruction, such as filter design, dictate that adjacent copies of the baseband are separated rather than contiguous for easier signal reconstruction. This implies that the tiling imposes an upper bound on the possible baseband size. This is analogous to the Nyquist rate in 1-D sampling. In the 1-D case, the Nyquist rate is the minimum sampling frequency. Extending this to 2-D and tilings yields the conclusion that the prototile specifies the minimum allowable sampling frequency for perfect signal reconstruction. The sampling frequency is related to the size of the prototile.

In summary, based on examining sampling from the tiling point of view, we can make the following observations. Firstly, the problem of uniformly sampling a 2-D signal is analogous to producing a periodic monohedral tiling of the plane. If we restrict ourselves to regular shapes then we only have three possibilities for the prototile, namely, a hexagon, square, or a triangle. Of these, the hexagon provides the greatest number of tiles/samples per unit area. Secondly, for signals which are band-limited, for efficient sampling and reconstruction it is best to fit the entire baseband spectrum into a prototile. The size and shape of the desired prototile is easy to determine if the signal has an isotropic baseband spectrum, i.e., the signal is circularly bandlimited. A hexagonal prototile is the best fit for such a spectrum. However, deriving a similar result for signals with complex baseband shapes is more complicated. Specifically, the equivalent 2-D Nyquist rate is difficult to compute, meaning that it is hard to sample the signal to give a simple reconstruction.

3.2 Addressing on hexagonal lattices

Related to the issue of how to sample the signal is the issue of how to address individual samples. Generally, addressing and storage are important issues, though often overlooked, when it comes to image processing. The specific storage mechanism can often drastically affect the performance of a system. In the human visual system (HVS), the lateral geniculate nucleus (LGN) serves to organise the information from the retina for use in the rest of the visual system. There are two sorts of arrangements that are apparent in the LGN. The first is spatiotopic where information from neighbouring sensors are stored near each other. The second is hierarchical organisation which selectively pools information from many sensors producing a variety of resolution-specific arrangements. The visual system manages to achieve these arrangements via the way in which it is organised. Such features are also desirable in computer vision systems, since neighbouring pixel information plays an important role in image manipulation/understanding and easy access to such information can impact system performance. We therefore take inspiration from the HVS and require that an addressing scheme for hexagonally sampled data has some or all of the following features:

- spatiotopic arrangement
- hierarchical
- computational efficiency

The solution as to how to achieve this can be found by examining tilings [102]. The details of this are now presented followed by a description of arithmetic operations.

3.2.1 Hexagonal addressing scheme

From the perspective of tilings, a hexagonally sampled image can be considered as a periodic monohedral tiling (see Section 3.1). By taking a group of hexagons, it is possible to make a more complex aggregate tile. The aggregation can be done in many ways. Figure 3.7(a) shows a circular arrangement where the tile is made of seven hexagons, six of which are arranged around a

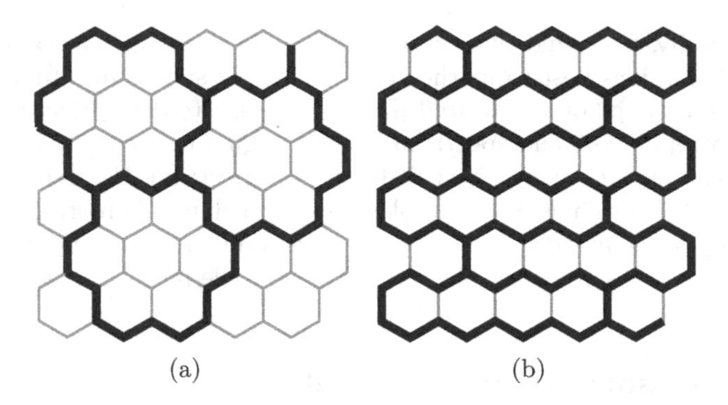

Fig. 3.7. Some possible hexagonal aggregate tiles: (a) circular (b) linear.

central hexagon. This is termed circular as the hexagons in the periphery are equidistant from the central hexagon. Figure 3.7(b)) consists of lines of four hexagons. Note that in addition to these two, there are numerous other methods in which the Euclidean plane can be tiled using hexagonal aggregate tiles. However, for the remainder of this section the discussion will be concerned with the circular aggregate tile illustrated in Figure 3.7(a). This is due to the fact that it contains more degrees of symmetry.

As discussed in Section 3.1, many important properties of tilings depend upon their symmetry. The four types of symmetry that can be considered, rotation, translation, reflection, and glide reflection, are illustrated in Figure 3.8 for the circular aggregate tile in Figure 3.7. However, due to the chosen vertical alignment of the individual hexagons, the rotational symmetry cannot be used in the addressing scheme. Thus, any of the other symmetries are permissible for the basis of the indexing scheme. For example, Figure 3.9 illustrates a possible hierarchical aggregation using just the translation property. Note, that other possible hierarchical aggregations are possible using different combinations of symmetries but this one is convenient and leads to a simple indexing scheme. Furthermore, there are several variations on this tiling, each with a different translation. However, what is important is that the tiles are placed in a consistent manner that will easily lead to the sort of hierarchical aggregation described here.

Until now, all the discussion of the aggregated tilings has been intuitive. However, if an appropriate addressing scheme is to be developed, a more rigorous mathematical approach is required. In Section 3.1, it was observed that the plane could be regularly sampled using integer multiples of a pair of basis vectors. Setting one vector to be parallel to the x axis and a second rotated 120° from the first, gives the basis vectors as:

$$B = \left\{ \begin{bmatrix} 1 \\ 0 \end{bmatrix}, \frac{1}{2} \begin{bmatrix} -1 \\ \sqrt{3} \end{bmatrix} \right\} \tag{3.2}$$

These vectors provide a convenient coordinate system which can be used to further develop the tiling with a view to producing an addressing scheme. In Figure 3.9 the original hexagonal tile (on the far left) can be considered the prototile. Let it be labelled as point $(0,0)$ relative to B. This is the zeroth level aggregate and the set containing all these points can be called A_0. The circular aggregate (or the first level aggregate), which is illustrated in the centre of the figure, can be considered as a set of seven points in this coordinate system. Continuing the naming convention, and ordering in an anti-clockwise direction starting from the horizontal, gives:

$$A_1 = \left\{ \begin{bmatrix} 0 \\ 0 \end{bmatrix}, \begin{bmatrix} 1 \\ 0 \end{bmatrix}, \begin{bmatrix} 1 \\ 1 \end{bmatrix}, \begin{bmatrix} 0 \\ 1 \end{bmatrix}, \begin{bmatrix} -1 \\ 0 \end{bmatrix}, \begin{bmatrix} -1 \\ -1 \end{bmatrix}, \begin{bmatrix} 0 \\ -1 \end{bmatrix} \right\} \tag{3.3}$$

Examination of these coordinates confirms the hierarchical nature of the tiling as A_0 is a member of A_1. A similar approach can be used to write all the points in the second level aggregate A_2, in Figure 3.9. It would be

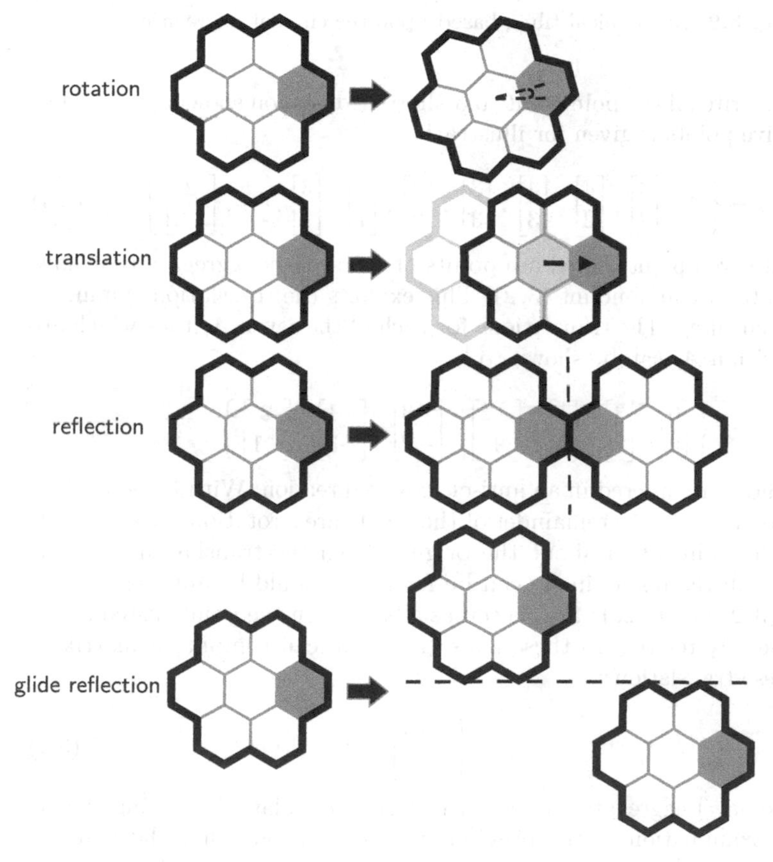

Fig. 3.8. Symmetries of a circular hexagonal aggregate.

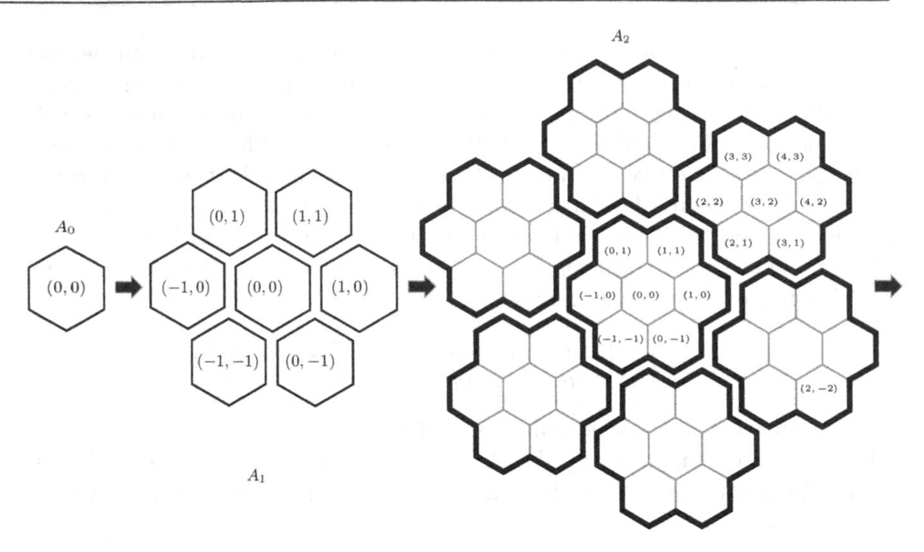

Fig. 3.9. Hierarchical tiling based upon the circular aggregate.

laborious to write all the points out so a shortened version showing only a few representative points is given for illustration:

$$A_2 = \left\{ A_1, \begin{bmatrix} 3 \\ 2 \end{bmatrix}, \begin{bmatrix} 4 \\ 2 \end{bmatrix}, \begin{bmatrix} 4 \\ 3 \end{bmatrix}, \begin{bmatrix} 3 \\ 3 \end{bmatrix}, \begin{bmatrix} 2 \\ 2 \end{bmatrix}, \begin{bmatrix} 2 \\ 1 \end{bmatrix}, \begin{bmatrix} 3 \\ 1 \end{bmatrix}, \cdots, \begin{bmatrix} 2 \\ -2 \end{bmatrix} \right\} \tag{3.4}$$

The first seven points listed are points of a first level aggregate that have been translated by an amount $(3, 2)$. This exploits the translation symmetry of the original tiling. The translations for each of the seven A_1 tiles which are contained within A_2 can be shown to be:

$$\left\{ \begin{bmatrix} 0 \\ 0 \end{bmatrix}, \begin{bmatrix} 3 \\ 2 \end{bmatrix}, \begin{bmatrix} 1 \\ 3 \end{bmatrix}, \begin{bmatrix} -2 \\ 1 \end{bmatrix}, \begin{bmatrix} -3 \\ -2 \end{bmatrix}, \begin{bmatrix} -1 \\ -3 \end{bmatrix}, \begin{bmatrix} 2 \\ -1 \end{bmatrix} \right\}$$

Again these are ordered in an anti-clockwise direction. With the exception of the first point $(0, 0)$ the remainder of the points are a rotation, by a multiple of $60°$, of the point $(3, 2)$ about the origin. Given the translations and the points in A_1 all the points in A_2 can be found. It should be apparent that if the points $(3, 2)$ and $(-2, 1)$ are chosen as a basis then the points listed above reduce to set A_1 relative to these axes. Thus, we can compute a matrix to perform these translations:

$$\mathbf{N}_1 = \begin{bmatrix} 3 & -2 \\ 2 & 1 \end{bmatrix} \tag{3.5}$$

The next level aggregate, A_3, is formed via a circular tiling using the A_2 aggregate. Examination of the previous two levels shows that the first tile is always placed so as to make a positive increase in angle from the previous aggregation. This means that there is only one location for the first tile. Thus, the first tile is centred on point $(5, 8)$ relative to the central tile using the basis B. The corresponding translation matrix is:

$$\mathbf{N}_2 = \begin{bmatrix} 5 & -8 \\ 8 & -3 \end{bmatrix} \tag{3.6}$$

Note that $\mathbf{N}_2 = \mathbf{N}_1\mathbf{N}_1$. This is no coincidence and comes as a natural consequence of the way in which the aggregate tiles are created. Generally, a level λ aggregate can be produced by a circular tiling of level $(\lambda - 1)$ aggregates. The translation matrix is given by:

$$\mathbf{N}_{\lambda-1} = \begin{bmatrix} 3 & -2 \\ 2 & 1 \end{bmatrix}^{\lambda-1} \tag{3.7}$$

Using this translation, and the previous sets, it is possible to generate a complete set of all possible points in a λ-level aggregate. These are:

$$A_\lambda = \mathbf{N}_{\lambda-1}A_1 + \cdots + \mathbf{N}_1 A_1 + A_1 \tag{3.8}$$

Here the $+$ operator concatenates the two sets together so that each element in the set on the left is added to all the elements in the set on the right. For example for A_2:

$$\begin{aligned}
A_2 &= \mathbf{N}_1 A_1 + A_1 \\
&= \begin{bmatrix} 3 & -2 \\ 2 & 1 \end{bmatrix} \left\{ \begin{bmatrix} 0 \\ 0 \end{bmatrix}, \cdots, \begin{bmatrix} 0 \\ -1 \end{bmatrix} \right\} + \left\{ \begin{bmatrix} 0 \\ 0 \end{bmatrix}, \cdots, \begin{bmatrix} 0 \\ -1 \end{bmatrix} \right\} \\
&= \left(\begin{bmatrix} 0 \\ 0 \end{bmatrix} + \left\{ \begin{bmatrix} 0 \\ 0 \end{bmatrix}, \cdots, \begin{bmatrix} 0 \\ -1 \end{bmatrix} \right\} \right) + \cdots + \left(\begin{bmatrix} 2 \\ -1 \end{bmatrix} + \left\{ \begin{bmatrix} 0 \\ 0 \end{bmatrix}, \cdots, \begin{bmatrix} 0 \\ -1 \end{bmatrix} \right\} \right) \\
&= \left\{ \begin{bmatrix} 0 \\ 0 \end{bmatrix}, \cdots, \begin{bmatrix} 0 \\ -1 \end{bmatrix} \right\} + \cdots + \left\{ \begin{bmatrix} 2 \\ -1 \end{bmatrix}, \cdots, \begin{bmatrix} 2 \\ -2 \end{bmatrix} \right\} \\
&= \left\{ \begin{bmatrix} 0 \\ 0 \end{bmatrix}, \cdots, \begin{bmatrix} 0 \\ -1 \end{bmatrix}, \cdots, \begin{bmatrix} 2 \\ -1 \end{bmatrix}, \cdots, \begin{bmatrix} 2 \\ -2 \end{bmatrix} \right\}
\end{aligned}$$

The aggregation process described has several important features. The first is that the number of points in a given level is seven times that of the previous level. This can be validated by examining Figure 3.9 and equation (3.8). In other words the number of points in a level λ aggregate is 7^λ. The second feature is that the angle between the horizontal, the first vector of the basis B, and the centre of each successive aggregate is increasing. The amount by which it increases for each aggregate is constant and is $\tan^{-1}\frac{\sqrt{3}}{2}$. For a λ-level aggregate this rotation is:

$$\theta_\lambda = (\lambda - 1)\tan^{-1}\frac{\sqrt{3}}{2}, \qquad \lambda > 0 \tag{3.9}$$

Another feature is that the Euclidean distance from the origin is increasing by a factor of $\sqrt{7}$ for each successive aggregate. This means that the locus of all the centres of each aggregated tile is a spiral. This locus can be computed by:

$$r = \exp(\theta \frac{\log \sqrt{7}}{\tan^{-1} \frac{\sqrt{3}}{2}}) \tag{3.10}$$

Where r and θ are the polar coordinates of a point on the spiral. Using these features, it is possible to return to the examination of the aggregates and assign an alternative addressing scheme for them. The key point to note in developing the alternative scheme is that the total number of tiles in each aggregate is a power of seven. Hence, the indexing scheme can exploit this by using base seven numbering. A simple addressing scheme that does this is as follows: Assign a base seven number to each of the points labelled in the aggregate, starting at 0. For example:

$$A_1 = \left\{ \begin{bmatrix} 0 \\ 0 \end{bmatrix}, \begin{bmatrix} 1 \\ 0 \end{bmatrix}, \begin{bmatrix} 1 \\ 1 \end{bmatrix}, \begin{bmatrix} 0 \\ 1 \end{bmatrix}, \begin{bmatrix} -1 \\ 0 \end{bmatrix}, \begin{bmatrix} -1 \\ -1 \end{bmatrix}, \begin{bmatrix} 0 \\ -1 \end{bmatrix} \right\}$$

yields:

$$G_1 = \{0, 1, 2, 3, 4, 5, 6\}$$

A more complicated example can be seen by looking at the second level aggregate:

$$A_2 = \left\{ A_1, \begin{bmatrix} 3 \\ 2 \end{bmatrix}, \begin{bmatrix} 4 \\ 2 \end{bmatrix}, \begin{bmatrix} 4 \\ 3 \end{bmatrix}, \begin{bmatrix} 3 \\ 3 \end{bmatrix}, \begin{bmatrix} 2 \\ 2 \end{bmatrix}, \begin{bmatrix} 2 \\ 1 \end{bmatrix}, \begin{bmatrix} 3 \\ 1 \end{bmatrix}, \cdots, \begin{bmatrix} 2 \\ -2 \end{bmatrix} \right\}$$

yielding:

$$G_2 = \{G_1, 10, 11, 12, 13, 14, 15, 16, \cdots, 66\}$$

The addressing for a second level aggregate is illustrated in Figure 3.10. In this scheme, the address of each individual hexagon also encodes the spatial location within a tile of a given level of aggregation. Consider the cell labelled 42 (the number is in base 7). The fact that it has two digits reveals that the hexagonal cell has its position within a second level aggregate. The first digit, 4, indicates that the address lies within the fourth, first-level aggregate and the second digit, 2, indicates that the tile is in the second position of the central (or zeroth-level) tile. The single index, multiple digit, addressing scheme outlined here turns out to be a modified form of the generalised balanced ternary [53–55]. The modification is in the way in which the cells are assigned with addresses. Instead of using a simple tiling construction, the generalised balanced ternary (GBT) assigns addresses to minimise the address difference between neighbours. The addressing outlined above is consistent with the way in which the space is tiled with the aggregate tiles. In order that a distinction is kept between our indexing scheme and the GBT, it will be referred to as HIP addressing.

The HIP addressing scheme that has been described is a radix 7 positional number system [57]. It can be defined via the rule:

$$(a_{\lambda-1} \cdots a_2 a_1 a_0)_7 = a_{\lambda-1}(7)^{\lambda-1} + \cdots + a_2 7^2 + a_1 7^1 + a_0 \qquad (3.11)$$

Here, the values of a_k are $0 \leq a_k < 7$. A numbering system based on radix 7 is also known as septenary. Obviously, it is possible to extend this definition to include numbers to the left of the radix point but these are not needed simply to address hexagonal cells. However, as this is a useful property we will return to it later (see Section 3.2.2). To distinguish the addresses from ordinary radix 7 numbers, bold italics will be used . For example, ***42*** is the cell highlighted in Figure 3.10. The set of all possible HIP addresses for a

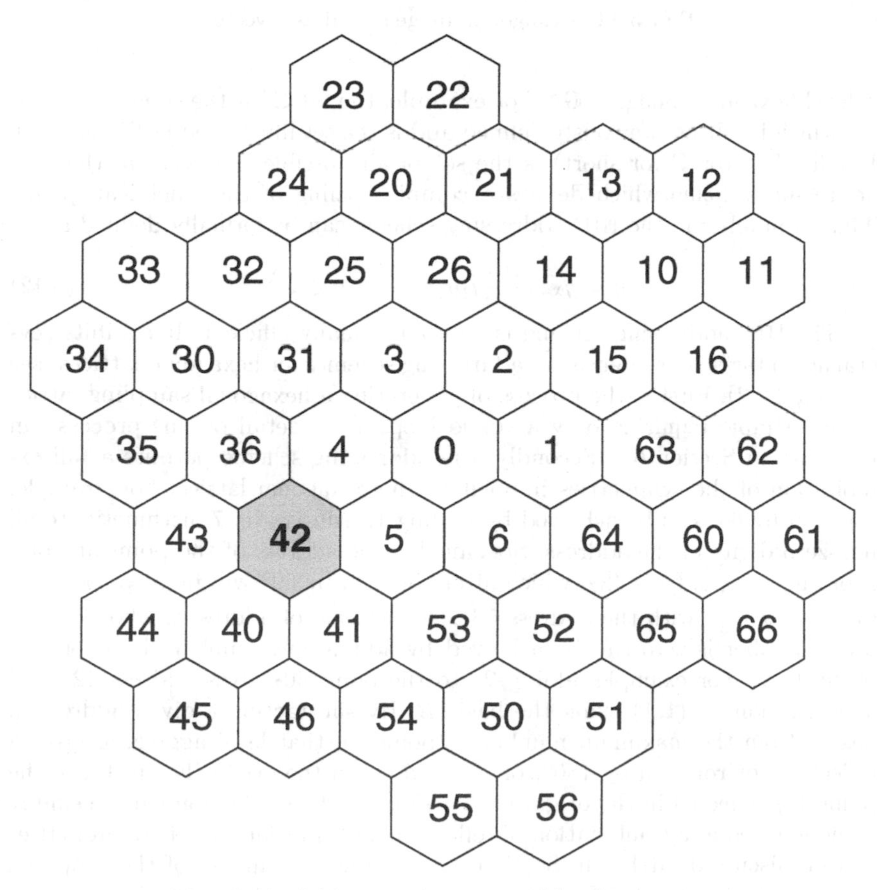

Fig. 3.10. Addressing for a second level aggregate.

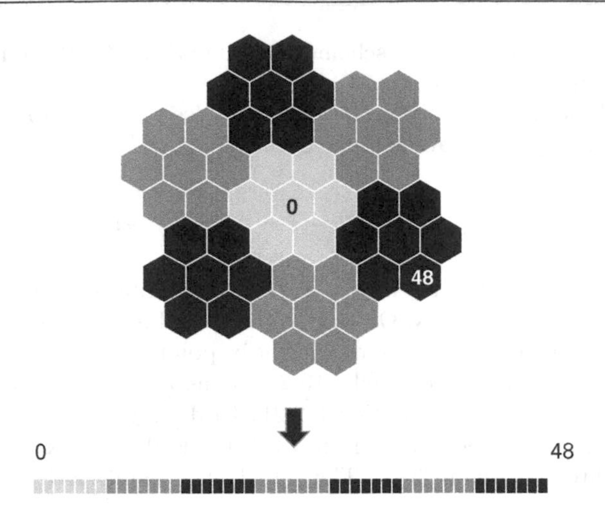

Fig. 3.11. Hexagonal image stored as a vector.

λ-level hexagonal image is $\mathbb{G}^\lambda$. For example, the set $\mathbb{G}^2$ is the same as the set G_2 which has been previously defined and is the set illustrated in Figure 3.10. Finally, $\mathbb{G}^\infty$ (or $\mathbb{G}$ for short) is the set of all possible addresses in the HIP addressing scheme which defines a complete tiling of the Euclidean plane. Thus, a number in the HIP addressing scheme can be formally defined as:

$$\boldsymbol{g} = g_{\lambda-1} \cdots g_2 g_1 g_0, \qquad \boldsymbol{g} \in \mathbb{G}^\lambda \tag{3.12}$$

The HIP addressing scheme is computationally efficient. It exhibits several advantages over alternative addressing schemes for hexagonal lattices (see Section 2.2.2). Firstly, the process of generating a hexagonal sampling lattice is very simple requiring only a single loop. More detail of this process can be found in Section 6.2. Secondly, the addressing scheme permits a full exploitation of the symmetries implicit in the hexagonal lattice. For example, rotation by 60° can be achieved by adding 1, using radix 7 arithmetic, to all non-zero digits in the address. Specifically, the address of the point at polar coordinates $(\sqrt{3}, \frac{\pi}{6})$ is 42_7, which after rotation by 60° will be at polar coordinates $(\sqrt{3}, \frac{\pi}{3})$ with the address 53_7. Translation of addresses whose lowest order number is zero can be achieved by adding any numbers between one to six to it. For example adding 2_7 to the point 40_7 gives address 42_7 and a translation by $(1, 1)$ using the basis B. By subtracting a given address in base 7 from the maximum number of points in that level aggregate gives a reflection, or rotation by 180°, of the address in the origin. For instance the point 42_7 reflected in the origin is 15_7. Finally, glide reflection can be simply demonstrated as a combination of reflection and translation. These properties will be discussed further in Section 3.4. Another advantage of the proposed addressing scheme is the fact that images can be stored using a single vector, which makes the scheme computationally efficient. This idea is illustrated in Figure 3.11. In the figure each square corresponds to an address in the original hexagonal image. The labels correspond to the first and last location in the image labelled using base 10. Note that neighbouring groups of pixels in the 2-D image remain neighbouring elements in the vector, a very useful feature in image processing operations.

The storage requirements for hexagonally sampled images using the HIP addressing scheme, depend upon two factors: the image resolution and the quantisation level of the individual sample points. For instance, a square sampled image of size $M \times N$ pixels with a quantisation depth of 24 bits will have a size of $3MN$ bytes. Similarly for a λ-level image the required storage space is $3 \times 7^{\lambda}$. Given a square sampled image, the number of layers for an equivalent hexagonal sampled image can be found by equating the numbers of points:

$$\lambda = \frac{\log M + \log N}{\log 7} \tag{3.13}$$

Due to the nature of the addressing scheme, the value of λ is also the number of digits in the corresponding addresses for the points in the image. For example if a square image of size 128×128 is to be examined, the formula gives $\lambda \approx 5$ as the number of layers in the equivalent hexagonal image. This structure contains 16807 points compared with a square array which contains 16384 points.

3.2.2 Arithmetic

The HIP addressing scheme is an example of a positional numbering system [57] using base (or radix) 7 and was formally defined in equation (3.12). The address itself indicates spatial location within the Euclidean plane relative to the origin. Each successive digit within an address gives an orientation and radius from the origin. This means that the address behaves as a vector quantity and arithmetic operations on the addresses can thus be defined with this in mind.

The first set of operations to examine are addition and subtraction. Due to the vectorial nature of the HIP address, the addition operation can be performed using vector addition as illustrated in Figure 3.12. The vectors corresponding to ***26*** and ***15*** are drawn as black lines and the resultant vectorial sum is a dashed line. This figure illustrates the well known parallelogram rule for vector addition. To generate a general rule for addition it is possible to compute the result of all possible additions of single digit HIP addresses as shown in Table 3.1. A detailed description of the generation of this table is given in Appendix B. Note that some of the additions result in a 2-digit HIP address. These occur when the angles are complementary and result in a net radius greater than 1. In these cases, the second digit can be considered as a *carry* into the next layer of the hexagonal image. Using this table, addition

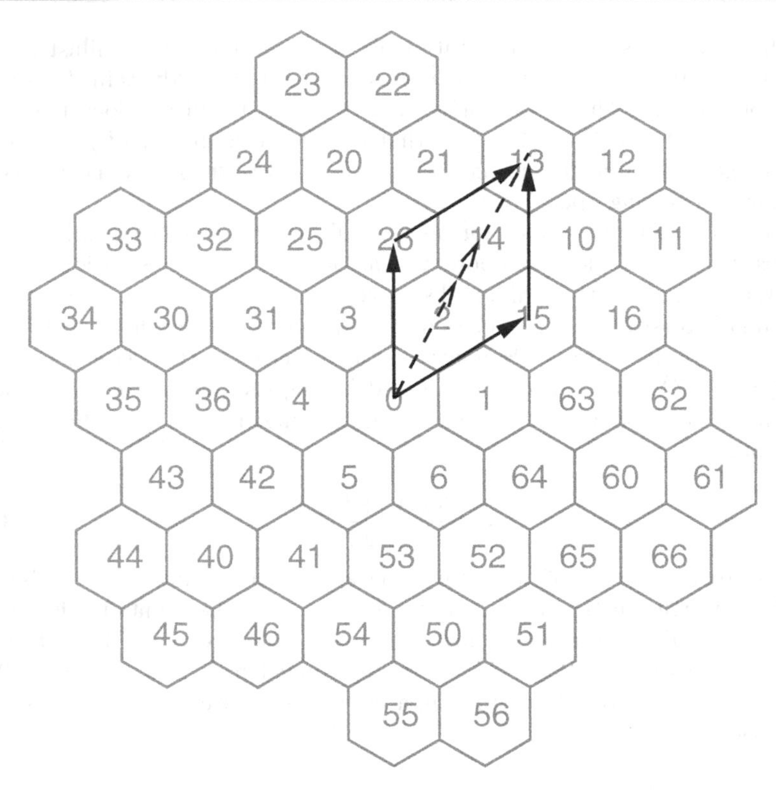

Fig. 3.12. Addition of ***26*** and ***15***.

can thus be performed in the same fashion as for ordinary arithmetic. For example, the vectorial sum of ***26*** and ***15*** is shown in Figure 3.12. This can alternatively be computed as:

$$\begin{array}{r} \boldsymbol{5} \\ \boldsymbol{2\ 6} \\ \oplus\ \boldsymbol{1\ 5} \\ \hline \boldsymbol{1\ 3} \\ \hline \end{array}$$

The operation proceeds as follows. First, the ***6*** and ***5*** are added to obtain ***53***. The ***3*** is written down and the ***5*** is carried. The ***5*** is then added to the ***2*** producing ***0***. This is then added to the ***1*** producing ***1***, which is also written down. The result is thus ***13***. This example is interesting as the ***5*** and ***2*** cancel each other out resulting in a net sum of ***0***. This is logical given that the addresses ***5*** and ***2*** correspond to vectors that are 180° apart.

The mathematical system described by HIP addressing and HIP addition, $(\mathbb{G}, \oplus)$, is a group as it satisfies the set of axioms given below:

A1. *Closure:* $\forall \boldsymbol{a}, \boldsymbol{b} \in \mathbb{G}$ applying the binary operation $\oplus$ yields $\boldsymbol{a} \oplus \boldsymbol{b} \in \mathbb{G}$.

Table 3.1. Addition table for HIP addresses.

⊕	*0*	*1*	*2*	*3*	*4*	*5*	*6*
0	*0*	*1*	*2*	*3*	*4*	*5*	*6*
1	*1*	*63*	*15*	*2*	*0*	*6*	*64*
2	*2*	*15*	*14*	*26*	*3*	*0*	*1*
3	*3*	*2*	*26*	*25*	*31*	*4*	*0*
4	*4*	*0*	*3*	*31*	*36*	*42*	*5*
5	*5*	*6*	*0*	*4*	*42*	*41*	*53*
6	*6*	*64*	*1*	*0*	*5*	*53*	*52*

A2. *Associative:* $\forall \boldsymbol{a}, \boldsymbol{b}, \boldsymbol{c} \in \mathbb{G}$ the following relationship holds $(\boldsymbol{a} \oplus \boldsymbol{b}) \oplus \boldsymbol{c} = \boldsymbol{a} \oplus (\boldsymbol{b} \oplus \boldsymbol{c})$.
A3. *Identity:* There is an element $\boldsymbol{e} \in \mathbb{G}$ such that, for every $\boldsymbol{a} \in \mathbb{G}, \boldsymbol{a} \oplus \boldsymbol{e} = \boldsymbol{a} = \boldsymbol{e} \oplus \boldsymbol{a}$. For HIP addition, $\boldsymbol{e} = \boldsymbol{0}$.
A4. *Inverse:* Given any $\boldsymbol{a} \in \mathbb{G}$ there exists a $\boldsymbol{b} \in \mathbb{G}$ such that $\boldsymbol{a} \oplus \boldsymbol{b} = \boldsymbol{0} = \boldsymbol{b} \oplus \boldsymbol{a}$.

Additionally, HIP addition is also commutative:

A5. *Commutative:* For all $\boldsymbol{a}, \boldsymbol{b} \in \mathbb{G}$ the following is true $\boldsymbol{a} \oplus \boldsymbol{b} = \boldsymbol{b} \oplus \boldsymbol{a}$.

Commutativity is a direct consequence of the vectorial nature of the addressing scheme. It can also be seen via the symmetry in Table 3.1. For example, the numbers ***26*** and ***15*** could be reversed in the addition example with no effect on the result. This extra property of commutativity makes $(\mathbb{G}, \oplus)$ an Abelian group.

The existence of a unique inverse for every address (axiom A4) makes it possible to define the negation of a number and thus a binary operation for subtraction, $\ominus$. To find the negation of any HIP address requires finding the negation of each digit within the address. For example the negation of ***6543210*** is ***3216540***. This can easily be validated by inspection and considering the vectorial nature of the addressing scheme. Subtraction can then proceed by adding a negated address. Repeating the previous addition example but using subtraction instead yields:

$$\begin{array}{r} \boldsymbol{2\ 6} \\ \ominus\, \boldsymbol{1\ 5} \\ \hline \boldsymbol{3\ 1} \\ \hline \end{array} \quad \longrightarrow \quad \begin{array}{r} \boldsymbol{2\ 6} \\ \oplus\, \boldsymbol{4\ 2} \\ \hline \boldsymbol{3\ 1} \\ \hline \end{array}$$

Due to the way in which it is defined, the system $(\mathbb{G}, \ominus)$ is also an Abelian group.

The addition operation (and subtraction, its inverse) examined above is in fact complex addition. The set of complex numbers $\mathbb{C}$ is an additive group, i.e., $(\mathbb{C}, +)$ is a group. Any lattice is a subset of the set of complex numbers, and it is well known that the lattice is an additive subgroup of complex numbers.

Since HIP addressing is just a particular form of representation of the points on a hexagonal lattice, the results obtained above are not surprising.

The next fundamental operation to examine is that of multiplication. Again this can be derived using the vectorial nature of the HIP addressing scheme. In this case it is more convenient to convert the address into a polar equivalent:

$$\begin{aligned} \boldsymbol{a} \in \mathbb{G}, \qquad & \boldsymbol{a} \to (r_a, \theta_a) \\ \boldsymbol{b} \in \mathbb{G}, \qquad & \boldsymbol{b} \to (r_b, \theta_b) \\ \boldsymbol{a} \otimes \boldsymbol{b} &= r_a r_b e^{\jmath(\theta_a + \theta_b)} \end{aligned}$$

Thus, the effect of multiplication of an address by another is to rotate and scale the original address. An example of the multiplication of *2* and *14* is illustrated in Figure 3.13. Note that although the original multiplicands are oriented in the same direction, the result is in another direction. As with addition, a general rule for multiplication can be found by computing all possible pairs of multiplications for single digit HIP addresses. The result is illustrated in Table 3.2 and its derivation is given in detail in Appendix B. Unlike addition, there are no carries in the multiplication table. This is due to the consistent connectivity of points in a hexagonal lattice. Specifically, all the points, by virtue of being in the same layer, are at a distance of 1 unit from the origin. Thus, the effect of multiplication is only to rotate the points in the anticlockwise direction.

Long multiplication can be carried out in a fashion similar to normal multiplication with the exception that HIP arithmetic is used for both the multiplication and addition steps. The example given in Figure 3.13 can also be computed as:

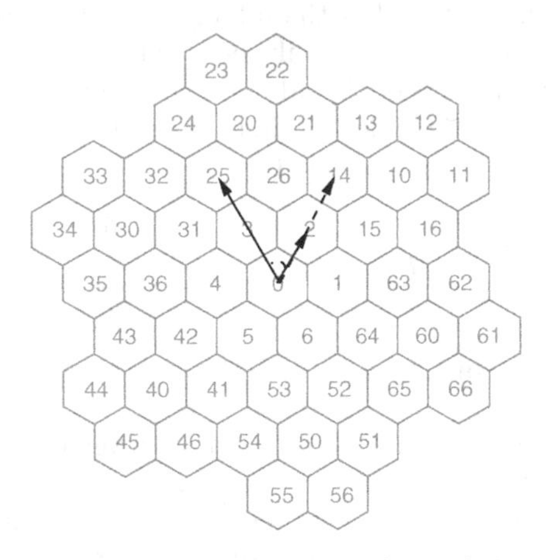

Fig. 3.13. Multiplication of *2* and *14*.

Table 3.2. Multiplication table for HIP addresses.

⊗	*0*	*1*	*2*	*3*	*4*	*5*	*6*
0	*0*	*0*	*0*	*0*	*0*	*0*	*0*
1	*0*	*1*	*2*	*3*	*4*	*5*	*6*
2	*0*	*2*	*3*	*4*	*5*	*6*	*1*
3	*0*	*3*	*4*	*5*	*6*	*1*	*2*
4	*0*	*4*	*5*	*6*	*1*	*2*	*3*
5	*0*	*5*	*6*	*1*	*2*	*3*	*4*
6	*0*	*6*	*1*	*2*	*3*	*4*	*5*

$$
\begin{array}{r}
\boldsymbol{14} \\
\otimes \quad \boldsymbol{2} \\
\hline
\boldsymbol{25} \\
\hline
\end{array}
$$

Examination of the physical meaning of this example shows that an address can be rotated (by 60°) via multiplication by ***2***. A more complicated example that illustrates both a scaling and rotation of the address is:

$$
\begin{array}{r}
\boldsymbol{21} \\
\otimes \quad \boldsymbol{44} \\
\hline
{}^{4}\boldsymbol{54} \\
\boldsymbol{540} \\
\hline
\boldsymbol{4224} \\
\hline
\end{array}
$$

In this example, the small superscript ***4*** indicates a carry caused by the operation of HIP addition. Multiplication by a integer scalar, or non HIP address, is permissible as a scalar multiplication is equivalent to:

$$\forall \boldsymbol{a} \in \mathbb{G}, k \in \mathbb{Z}, \qquad k \otimes \boldsymbol{a} = \boldsymbol{a} \oplus \cdots \oplus \boldsymbol{a}$$

This leads to a general definition for scalar multiplication, namely:

$$\forall \boldsymbol{a} \in \mathbb{G}, k \in \mathbb{Z}, \qquad k \otimes \boldsymbol{a} = \left(\sum_{k=0}^{k-1} \boldsymbol{1} \right) \otimes \boldsymbol{a}$$

The expression in the brackets replaces the integer scale quantity. It is equivalent to a HIP address with a magnitude corresponding to the scalar, but with an angle of zero. All such points lie on a horizontal line that runs through points ***1*** and ***0***. For the integer scalar k, the effect of scalar multiplication is to produce a new address which is k times further away from the origin than the original address.

Next we turn to the question of the inverse of multiplication. The requirement for an inverse for multiplication is as follows: given any $\boldsymbol{a} \in \mathbb{G}$ there should exist a $\boldsymbol{b} \in \mathbb{G}$ such that $\boldsymbol{a} \otimes \boldsymbol{b} = \boldsymbol{1} = \boldsymbol{b} \otimes \boldsymbol{a}$. From Table 3.2, we see

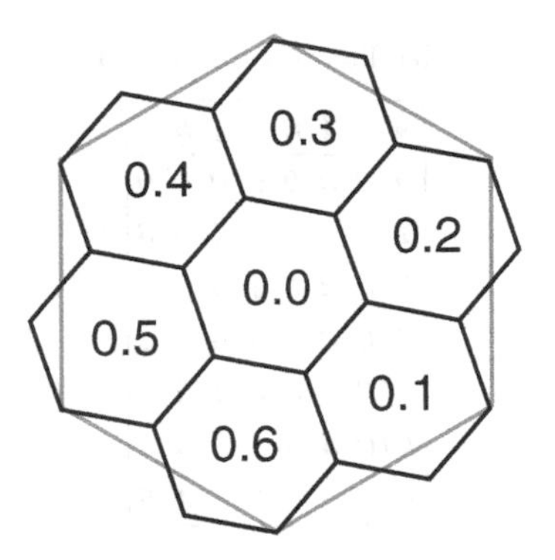

Fig. 3.14. Illustration of fractional addresses.

such an inverse exists for all numbers except the ***0***. For example, ***6*** and ***2*** are inverses of each other, while ***1*** and ***4*** are self inverses.The preceding discussion on inverse of multiplication signals a difficulty in defining the division operation. If we use the complex division notion, using polar coordinates we have:

$$\begin{aligned} \boldsymbol{a} \in \mathbb{G}, \qquad & \boldsymbol{a} \rightarrow (r_a, \theta_a) \\ \boldsymbol{b} \in \mathbb{G}, \qquad & \boldsymbol{b} \rightarrow (r_b, \theta_b) \\ \boldsymbol{a} \oslash \boldsymbol{b} &= \frac{r_a}{r_b} e^{\jmath(\theta_a - \theta_b)} \end{aligned}$$

It is quite possible that this ratio $\frac{r_a}{r_b}$ will not be an integer. This means that the result of division may yield a point which is not on the hexagonal lattice and hence has no HIP address. One solution to this problem is cell subdivision and extension of the addressing scheme to include fractional addresses. These fractional addresses are analogous to the integer addresses but each subdivision reduces the size of the cell by $\sqrt{7}$ and rotates by $\tan^{-1}\frac{\sqrt{3}}{2}$ clockwise. An example of fractional addressing is given in Figure 3.14. For two HIP addresses $\boldsymbol{g_1}$ and $\boldsymbol{g_2}$, such that $\boldsymbol{g_2} > 0$, division can be computed as follows:

$$\begin{aligned} \boldsymbol{g_1} \oslash \boldsymbol{g_2} &= \boldsymbol{g_1} \otimes \boldsymbol{g_2}^{-1} \\ &= \boldsymbol{g_1} \otimes (\boldsymbol{1} \oslash \boldsymbol{g_2}) \end{aligned}$$

The bracketed expression is the inverse of the HIP address as described earlier. It can be computed using long division with the exception that HIP arithmetic operators are employed. For example, to compute $\frac{\boldsymbol{1}}{\boldsymbol{14}}$:

$$\begin{array}{r|l} & \mathit{0.1}\cdots \\ \hline \mathit{14} & \mathit{1\ 0} \\ & \mathit{1\ 4} \\ \hline & \ \ \mathit{1\ 0} \end{array}$$

Placing the radix point correctly gives $\mathit{1} \oslash \mathit{14} = \mathit{0.111}\cdots$. Next, to compute $\mathit{25} \oslash \mathit{14} = \mathit{25} \otimes \mathit{0.111}\cdots$ requires a multiplication, namely:

$$\begin{array}{r} \mathit{2\ 5} \\ \otimes\ \mathit{1\ 1\ 1} \\ \hline \mathit{2\ 5} \\ \mathit{2\ 5\ 0} \\ \mathit{2\ 5\ 0\ 0} \\ \hline \mathit{2\ 0\ 0\ 5} \\ \hline \end{array}$$

Truncation of the inverse to three significant figures has resulted in a numerical error. This is why the result is $\mathit{2.005}$ rather than $\mathit{2}$. This is however, not a problem as all fractional parts of an address lie within the address itself. This means that rounding can be performed by truncation of the address to the required number of significant figures. This property is a specific feature of the HIP addressing scheme as it is a balanced ternary numbering scheme [57].

Based on the arithmetic that has been defined upon the HIP addressing scheme, it is possible to revisit the definition on an arbitrary number within the scheme as seen in equation (3.12). This requires the introduction of a few definitions:

i. $\mathit{10}^n = \mathit{10} \otimes \cdots \otimes \mathit{10}$ or address $\mathit{1}$ followed by n zeros.
ii. $\forall \boldsymbol{m} \in \mathbb{G}^1, \boldsymbol{m} \otimes \mathit{10} = \boldsymbol{m}\mathit{0}$
iii. $\forall \boldsymbol{m}, \boldsymbol{n} \in \mathbb{G}^1, \boldsymbol{m}\mathit{0} \oplus \boldsymbol{n} = \boldsymbol{mn}$

Definition (i) can be validated by performing repeated multiplication by $\mathit{10}$. The second and third definitions are a consequence of $\mathit{1}$ being the identity for multiplication and $\mathit{0}$ being the identity for addition. We can generalise this and say that for an address $\boldsymbol{g}$ which belongs to $\mathbb{G}^\lambda$, the following holds:

$$\begin{aligned} \boldsymbol{g} &= g_{\lambda-1} \cdots g_2 g_1 g_0 \\ &= \boldsymbol{g_{\lambda-1}} \otimes \mathit{10}^{\lambda-1} \oplus \cdots \oplus \boldsymbol{g_2} \otimes \mathit{10}^2 \oplus \boldsymbol{g_1} \otimes \mathit{10} \oplus \boldsymbol{g_0} \end{aligned} \tag{3.14}$$

3.2.3 Closed arithmetic

It is clear from the above discussion that given a λ-level hexagonal image, some of the arithmetic operations on it will result in an address outside the image. A solution to this is to make the image closed by using closed arithmetic. Such an approach was first explored by Sheridan [58] and is explored next. Closure requires the arithmetic operations defined in Section 3.2.2 to be modified using a modulo-n operation. The modulo-n operation yields the remainder after division by n. As an example, 10 mod 7 is 3. Hence, in order to keep a λ-level hexagonal image closed under arithmetic operations, the use of modulo-7^λ is required. For example, if in a two-level hexagonal image, the arithmetic results in an address ***342*** then ***342*** mod ***100*** = ***42*** will give a valid address. It is now possible to redefine the arithmetic operators using the modulo operator. These will be examined in the same order as in Section 3.2.2.

Addition of two HIP addresses using the modulo operator can be defined as:

$$\boldsymbol{a} \oplus_\lambda \boldsymbol{b} = (\boldsymbol{a} \oplus \boldsymbol{b}) \mod \boldsymbol{10}^\lambda \tag{3.15}$$

As subtraction is defined as the negation of the addition operator then the following also holds:

$$\boldsymbol{a} \ominus_\lambda \boldsymbol{b} = (\boldsymbol{a} \ominus \boldsymbol{b}) \mod \boldsymbol{10}^\lambda \tag{3.16}$$

The closure of both $\oplus_\lambda$ and $\ominus_\lambda$ can be ascertained by observing the result of varying $\boldsymbol{a}$ from $\boldsymbol{0}$ and $\boldsymbol{10}^{\lambda-1}$ while keeping $\boldsymbol{b}$ constant. This will visit every point in the set $\mathbb{G}^\lambda$ once and once only though not necessarily in numerical order. As an example:

$$\boldsymbol{11} \oplus_2 \boldsymbol{15} = \boldsymbol{36} \qquad \boldsymbol{14} \oplus_2 \boldsymbol{15} = \boldsymbol{12}$$
$$\boldsymbol{12} \oplus_2 \boldsymbol{15} = \boldsymbol{30} \qquad \boldsymbol{15} \oplus_2 \boldsymbol{15} = \boldsymbol{11}$$
$$\boldsymbol{13} \oplus_2 \boldsymbol{15} = \boldsymbol{34} \qquad \boldsymbol{16} \oplus_2 \boldsymbol{15} = \boldsymbol{43}$$

Obviously, the repeat length for the modulo operation is 7^λ. The properties of commutativity and associativity are preserved under closed addition and subtraction. The effect of the closed addition operator is illustrated in Figure 3.15. There is no need to illustrate the closed subtraction operator as it

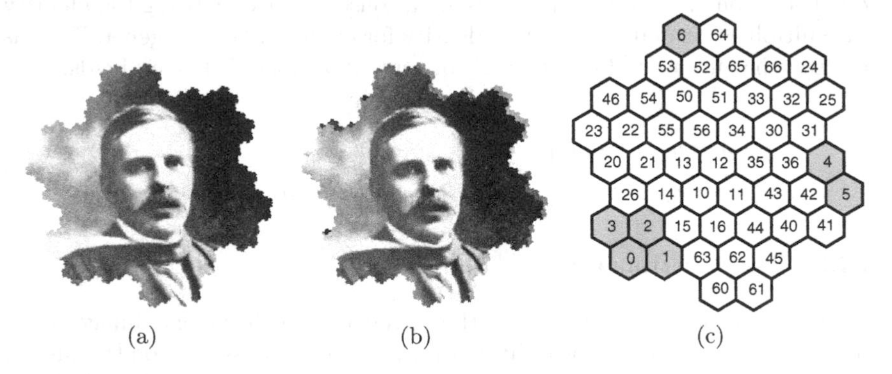

Fig. 3.15. The effect of the $\oplus_\lambda$ operator: (a) original image (© The Nobel Foundation) (b) image with ***12*** added to the addresses within it (c) addresses with ***12*** added to them.

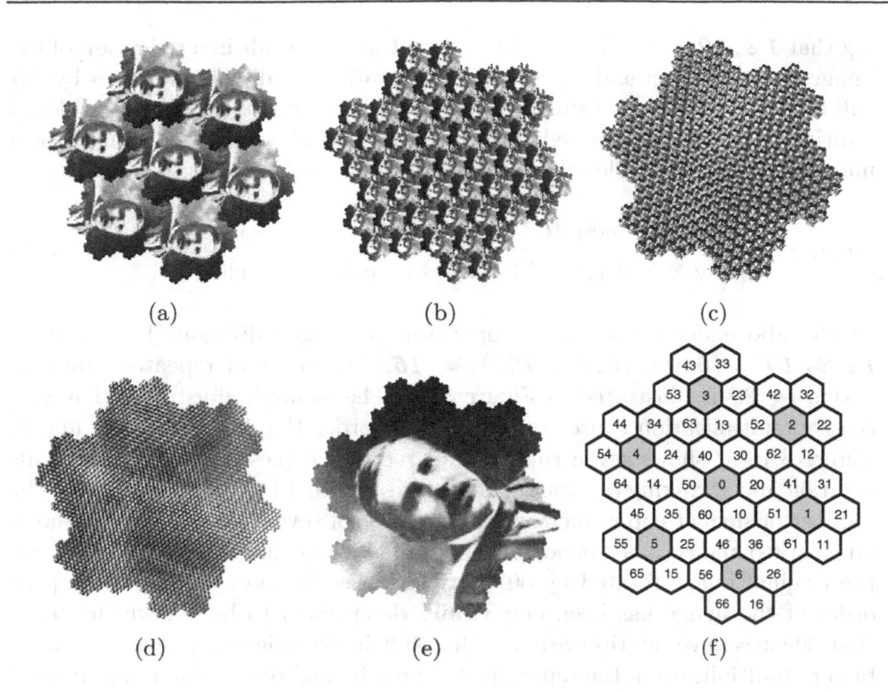

Fig. 3.16. The effect of the $\circledast_\lambda$ operator: Multiplication (a) by ***20*** (b) by ***20***^{2} (c) by ***20***^{3} (d) by ***20***^{4} (e) by ***20***^{5} (f) addresses in a 2-layer image after multiplication by ***20***.

can be performed using negation and closed addition. In the figure, the image in the middle shows the effect of adding ***12*** to all the addresses in the original image and re-plotting the result. Notice that the image appears to have been shifted downwards and to the left. Specifically, the centre of the image has translated by $(\frac{5}{2}, \frac{3\sqrt{3}}{2})$, which is the Cartesian coordinate corresponding to ***12***. Further, some of the information can be seen to wrap from the right to the left side of the image. This is a consequence of the modulo operator. Figure 3.15(c) shows how the addresses of a smaller two-level image are redistributed under the same addition operation. This again illustrates the translation effect. The fringing effect is seen with the addresses particularly ***0*** to ***6*** being spread across several sides of the hexagonal image. This operation is extremely useful as it allows the data to be translated with a simple manipulation of the addresses.

Multiplication can also be redefined using the modulo operator. HIP multiplication results in both a scaling and a rotation of the address. The scaling is the primary cause of the address going outside the current layer. In contrast to closed addition, closed multiplication is more complex. For example, multiplication by multiples of ***10*** will result in decimation of the number of addresses if the modulo operation is applied naïvely. This can seen by observing that $\boldsymbol{1} \otimes_2 \boldsymbol{10} = \boldsymbol{11} \otimes_2 \boldsymbol{10} = \boldsymbol{10}$. In fact this will result in a reduction of the image data by seven and repeated multiplication of all the addresses by $\boldsymbol{10}$ will eventually result in a single point at $\boldsymbol{0}$. To avoid this problem, the closed multiplication needs to be redefined for the cases where the multiplicand is a multiple of $\boldsymbol{10}$. Thus, closed multiplication can be defined as:

$$\boldsymbol{a} \otimes_\lambda \boldsymbol{b} = \begin{cases} (\boldsymbol{a} \otimes \boldsymbol{b}) \mod \boldsymbol{10}^\lambda & \text{if } \boldsymbol{b} \neq \boldsymbol{k} \otimes \boldsymbol{10} \\ \left((\boldsymbol{a} \otimes \boldsymbol{b}) \oplus ((\boldsymbol{a} \otimes \boldsymbol{b}) \div \boldsymbol{10}^\lambda)\right) \mod \boldsymbol{10}^\lambda & \text{else} \end{cases} \tag{3.17}$$

In the above definition, the $\div$ operator is integer division. For example $\boldsymbol{61} \otimes_2 \boldsymbol{10} = (\boldsymbol{610} \oplus (\boldsymbol{610} \div \boldsymbol{100})) = \boldsymbol{16}$. The effect of repeated multiplication by $\boldsymbol{20}$ is illustrated in Figure 3.16. The example illustrates that successive multiplication causes the image to reorder the data. The first image, Figure 3.16(a) shows seven rotated and reduced copies of the original. This rotation and shrinking continues through to Figure 3.16(c). In each figure, the number of smaller copies increases by a factor of seven. Figure 3.16(d) shows an enlarged and rotated version of the original image and Figure 3.16(e) shows the original image rotated by 60° anticlockwise. In the example, the repeat order of the image has been empirically determined to be 30, which means that 30 successive multiplications will result in the original image. For an arbitrary multiplication, the repeat factor may be different, from a minimum of 5 (multiplication by $\boldsymbol{10}$) to a maximum of $7^{\lambda-1}$ (multiplication by $\boldsymbol{12}$). The repetition of the original image is guaranteed due to the closed nature of the multiplication operation. Figure 3.16(f) shows the effect of multiplication by $\boldsymbol{20}$ on the addresses for a two-layer image. This clearly illustrates that closed multiplication is nothing but down-sampling of the image with unequal offsets. Closed division is not discussed as it is analogous to closed multiplication.

3.3 Conversion to other coordinate systems

As reported in Chapter 2, three different coordinate systems have been used for hexagonal image processing. The first is the three-coordinate scheme of Her [20, 51] which has been discussed in Section 2.2.2. The next is a two-coordinate system based on skewed axes. There are various interpretations of this (see Section 2.2.2) but only the scheme defined as B_h (see Section 6.1, equation (6.3)) will be discussed. The last scheme is the standard Cartesian coordinate scheme. It is possible to convert from the HIP addressing scheme to these coordinate systems and the methodology for conversion is the same.

The HIP address for a point in a λ-layer hexagonal image can be written as:

$$\boldsymbol{g_{\lambda-1} \cdots g_2 g_1 g_0} = \boldsymbol{g_{\lambda-1}} \otimes \boldsymbol{10}^{\lambda-1} \oplus \cdots \oplus \boldsymbol{g_2} \otimes \boldsymbol{10}^2 \oplus \boldsymbol{g_1} \otimes \boldsymbol{10} \oplus \boldsymbol{g_0} \tag{3.18}$$

Table 3.3. Equivalent rotation matrices for multiplication by various HIP indices.

index	angle	f_2	f_3
0	–	$\begin{bmatrix} 0 & 0 \\ 0 & 0 \end{bmatrix}$	$\begin{bmatrix} 0 & 0 & 0 \\ 0 & 0 & 0 \\ 0 & 0 & 0 \end{bmatrix}$
1	0°	$\begin{bmatrix} 1 & 0 \\ 0 & 1 \end{bmatrix}$	$\begin{bmatrix} 1 & 0 & 0 \\ 0 & 1 & 0 \\ 0 & 0 & 1 \end{bmatrix}$
2	60°	$\begin{bmatrix} 1 & -1 \\ 1 & 0 \end{bmatrix}$	$\begin{bmatrix} 0 & -1 & 0 \\ 0 & 0 & -1 \\ -1 & 0 & 0 \end{bmatrix}$
3	120°	$\begin{bmatrix} 0 & -1 \\ 1 & -1 \end{bmatrix}$	$\begin{bmatrix} 0 & 0 & 1 \\ 1 & 0 & 0 \\ 0 & 1 & 0 \end{bmatrix}$
4	180°	$\begin{bmatrix} -1 & 0 \\ 0 & -1 \end{bmatrix}$	$\begin{bmatrix} -1 & 0 & 0 \\ 0 & -1 & 0 \\ 0 & 0 & -1 \end{bmatrix}$
5	240°	$\begin{bmatrix} -1 & 1 \\ -1 & 0 \end{bmatrix}$	$\begin{bmatrix} 0 & 1 & 0 \\ 0 & 0 & 1 \\ 1 & 0 & 0 \end{bmatrix}$
6	300°	$\begin{bmatrix} 0 & 1 \\ -1 & 1 \end{bmatrix}$	$\begin{bmatrix} 0 & 0 & -1 \\ -1 & 0 & 0 \\ 0 & -1 & 0 \end{bmatrix}$

The above equation shows that the conversion process can be carried out by $\lambda - 1$ additions using the HIP addition operator. Each term will consist of a multiplication of one digit of the original HIP address and a power of the address ***10***. Since HIP multiplication typically results in a rotation and scaling of the address, and the required multiplications will be by a single digit (***0*** to ***6***) the result will simply be a rotation. This is always true with the exception that a zero digit will result in the term being ignored. Thus to map the HIP address to a different coordinate system requires finding two matrices. The first is a matrix that corresponds to $\mathbf{10}^\lambda$ and the second is a matrix corresponding to the rotation due to a digit within the index. For the B_h coordinate scheme $\mathbf{10}^\lambda$ can be replaced by:

$$\mathbf{10}^\lambda \equiv \begin{bmatrix} 3 & -2 \\ 2 & 1 \end{bmatrix}^\lambda$$

For Her it can be replaced by:

$$\mathbf{10}^\lambda \equiv \begin{bmatrix} 3 & 0 & 2 \\ 2 & 3 & 0 \\ 0 & 2 & 3 \end{bmatrix}^\lambda$$

The rotations can be defined by observing how much rotation is due to multiplication by a particular HIP digit and devising an equivalent rotation matrix. The cases for both Her and B_h are illustrated in Table 3.3. Notice the large degree of symmetry in this table.

It is now possible to define a couple of mappings which perform the same function as Table 3.3. For the two-coordinate scheme, the mapping function is $f_2 : \mathbb{G}^1 \to \mathbb{Z}^{2\times 2}$ and, for Her's three-coordinate scheme, the mapping function is $f_3 : \mathbb{G}^1 \to \mathbb{Z}^{3\times 3}$. Thus the coordinates of a HIP address in terms of the basis B_h are:

$$\begin{bmatrix} b_1 \\ b_2 \end{bmatrix} = \sum_{i=0}^{\lambda-1} f_2(g_i) \begin{bmatrix} 3 & -2 \\ 2 & 1 \end{bmatrix}^i \begin{bmatrix} 1 \\ 0 \end{bmatrix} \tag{3.19}$$

For Her's three-coordinate scheme, the coordinates can be written as:

$$\begin{bmatrix} b_1 \\ b_2 \\ b_3 \end{bmatrix} = \sum_{i=0}^{\lambda-1} f_3(g_i) \begin{bmatrix} 3 & 0 & 2 \\ 2 & 3 & 0 \\ 0 & 2 & 3 \end{bmatrix}^i \begin{bmatrix} 1 \\ 0 \\ -1 \end{bmatrix} \tag{3.20}$$

These two equations produce a unique set of coordinates for an arbitrary HIP address. To convert from either of these coordinate schemes to Cartesian coordinates requires a further multiplication by a matrix. For the two-coordinate scheme, this matrix is:

$$C_{2e} = \frac{1}{2} \begin{bmatrix} 2 & -1 \\ 0 & \sqrt{3} \end{bmatrix}$$

The corresponding conversion matrix for the three-coordinate scheme is:

$$C_{3e} = \frac{1}{2\sqrt{3}} \begin{bmatrix} \sqrt{3} & 0 & -\sqrt{3} \\ -1 & 2 & -1 \end{bmatrix}$$

3.4 Processing

We now turn to examining how hexagonally sampled images can be processed. Specifically, we describe how the HIP addressing scheme can be exploited to perform simple image processing operations. Before delving into specific spatial or frequency domain operations however, it is essential to examine some mathematical preliminaries.

3.4.1 Boundary and external points

The addressing scheme, outlined in Section 3.2 covers an infinite space permitting addressing of all points on a hexagonal lattice defined in the Euclidean plane. Images however, are finite in extent. Hence, it is necessary to consider

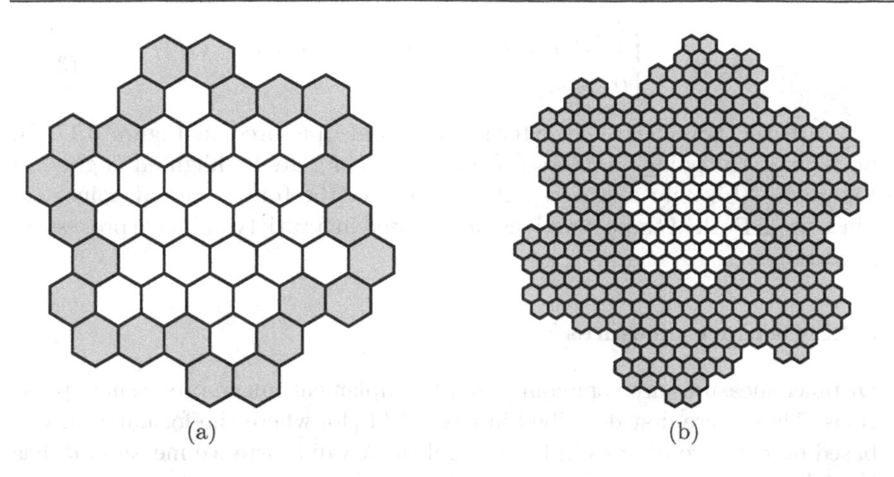

Fig. 3.17. Illustration of (a) boundary points and (b) external points for a 2-layer hexagonal image.

a hexagonal image of a finite number of layers when considering image processing applications. Pixels in such an image will have addresses belonging to the set $\mathbb{G}^\lambda$ for a λ-level image. However, given an image of finite extent, operations often result in pixel addresses that lie outside the original image. There are two ways to deal with this problem. The first is to use closed operations as described in Section 3.2.3 and the second is to treat such pixels as boundary pixels. The latter will now be discussed.

For a given λ-level hexagonal image there are a total of 7^λ pixels which are numbered from 0 to $7^\lambda - 1$. External pixels start with 7^λ and extend to infinity. Any external pixel can be considered, somewhat naively, to lie on the boundary of the image. For an address, $\boldsymbol{x}$, to determine if a pixel is external to the image can be found via a simple comparison:

$$e_\lambda(\boldsymbol{x}) = \begin{cases} 0 & \text{if } 0 \leq \boldsymbol{x} < (7^\lambda)_7 \\ 1 & \text{if } \boldsymbol{x} \geq (7^\lambda)_7 \end{cases} \tag{3.21}$$

In the above, the term $(7^\lambda)_7$ implies that the number is computed using radix 7. For instance, a two level HIP image gives a value of $\boldsymbol{100} = 100_7$ as the threshold. Simply stated, this number is one *1* followed by λ zeros for a λ-level image. The result of the $e_\lambda(\boldsymbol{x})$ is 1 if the point is external and 0 otherwise. The external points are not the best method to define the image's boundary. However, by contemplation of e_λ it is possible to define the boundary points of the image as follows. The boundary points are points for which one of the surrounding six points are external to the image. For an address $\boldsymbol{x}$ the decision as to whether it is a boundary point is:

$$b_\lambda(\boldsymbol{x}) = \begin{cases} 1 & \text{if } \boldsymbol{x} \oplus \boldsymbol{1} \cup \boldsymbol{x} \oplus \boldsymbol{2} \cup \cdots \cup \boldsymbol{x} \oplus \boldsymbol{6} \geq (7^\lambda)_7 \\ 0 & \text{else} \end{cases} \tag{3.22}$$

Both the boundary and external points are pictured in Figure 3.17. In both cases, the points of interest, $b_\lambda(\boldsymbol{x})$ and $e_\lambda(\boldsymbol{x})$, are highlighted in gray. In Figure 3.17(b), only a representative subset of the total external points are illustrated. Both of these functions can be used in a variety of image processing operations.

3.4.2 Distance measures

Distance measures are commonly used to implement image processing operations. These were first described in Rosenfeld [46] where the formulation was based on distance metrics in linear algebra. A valid distance measure, d, has the following requirements:

i. *Positive Definiteness:* The distance between two HIP addresses $d(\boldsymbol{a}, \boldsymbol{b}) \geq 0$. Equality only holds if $\boldsymbol{a} = \boldsymbol{b}$.
ii. *Symmetry:* For all $\boldsymbol{a}, \boldsymbol{b} \in \mathbb{G}$ then $d(\boldsymbol{a}, \boldsymbol{b}) = d(\boldsymbol{b}, \boldsymbol{a})$.
iii. *Triangle Inequality:* For all $\boldsymbol{a}, \boldsymbol{b}, \boldsymbol{c} \in \mathbb{G}$ then $d(\boldsymbol{a}, \boldsymbol{b}) \leq d(\boldsymbol{a}, \boldsymbol{c}) + d(\boldsymbol{c}, \boldsymbol{b})$.

In the HIP framework these are fulfilled due to the vector-like nature of the addresses. Distances are usually defined in terms of the p-norm of two vectors, $\mathbf{a}, \mathbf{b} \in \mathbb{R}^n$. This can be defined as:

$$\|\mathbf{a} - \mathbf{b}\|_p = \left(\sum_{i=0}^{n-1} |a_i - b_i|^p \right)^{\frac{1}{p}} \tag{3.23}$$

In the above a_i and b_i are the components of the individual vectors. For the special case when $p \to \infty$ the above reduces to:

$$\|\mathbf{a} - \mathbf{b}\|_\infty = \max |a_i - b_i| \tag{3.24}$$

Usually, distance measures are defined with respect to a set of axes. Since the HIP addressing scheme uses positional numbering it is difficult to define a valid distance measure directly. Distance measures will instead be defined with respect to other coordinate schemes. The first distance measure that will be examined, is in terms of Her's 3-coordinate scheme. Assume that the coordinates of two HIP addresses, $\boldsymbol{a}$ and $\boldsymbol{b}$, have been converted into coordinates (a_1, a_2, a_3) and (b_1, b_2, b_3) respectively. These coordinates can be used to compute the 1-norm to be:

$$d_1(\boldsymbol{a}, \boldsymbol{b}) = |a_1 - b_1| + |a_2 - b_2| + |a_3 - b_3| \tag{3.25}$$

This metric is also known in the literature [46] as the Manhattan or city block distance. The minimum non-zero value of this measure is two so it is

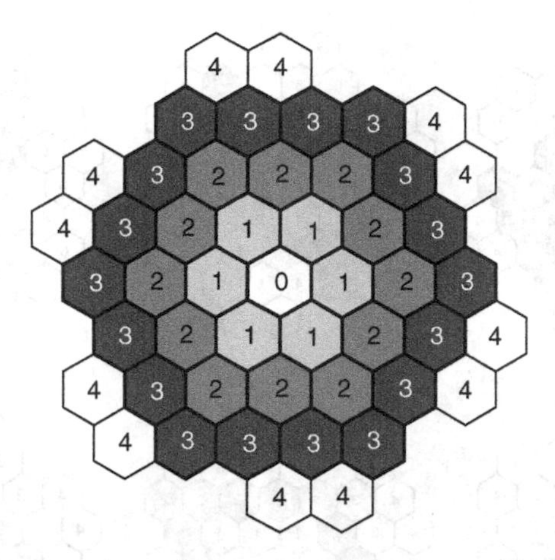

Fig. 3.18. The distance from the centre measured using Her's symmetric axes.

possible to introduce another measure $d_1' = \frac{1}{2}d_1$. This is still a valid distance measure. The ∞-norm is:

$$d_2(\boldsymbol{a}, \boldsymbol{b}) = \max(|a_1 - b_1|, |a_2 - b_2|, |a_3 - b_3|) \tag{3.26}$$

This is the familiar chessboard distance. Due to the redundancy in the original coordinate system it turns out that this distance measure gives the same result as d_1'. This is because Her's coordinate scheme has as a requirement that the elements of an individual coordinate sum to 0. Figure 3.18 illustrates the metrics d_1' and d_2 superimposed on a second layer HIP data structure. In the figure, different colours represent hexagonal cells with equivalent distances from the origin. Notice that the shape of the region is hexagonal. This is primarily due to the symmetric nature of using three coordinates to represent points in the hexagonal lattice.

It is also possible to define distance measures based on the different skewed coordinate systems (see Figure 2.8 in Section 2.2.2). For the skewed axes, there are three unique possibilities: (i) one axis aligned with the horizontal and the second at 120° to the first, (ii) one axis aligned with the horizontal and the second axis at 60° to the first, and finally (iii) one at 60° to the horizontal and the second axis at 120° to the horizontal. Both the norms described above can be expressed relative to each of these three skewed axes. Taking them in order and computing the city block distance, gives:

$$d_3(\boldsymbol{a}, \boldsymbol{b}) = |b_3 - a_3| + |a_2 - b_2| \tag{3.27}$$

$$d_4(\boldsymbol{a}, \boldsymbol{b}) = |b_2 - a_2| + |a_1 - b_1| \tag{3.28}$$

$$d_5(\boldsymbol{a}, \boldsymbol{b}) = |b_3 - a_3| + |a_1 - b_1| \tag{3.29}$$

Using the ∞-norm, the distance metrics on each of the three skewed axes become:

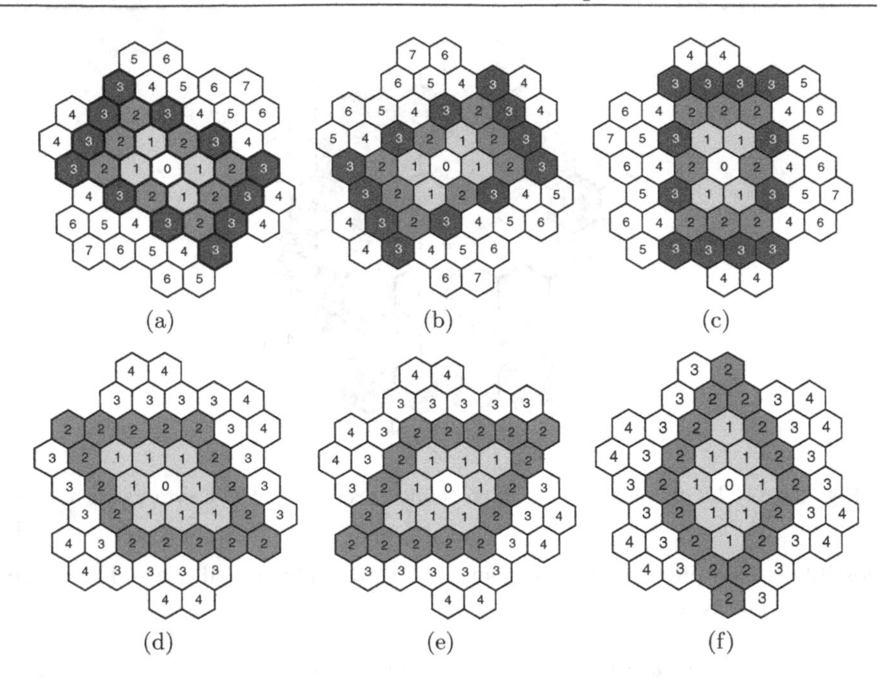

Fig. 3.19. Different distance measures using skewed axes: (a) d_3 (b) d_4 (c) d_5 (d) d_6 (e) d_7 (f) d_8.

$$d_6(\boldsymbol{a}, \boldsymbol{b}) = \max(|b_3 - a_3|, |a_2 - b_2|) \tag{3.30}$$

$$d_7(\boldsymbol{a}, \boldsymbol{b}) = \max(|b_2 - a_2|, |a_1 - b_1|) \tag{3.31}$$

$$d_8(\boldsymbol{a}, \boldsymbol{b}) = \max(|b_3 - a_3|, |a_1 - b_1|) \tag{3.32}$$

Figure 3.19 illustrates the various distance measures on two skewed axes overlaid upon a second level HIP structure. Different shades are used to represent regions which have equivalent distances from the structure's centre. The three distance measures computed based upon the 1-norm give rectangular shapes whereas the ones based upon the ∞-norm give rhombuses. There are two important features that can be observed in these figures. The first is that in all cases, the obtained shape, whether rectangular or rhombus, is aligned to the skewed axes. The second is that for a class of shape, the three different distance measures are just rotations of one another.

Another commonly used distance function is the Euclidean distance. This is the distance expressed in terms of Cartesian coordinates. The coordinates of a HIP address can easily be converted to a pair of Cartesian coordinates. Thus for two HIP addresses $\boldsymbol{a}$ and $\boldsymbol{b}$, with Cartesian coordinates (x_a, y_a) and (x_b, y_b) respectively, the Euclidean distance is defined as:

$$d_9(\boldsymbol{a}, \boldsymbol{b}) = \sqrt{(x_a - x_b)^2 + (y'_a - y'_b)^2} \tag{3.33}$$

This distance measure will yield a shape that is roughly circular in nature. All of the distance measures that have been mentioned require conversion from HIP addresses to some other coordinate scheme.

3.4.3 HIP neighbourhood definitions

Neighbourhood of a pixel is one of the basic relationships of interest in image processing, and it is used in many processing techniques. Two cells in a lattice can be regarded as neighbours when they have a common edge or corner. Examination of a hexagonally tiled plane shows that each hexagonal cell only has neighbours with a common edge and that for each cell there are six such neighbours. Each of these neighbours is equidistant from the central cell. This collection of seven hexagons (the central plus six neighbours) is the smallest single neighbourhood that can be defined for a hexagonal lattice. This smallest neighbourhood is analogous to the first level HIP structure. Consequently, the addresses of the individual cells are given by $\mathbb{G}^1$. To find the nearest neighbourhood of an arbitrary point, $\boldsymbol{x}$, then just requires the HIP addresses of the points in $\mathbb{G}^1$ to be added to $\boldsymbol{x}$.

It is possible to define different neighbourhood shapes. The most intuitive one is analogous to the case for square neighbourhoods. This can be defined by walking at most a fixed distance, using distance measure d_1', from the central tile. In this case the first neighbourhood N_1 of a point $\boldsymbol{x}$ can be defined to be the set of addresses as follows:

$$N_1(\boldsymbol{x}) = \{\boldsymbol{x}, \boldsymbol{x} \oplus \boldsymbol{1}, \boldsymbol{x} \oplus \boldsymbol{2}, \cdots, \boldsymbol{x} \oplus \boldsymbol{6}\} \qquad \boldsymbol{x} \in \mathbb{G}, N_1 \subset \mathbb{G} \tag{3.34}$$

This neighbourhood is a set of seven points. The second neighbourhood can be defined using a variety of alternatives. The easiest definition, however, is to use the already existing N_1 definition. The second neighbourhood is then the set of unique points in N_1 together with its neighbours. Thus:

$$\begin{aligned} N_2(\boldsymbol{x}) =& N_1(\boldsymbol{x}) \cup N_1(\boldsymbol{x} \oplus \boldsymbol{1}) \cup N_1(\boldsymbol{x} \oplus \boldsymbol{2}) \cup N_1(\boldsymbol{x} \oplus \boldsymbol{3}) \cup N_1(\boldsymbol{x} \oplus \boldsymbol{4}) \\ & \cup N_1(\boldsymbol{x} \oplus \boldsymbol{5}) \cup N_1(\boldsymbol{x} \oplus \boldsymbol{6}) \end{aligned} \tag{3.35}$$

If $\boldsymbol{x}$ is defined to be the origin, this defines a set of points:

$$N_2(\boldsymbol{0}) = \{\boldsymbol{0}, \boldsymbol{1}, \boldsymbol{2}, \boldsymbol{3}, \boldsymbol{4}, \boldsymbol{5}, \boldsymbol{6}, \boldsymbol{14}, \boldsymbol{15}, \boldsymbol{25}, \boldsymbol{26}, \boldsymbol{36}, \boldsymbol{31}, \boldsymbol{41}, \boldsymbol{42}, \boldsymbol{52}, \boldsymbol{53}, \boldsymbol{63}, \boldsymbol{64}\}$$

N_2 contains 19 points which includes the original point, the six immediate neighbouring points (N_1), and 12 further neighbours which are the unique neighbours of the N_1 points. Generally, an arbitrary order neighbourhood can be recursively defined to be:

$$N_n(\boldsymbol{x}) = N_{n-1}(\boldsymbol{x}) \cup N_{n-1}(\boldsymbol{x} \oplus \boldsymbol{1}) \cup \cdots \cup N_{n-1}(\boldsymbol{x} \oplus \boldsymbol{6}) \tag{3.36}$$

This definition can be used without loss of generality as it is uncommon for a neighbourhood to be defined which only includes the points on the boundary. However, it is possible to find the boundary points by subtracting two such sets. Thus $N_n - N_{n-1}$ gives the new points added to the (n-1)th neighbourhood to achieve the nth neighbourhood. These additional points are the N_1 neighbourhood of the points in N_{n-1}. The cardinality, or number of elements, of the sets that correspond to the neighbourhoods is as follows:

$$\begin{aligned} \text{card}(N_1) &= 7 \\ \text{card}(N_2) &= 19 \\ &\cdots \\ \text{card}(N_n) &= 3n^2 + 3n + 1 \end{aligned}$$

The inherent aggregation of the HIP addressing scheme makes it possible to define a second neighbourhood. In the first level, this new neighbourhood, N_1^h, is identical to N_1. The second level aggregate can be defined in several ways. The first is to use the methodology described in Section 3.2. However, it is more convenient to use the idea that the neighbourhood can be defined using the existing N_1^h neighbourhood. This yields:

$$N_2^h(\boldsymbol{x}) = N_1^h(\boldsymbol{x}) \cup N_1^h(\boldsymbol{x} \oplus \boldsymbol{10}) \cup \cdots \cup N_1^h(\boldsymbol{x} \oplus \boldsymbol{60}) \tag{3.37}$$

For the origin, this neighbourhood gives the following points:

$$N_2^h(\boldsymbol{0}) = \{\boldsymbol{0}, \boldsymbol{1}, \boldsymbol{2}, \boldsymbol{3}, \boldsymbol{4}, \boldsymbol{5}, \boldsymbol{6}, \boldsymbol{10}, \boldsymbol{11}, \cdots \boldsymbol{66}\}$$

As expected, N_2^h contains exactly 49 points. This includes the original point and its six neighbours, plus the six second-level aggregate tiles consisting of 42 points. This neighbourhood can also be defined recursively:

$$N_n^h(\boldsymbol{x}) = N_{n-1}^h(\boldsymbol{x}) \cup N_{n-1}^h(\boldsymbol{x} \oplus \boldsymbol{10}^{n-1}) \cup \cdots \cup N_{n-1}^h(\boldsymbol{x} \oplus \boldsymbol{6} \otimes \boldsymbol{10}^{n-1}) \tag{3.38}$$

Since this neighbourhood is based upon the aggregation process, the number of points in a given set is the same as for the number of addresses at a particular level of aggregation, viz.:

$$\begin{aligned} \text{card}(N_1^h) &= 7 \\ \text{card}(N_2^h) &= 49 \\ &\cdots \\ \text{card}(N_n^h) &= 7^n \end{aligned}$$

The two different neighbourhood definitions are visually compared in Figure 3.20. The immediate difference is that the first neighbourhood definition is hexagonal in shape whilst the neighbourhood definition based on aggregates has a unusual snowflake like shape. Both cases approximate a circular

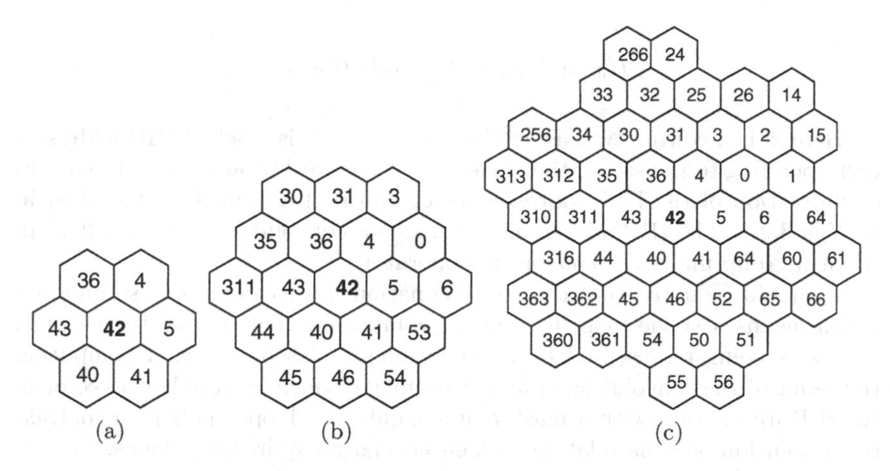

Fig. 3.20. Examples of the different neighbourhood definitions for the point *42* a) $N_1 = N_1^h$ b) N_2 c) N_2^h.

shape well. Due to the uniform connectivity of the hexagonal lattice, it should be expected that circular neighbourhoods should be implemented easily and accurately using either of these neighbourhoods.

It is possible to exploit different distance measures to generate alternative neighbourhoods. These neighbourhoods have shapes like those in Figure 3.19. The definition process is similar to that given for N_i. Using distance measure d_3 (equation (3.27)) as an example, the first level neighbourhood, N_1^r, about an arbitrary HIP address $\boldsymbol{x}$ is:

$$N_1^r(\boldsymbol{x}) = \{\boldsymbol{x}, \boldsymbol{x} \oplus \boldsymbol{1}, \boldsymbol{x} \oplus \boldsymbol{3}, \boldsymbol{x} \oplus \boldsymbol{4}, \boldsymbol{x} \oplus \boldsymbol{6}\}$$

This can also be used to define a neighbourhood recursively as follows:

$$N_n^r(\boldsymbol{x}) = N_{n-1}^r(\boldsymbol{x}) \cup N_{n-1}^r(\boldsymbol{x} \oplus \boldsymbol{1}) \cup N_{n-1}^r(\boldsymbol{x} \oplus \boldsymbol{3}) \cup N_{n-1}^r(\boldsymbol{x} \oplus \boldsymbol{4}) \cup N_{n-1}^r(\boldsymbol{x} \oplus \boldsymbol{4})$$

The cardinality of this set is $\text{card}(N_n^r) = 2n^2+2n+1$. Using these methods, neighbourhoods can be defined for arbitrary distance measures. For instance, neighbourhoods based upon rectangles and rhombuses are often used in morphological operations.

3.4.4 Convolution

Convolution is a widely used operation in image processing. It is a neighbourhood operation where a given pixel is replaced by the weighted sum of pixels in its neighbourhood. The neighbouring pixels can be found using either of the neighbourhood definitions, N_n and N_n^h. Thus the convolution of an image I with a λ-level mask M is defined as:

$$M(\boldsymbol{x}) \circledast I(\boldsymbol{x}) = \sum_{\boldsymbol{k}\in R} M(\boldsymbol{k}) I(\boldsymbol{x} \ominus \boldsymbol{k}) \tag{3.39}$$

Here $\circledast$ is a convolution using HIP addresses. R is a set of HIP addresses corresponding to the size of the mask, M. The convolution operation requires a subtraction of the HIP addresses which can be performed as described in Section 3.2.2. The HIP addressing scheme is a convolution ring, i.e., it is an Abelian group under the convolution operator [56].

It should be noted that the above equation involves only one summation which means a single loop in terms of computing. This is due to the pixel addresses being represented by a single index. This means that computing the result of the convolution of an entire image, which is stored as a vector in the HIP framework, with a mask requires only two loops. This is in contrast to convolution on square lattices which normally requires four loops.

3.4.5 Frequency Domain processing

Finally, we turn to fundamental operations pertaining to the frequency domain using the HIP addressing scheme. We consider only an orthogonal transform for mapping the spatial domain to the frequency domain, namely, the discrete Fourier transform (DFT), due to its important role in image processing. We start with the development of a reciprocal sampling lattice and the corresponding addressing scheme for the frequency domain. The hexagonal DFT is then defined based on these developments.

Given an orthogonal mapping function that maps the spatial domain to the frequency domain, the underlying lattices in the two domains are said to be duals or reciprocals of one another [110]. This means that the axes used for the coordinates in the spatial and frequency domains are orthogonal to another. The spatial domain coordinates can be represented by a matrix, $\mathbf{V}$, which has as its columns the basis vectors for the spatial domain lattice. This matrix is called the spatial domain sampling matrix. Likewise a frequency domain sampling matrix can be defined as $\mathbf{U}$. The criteria of orthogonality imposes a relationship between these two matrices:

$$\mathbf{V}^T\mathbf{U} = \mathbf{I} \tag{3.40}$$

Hence, the frequency domain matrix can be computed from the spatial domain matrix, viz. $\mathbf{U} = (\mathbf{V}^T)^{-1}$. The spatial domain matrix for HIP addressing was defined in Section 3.2 using the basis vector set B_h. We repeat that here for convenience.

$$\mathbf{V} = \begin{bmatrix} 1 & -\frac{1}{2} \\ 0 & \frac{\sqrt{3}}{2} \end{bmatrix} \tag{3.41}$$

From this we derive $\mathbf{U}$ as follows:

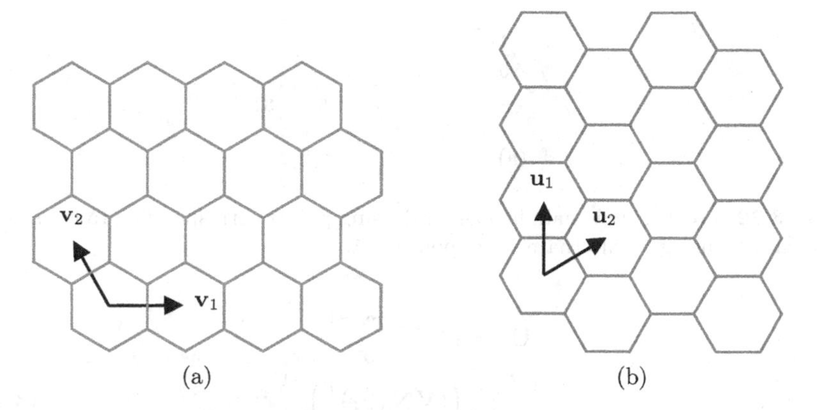

Fig. 3.21. Spatial and frequency domain lattices defined by (a) **V** and (b) **U**.

$$\mathbf{U} = \begin{bmatrix} 1 & 0 \\ \frac{1}{\sqrt{3}} & \frac{2}{\sqrt{3}} \end{bmatrix} \tag{3.42}$$

The spaces that are generated by these two matrices are illustrated in Figure 3.21. It should be noted that they are not drawn to scale. Consistent to the orthogonality requirement, the orientation of the axes are seen to have opposite sense in the frequency domain. In order to generate the frequency domain HIP indexing scheme, we wish to extend the hierarchical nature of the addressing scheme to the frequency domain as well. Due to the hierarchical nature of HIP images, a particular λ-layer image will be periodic in the $(\lambda+1)$-layer image. For instance, a first-layer HIP structure can be seen repeated in the second-layer HIP structure a total of seven times. A matrix which describes the translation from a λ-layer image to a $(\lambda+1)$-layer image is called the periodicity matrix. In the case of spatial HIP addressing, this matrix is equivalent to the translation matrix defined in Section 3.2. Thus the periodicity matrix for the λ-layer HIP image is:

$$\mathbf{N}_{\lambda-1} = \begin{bmatrix} 3 & -2 \\ 2 & 1 \end{bmatrix}^{\lambda-1}$$

An arbitrary layer image can be found by repeated multiplications of this periodicity matrix as per equation (3.8). The spatial sampling matrix for a λ-layer HIP image can also be modified to achieve this as:

$$\mathbf{V}'_\lambda = \mathbf{V}\mathbf{N}_{\lambda-1}$$

Using the orthogonality criteria given in equation (3.40) yields a new frequency sampling matrix:

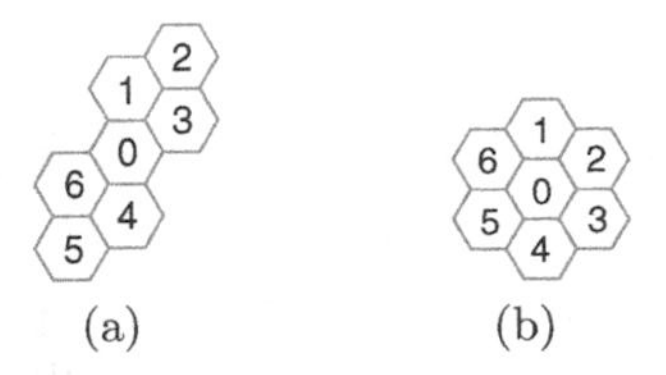

Fig. 3.22. Frequency domain indexing (a) using points corresponding to the spatial domain (b) using an alternate set of points.

$$\begin{aligned} \mathbf{U}'_\lambda &= \left((\mathbf{V}'_\lambda)^T \right)^{-1} \\ &= \left((\mathbf{V}\mathbf{N}_{\lambda-1})^T \right)^{-1} \end{aligned} \tag{3.43}$$

As an example, the frequency sampling matrix for a two-layer HIP image is:

$$\mathbf{U}'_2 = \frac{1}{7\sqrt{3}} \begin{bmatrix} \sqrt{3} & -2\sqrt{3} \\ 5 & 4 \end{bmatrix}$$

The orthogonality criterion relating the sampling matrices also requires their determinants to be inverses of each other. Consequently, the frequency domain image is scaled by a factor proportional to the determinant of $\mathbf{V}_\lambda$ squared or $\det|\mathbf{V}_\lambda|^2$. For example, a two-layer HIP image has $\det|\mathbf{V}|^2 = \frac{49\times3}{4} = \frac{147}{4}$. The reduction (scaling) between the spatial and frequency domain is the inverse of this value, or $\frac{4}{147}$. Generally, for a λ-layer spatial image the reduction is by $\frac{3\times7^\lambda}{4}$. The opposite sense, illustrated in Figure 3.21, is maintained for the frequency HIP image through an extra rotation by $-\tan^{-1}\frac{\sqrt{3}}{2}$ introduced for each successive layer.

Now that the frequency sampling matrix has been defined, we can define the frequency domain HIP coordinates. In the spatial domain, the coordinates of the points in the first-level aggregate were given in equation (3.3). Using the frequency domain sampling matrix derived in equation (3.43) and HIP addresses instead of coordinates, we obtain a scheme for a first-layer HIP image as illustrated in Figure 3.22(a). The envelope of the shape can be seen to be two linked diamonds. This will result in an elongated shape for the frequency domain HIP image when this shape is repeatedly tiled. This is undesirable and ideally a shape equivalent to the spatial domain image is required. Hence, an alternate set of coordinates is used which is shown in Figure 3.22(b). This choice is valid since, given the periodic nature of the HIP structure, any set of points could be used so long as each point is unique within a single period. Simply stated, any unique group of seven points could be employed to be the generating tile for the next layer of the HIP frequency image. The coordinates of the points in the set illustrated in Figure 3.22 correspond to:

$$A'_1 = \left\{ \begin{bmatrix} 0 \\ 0 \end{bmatrix}, \begin{bmatrix} 1 \\ 0 \end{bmatrix}, \begin{bmatrix} 0 \\ 1 \end{bmatrix}, \begin{bmatrix} -1 \\ 1 \end{bmatrix}, \begin{bmatrix} -1 \\ 0 \end{bmatrix}, \begin{bmatrix} 0 \\ -1 \end{bmatrix}, \begin{bmatrix} 1 \\ -1 \end{bmatrix} \right\} \tag{3.44}$$

The points in a λ layer HIP frequency image can be found in a fashion similar to the spatial case using the periodicity matrix for the frequency domain $\mathbf{N}_\lambda^T$. Thus equation (3.8) can be rewritten as:

$$A'_\lambda = \mathbf{N}_{\lambda-1}^T A'_1 + \cdots + \mathbf{N}_1^T A'_1 + A'_1 \tag{3.45}$$

Analogous to the spatial domain case, HIP addressing can be used to label them in order, using base 7 numbers, to obtain frequency domain addresses. For example, a second layer HIP frequency image contains the following points:

$$A'_2 = \left\{ A'_1, \begin{bmatrix} 2 \\ 1 \end{bmatrix}, \begin{bmatrix} 2 \\ 2 \end{bmatrix}, \begin{bmatrix} 3 \\ 1 \end{bmatrix}, \begin{bmatrix} 3 \\ 0 \end{bmatrix}, \begin{bmatrix} 2 \\ 0 \end{bmatrix}, \begin{bmatrix} 1 \\ 1 \end{bmatrix}, \begin{bmatrix} 1 \\ 2 \end{bmatrix}, \cdots \begin{bmatrix} -2 \\ 4 \end{bmatrix} \right\} \tag{3.46}$$

This set is illustrated in Figure 3.23 to aid in visualisation. With reference to the spatial domain image, there is a reversal in the sense of the addressing along with a rotation of the starting point. All of the properties previously attributed to the HIP addressing scheme are equally applicable to frequency domain HIP addressing. The definition of $\mathbb{G}$ also holds for the frequency domain version of the HIP addressing scheme. In fact the set illustrated in Figure 3.23 is equivalent to $\mathbb{G}^2$. In general, an arbitrary HIP address could be in either the spatial or the frequency domain as it is independent of both. The difference is the physical significance of the address and what it means in either the spatial or frequency domain.

Next, we define the hexagonal discrete Fourier transform (HDFT). The methodology described here follows Mersereau [111] whose method generally extends to periodic functions with arbitrary dimensionality. As previously stated, HIP images are periodic with periodicity matrix $\mathbf{N}_\lambda$. The images are periodic in an image of order one higher than the original image. For a given HIP image x of order λ the following holds:

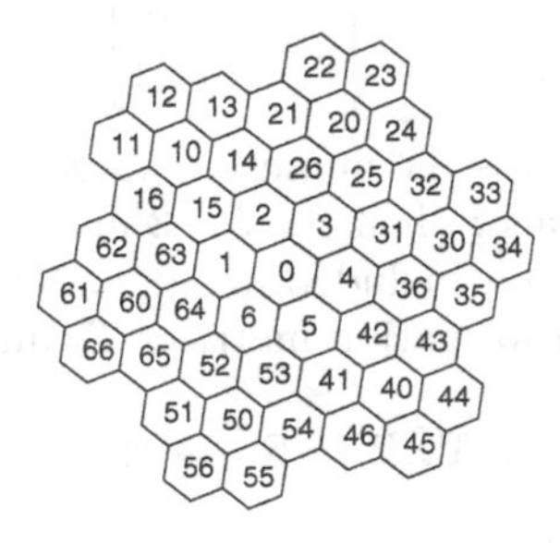

Fig. 3.23. Addresses of a two-layer HIP image in the frequency domain.

$$x(\mathbf{n}) = x(\mathbf{n} + \mathbf{N}_\lambda \mathbf{r})$$

$\mathbf{n}$ and $\mathbf{r}$ are integer vectors (in $\mathbb{Z}^2$) aligned with basis B_h. The number of sample points in a period is equal to $\det \mathbf{N}_\lambda$ which is 7^λ. Let us denote the set of the points $I_{\mathbf{N}_\lambda}$ as the spatial domain samples and $J_{\mathbf{N}_\lambda}$ as the frequency domain samples. The sequence $x(\mathbf{n})$ can be expanded as a Fourier series with coefficients denoted as $X(\mathbf{k})$, where:

$$X(\mathbf{k}) = \sum_{\mathbf{n} \in I_{\mathbf{N}_\lambda}} x(\mathbf{n}) \exp\left[-2\pi \mathrm{j} \mathbf{k}^T {\mathbf{N}_\lambda}^{-1} \mathbf{n}\right] \tag{3.47}$$

$$x(\mathbf{n}) = \frac{1}{|\det \mathbf{N}_\lambda|} \sum_{\mathbf{k} \in J_{\mathbf{N}_\lambda}} X(\mathbf{k}) \exp\left[2\pi \mathrm{j} \mathbf{k}^T {\mathbf{N}_\lambda}^{-1} \mathbf{n}\right] \tag{3.48}$$

These two equations are the well known discrete Fourier transform (DFT) pair. Both x and X are unique for the periods in which they are defined ($I_{\mathbf{N}_\lambda}$ and $J_{\mathbf{N}_\lambda}$), beyond which they repeat periodically. The formulation of the discrete Fourier transform, as described, requires the computation of a vector inner product defined by $\mathbf{k}^T {\mathbf{N}_\lambda}^{-1} \mathbf{n}$. Since HIP addressing uses aggregation, the computation of the inner product will be done by introducing an intermediate coordinate system. In this formulation, the three-coordinate scheme proposed by Her [51] is used, though, there is no reason why other schemes could not be employed. Her's scheme was chosen due to the fact that it exploits the symmetry of the hexagonal lattices and so should be applicable to both the spatial and frequency domain versions of the HIP addressing scheme. As a first step, a function $c : \mathbb{G}^1 \to \mathbb{R}^3$ is defined, the purpose of which is to convert a single digit HIP address to a 3-tuple coordinate as described by Her. The function is defined by:

$$c(\boldsymbol{n}) = \begin{cases} \begin{bmatrix} 0 \\ 0 \\ 0 \end{bmatrix} & \text{if } \boldsymbol{n} = \boldsymbol{0} \\ \begin{bmatrix} 0 & 0 & -1 \\ -1 & 0 & 0 \\ 0 & -1 & 0 \end{bmatrix}^{(7-n)} \begin{bmatrix} 1 \\ 0 \\ -1 \end{bmatrix} & \text{if } \boldsymbol{1} \le \boldsymbol{n} < \boldsymbol{6} \end{cases} \tag{3.49}$$

In the above equation, n is the numerical value of the HIP index $\boldsymbol{n}$ in base 10. Note that when n equals 1 then the 3×3 matrix becomes the identity. A derivation for function $c(\boldsymbol{n})$ is included in Appendix A.3. It is now possible to define a pair of linear transformation matrices $\mathbf{T}_s$ and $\mathbf{T}_f$ using this function:

$$\mathbf{T}_s, \mathbf{T}_f : \mathbb{R}^3 \to \mathbb{R}^2$$

These can be defined as:

$$\mathbf{T}_s = \frac{1}{3}\begin{bmatrix} 1 & 1 & -2 \\ -1 & 2 & -1 \end{bmatrix} \tag{3.50}$$

$$\mathbf{T}_f = \frac{1}{3}\begin{bmatrix} -1 & 2 & -1 \\ 2 & -1 & -1 \end{bmatrix} \tag{3.51}$$

These matrices can convert from three coordinates into the previously defined spatial coordinates (using $\mathbf{T}_s$) or frequency coordinates (using $\mathbf{T}_f$). As an example when $\boldsymbol{n} = \mathit{2}$:

$$c(\mathit{2}) = \begin{bmatrix} 0 \\ 1 \\ -1 \end{bmatrix} \qquad \mathbb{G}^1 \to \mathbb{R}^3$$

$$\left.\begin{aligned} \mathbf{T}_s c(\mathit{2}) &= \begin{bmatrix} 1 \\ 1 \end{bmatrix} \\ \mathbf{T}_f c(\mathit{2}) &= \begin{bmatrix} 1 \\ 0 \end{bmatrix} \end{aligned}\right\} \qquad \mathbb{R}^3 \to \mathbb{R}^2$$

Whilst this only works for single digit HIP addresses, it is a simple exercise to extend to arbitrary HIP addresses. The method employed exploits the use of the periodicity matrix. First we define a pair of functions which perform the appropriate conversion:

$$h(\boldsymbol{g}), H(\boldsymbol{g}) : \mathbb{G}^\lambda \to \mathbb{R}^2$$

Now the mapping function, $h(\boldsymbol{g})$, for the spatial domain is defined as:

$$h(\boldsymbol{g}) = \sum_{i=0}^{\lambda-1} \mathbf{N}_i \mathbf{T}_s c(g_i) \tag{3.52}$$

and the frequency domain mapping function, $H(\boldsymbol{g})$, is defined as:

$$H(\boldsymbol{g}) = \sum_{i=0}^{\lambda-1} (\mathbf{N}_i)^T \mathbf{T}_f c(g_i) \tag{3.53}$$

In the above definitions, g_i is the i-th digit of a HIP address. Using the above functions, the inner product in equations (3.54) and (3.55) can be rewritten as:

$$\mathbf{k}^T \mathbf{N}_\lambda^{-1} \mathbf{n} = H(\boldsymbol{k})^T \mathbf{N}_\lambda^{-1} h(\boldsymbol{n})$$

where $\boldsymbol{k}$ and $\boldsymbol{n}$ are HIP indices. Using this relationship makes it possible to rewrite equations (3.47) and (3.48) as:

$$X(\boldsymbol{k}) = \sum_{\boldsymbol{n} \in \mathbb{G}^\lambda} x(\boldsymbol{n}) \exp\left[-2\pi \mathrm{j} H(\boldsymbol{k})^T \mathbf{N}_\lambda^{-1} h(\boldsymbol{n})\right] \tag{3.54}$$

$$x(\boldsymbol{n}) = \frac{1}{|\det \mathbf{N}_\lambda|} \sum_{\boldsymbol{k} \in \mathbb{G}^\lambda} X(\boldsymbol{k}) \exp\left[2\pi \mathrm{j} H(\boldsymbol{k})^T \mathbf{N}_\lambda^{-1} h(\boldsymbol{n})\right] \tag{3.55}$$

These equations are the HIP discrete Fourier transform (or HDFT) pair. The equations are purely in terms of HIP addresses and the periodicity matrix. Straightforward computation of the HDFT requires a large amount of effort which is also the case with the DFT computation on square grids. Methods to speed up the computation are examined in Section 4.2.1.

The ordinary DFT has many properties which are commonly exploited in applications. These properties also hold for the HDFT. Some of those which are exploited in the later part of this book are:

i. *Linearity:* Given two HDFT pairs $x(\boldsymbol{n}) \leftrightarrow X(\boldsymbol{k}), y(\boldsymbol{n}) \leftrightarrow Y(\boldsymbol{k})$ then for $a, b \in \mathbb{C}$ the following holds $ax(\boldsymbol{n}) + by(\boldsymbol{n}) \leftrightarrow aX(\boldsymbol{k}) + bY(\boldsymbol{k})$.
ii. *Spatial Shift:* Given $x(\boldsymbol{n}) \leftrightarrow X(\boldsymbol{k})$ and a HIP address $\boldsymbol{a}$ then $x(\boldsymbol{n} \ominus \boldsymbol{a}) \leftrightarrow X(\boldsymbol{k}) \exp\left[-2\pi \mathrm{j} H(\boldsymbol{k})^T \mathbf{N}_\lambda^{-1} h(\boldsymbol{a})\right]$.
iii. *Modulation:* Given $x(\boldsymbol{n}) \leftrightarrow X(\boldsymbol{k})$ and a HIP address $\boldsymbol{a}$ then $x(\boldsymbol{n}) \exp\left[2\pi \mathrm{j} H(\boldsymbol{a})^T \mathbf{N}_\lambda^{-1} h(\boldsymbol{n})\right] \leftrightarrow X(\boldsymbol{k} \ominus \boldsymbol{a})$.
iv. *Convolution:* Given two HDFT pairs $x(\boldsymbol{n}) \leftrightarrow X(\boldsymbol{k}), y(\boldsymbol{n}) \leftrightarrow Y(\boldsymbol{k})$ then $x(\boldsymbol{n}) \circledast y(\boldsymbol{n}) \leftrightarrow X(\boldsymbol{k}) Y(\boldsymbol{k})$. The converse also holds $X(\boldsymbol{k}) \circledast Y(\boldsymbol{k}) \leftrightarrow x(\boldsymbol{n}) \circledast y(\boldsymbol{n})$.

Proofs of these properties are described in Section A.4. There is one property that is missing from the above list and this is separability. Unlike the DFT for square grids, the HDFT is not separable. This should be evident by looking at the HDFT pair as given by equations (3.54) and (3.55). A fundamental reason for this is that the basis vector set, B_h, which spans the hexagonal lattice consists of vectors which are not mutually orthogonal. The lack of separability makes the HDFT even more computationally expensive to compute than the square equivalent. This has been addressed later by deriving a fast algorithm for the HDFT computation in Section 4.2.1 which exploits, among other things, the shifting and modulation properties.

3.5 Concluding remarks

This chapter presented a new representation for hexagonal images and used it to develop a framework, called the HIP framework, for processing hexagonal images. Since, the intention behind the development of this framework is to have an efficient test bed for conducting experiments on hexagonal image processing, the chapter included details of carrying out basic operations in the spatial and frequency domains within this framework.

The specific addressing scheme used in the HIP framework represents a point in the lattice as a single, radix 7, multiple-digit index rather than a pair or triple of coordinates as are generally used. As each point in Euclidean space is represented by a single address, an entire image can now be stored using a vector. This representation comes as a natural consequence of the addressing scheme rather than as a by-product of a matrix manipulation through row or column ordering as for two- and three-coordinate schemes. Furthermore, the addressing scheme generates an image shape which is hexagonal (or roughly like a snowflake) and consequently the addresses exhibit the symmetries of the hexagon. Due to the construction of the addresses based on aggregate tilings, the addressing scheme is hierarchical in nature. Taken together, these factors mean that the HIP addressing scheme outlined in this chapter is computationally efficient.

A number of fundamental operations in image processing were considered for implementation within the HIP framework. These can be used to build more complex image processing algorithms. Address manipulation is used in many image processing techniques such as neighbourhood operations, down/up-sampling, etc. This was considered first and it required the definition of operations on HIP addresses using modulo-7 arithmetic. Address manipulation using closed arithmetic was found to lead to interesting results which could be used in image pyramid generation. Secondly, fundamental concepts such as image boundary, external points, distance measures, and neighbourhoods were examined within the framework and definitions for the same were developed. The neighbourhood definition was used to define the convolution operation. Convolution within HIP involves processing two 1-D matrices namely, the mask and a given image(now a vector) and hence is more efficient. Finally, operations in the frequency domain were described. This began with the derivation of a frequency domain analog to the spatial HIP addressing scheme previously outlined. This was based upon two properties: the hierarchical nature of the addressing scheme, and the reciprocal relationship of spatial and frequency domains. The discrete Fourier transform for the HIP framework (HDFT) was also derived to enable frequency domain operations where the transform is completely defined in terms of HIP addresses. Thus, we are all set for studying in detail the processing of hexagonally sampled images both in the spatial and frequency domains.

4

Image processing within the HIP framework

The HIP framework provides a test bed for studying the performance of various processing techniques on images sampled on a hexagonal grid. In this chapter, we examine how some of the basic image processing techniques can be implemented within the HIP framework. Since the material in this chapter is also intended to inform on the utility of the HIP framework, problems have been selected to be representative and hence cover processing techniques in the spatial and frequency domains, multiresolution representations and morphological operations.

4.1 Spatial domain processing

Spatial domain methods are often the natural choice for processing and extracting pertinent information from an image. The direct access to the raw intensity information can aid in the design of computationally efficient analysis and processing. Furthermore, it is often more intuitive to design operations to be performed directly upon the raw image data. This section examines two distinct operations in the spatial domain. The first is edge detection and the second is skeletonisation. These two methodologies are considered to be complementary.

4.1.1 Edge detection

Edge detection is an important operation in both biological and computer vision. In biological vision, there is significant evidence [112] that the primary visual cortex serves to spatially arrange the visual stimuli into maps of oriented edges. There is also weaker evidence for edge preference in the retina and the LGN [3]. In computer vision, edge detection is a pre-processing step in many applications such as object recognition, boundary extraction, and segmentation. The basic assumption used in computer vision is that edges are characterised by significant (step) changes in intensity. Hence, at the location of an edge, the first derivative of the intensity function should be a maximum or the second derivative should have a zero-crossing. This was the basis for the design of basic edge detection techniques. In real images however, edges are also often marked by subtle changes in intensity which is noted and addressed by more advanced edge detection techniques.

This section examines three commonly used techniques for edge detection based on the derivative operation. The three techniques are the Prewitt edge operators, the Laplacian of Gaussian and the Canny edge detector. The Prewitt edge operators are first derivative operators while the Laplacian of Gaussian (LoG) is a second derivative operator. The Canny edge detector is a good example of a near optimal edge detector combining the features of the Prewitt and LoG operators. In terms of directional sensitivity, the Prewitt operator is maximally sensitive to edges in horizontal and vertical directions while the LoG operator is isotropic. The isotropic nature of the LoG operator is due to the Gaussian smoothing function employed to reduce the noise sensitivity of the second derivative operation.

Prewitt edge detector

The Prewitt edge detector [113] is an example of a gradient based edge detector. It approximates the gradient operation by a pair of 3×3 masks. These masks are illustrated in Figure 4.1(a) as s_1 and s_2. The masks are aligned in the horizontal (s_1) and vertical directions (s_2). Operation of the Prewitt edge detector involves computing a pair of gradient images by first convolving the image with each of the masks. Each point in the two gradient images are then combined either using a sum of squares or a sum of absolute values to generate a candidate edge map for the original image. As a final step, the edge map is thresholded to yield a final edge map. Generally, the Prewitt operator is considered a poor edge detector for square images. The reasons for this are twofold. Firstly, the masks are a weak approximation to the gradient operation and secondly the approximation fails to consider the disparate distances between the eight neighbouring pixels and the centre pixel in the mask. However, due to the ease of implementation and low computational cost, the Prewitt edge detector is often employed.

On a square lattice, the Prewitt operator is designed to approximate the gradient computation in two orthogonal directions which are the two principal axes of symmetry for a square. For a hexagonal lattice, a simple approach would be to compute gradients aligned along each of the three axes of symmetry of the lattice [24]. This gives the three masks illustrated in Figure 4.1(b). However, one mask is redundant as it can be written as a combination of the other two. For instance, the mask aligned along 0° can be obtained by taking the difference between masks at 60° and 120°. Generally, edge detection algorithms on square images are often employed to find both edge strength and direction. For this reason the masks are computed in orthogonal directions.

$$s_1 = \begin{bmatrix} 1 & 0 & -1 \\ 1 & 0 & -1 \\ 1 & 0 & -1 \end{bmatrix} \quad s_2 = \begin{bmatrix} 1 & 1 & 1 \\ 0 & 0 & 0 \\ -1 & -1 & -1 \end{bmatrix}$$

(a)

$$h_1 = \begin{bmatrix} & 1 & & 1 & \\ 0 & & 0 & & 0 \\ & -1 & & -1 & \end{bmatrix} \quad h_2 = \begin{bmatrix} & 0 & & 1 & \\ -1 & & 0 & & 1 \\ & -1 & & 0 & \end{bmatrix} \quad h_3 = \begin{bmatrix} & 1 & & 0 & \\ 1 & & 0 & & -1 \\ & 0 & & -1 & \end{bmatrix}$$

(b)

Fig. 4.1. The masks used in the Prewitt edge detector implementation (a) square (b) hexagonal.

Using the redundancy, the orthogonal gradients can be computed as $G_x = h_1$ and $h_2 - h_3$.

From these we can compute the gradient magnitude, $M = \sqrt{h_2^2 + h_3^3 + h_2 h_3}$, and direction, $\theta = \tan^{-1} \frac{h_2 + h_3}{h_2 - h_3}$. The size of the mask is determined by the neighbourhood of a pixel on the given lattice. Hence, the hexagonal masks are specified by seven weights as compared to nine weights for the square masks.

We now examine how the Prewitt operator can be implemented using the HIP framework. Recalling that an image is stored as a vector in this framework, the task at hand is to first convolve a 7-long or $\lambda = 1$ layer mask vector with a given λ-layer image $f(\boldsymbol{x})$ and then follow it by gradient magnitude computation at every point:

for all $\boldsymbol{x} \in \mathbb{G}^{\lambda}$ **do**
$f_2(\boldsymbol{x}) = f(\boldsymbol{x}) \circledast h_2(\boldsymbol{x})$
$f_3(\boldsymbol{x}) = f(\boldsymbol{x}) \circledast h_3(\boldsymbol{x})$
$M(\boldsymbol{x}) = \sqrt{f_2(\boldsymbol{x})^2 + f_3(\boldsymbol{x})^2 - f_2(\boldsymbol{x}) f_3(\boldsymbol{x})}$
end for

The above approach can be generalised to derive gradient operators for other directions. Such operators are popularly known as compass operators as they compute gradients in the eight compass directions in the square lattice. Analogously, in the hexagonal case, we can compute derivatives in six directions by designing operators at $k \times 60°$ where k = 0,1,2,3,4,5. The corresponding masks are obtained by simply rotating the h_1 successively by 60°. Since derivatives enhance noise, special operators are designed for square lattices which combine smoothing with gradient computation. The Sobel and Frei-Chen operators are examples. Here, the weights are assigned to pixels according to their distance from the centre pixel unlike the Prewitt operator. There is no equivalent for Sobel or Frei-Chen operators in the hexagonal case as all pixels are equidistant from the centre in the one-layer mask.

The calculation of the masking operation is the largest inhibition in computational performance for the Prewitt edge detector. This requires distinct masking operations for each of the required directions. For the Prewitt operator, since the mask size is seven, computing each point in the output image

$$L_s = \begin{bmatrix} 0 & 0 & -1 & -1 & -1 & 0 & 0 \\ 0 & -1 & -5 & -6 & -5 & -1 & 0 \\ -1 & -5 & -3 & 10 & -3 & -5 & -1 \\ -1 & -6 & 10 & 49 & 10 & -6 & -1 \\ -1 & -5 & -3 & 10 & -3 & -5 & -1 \\ 0 & -1 & -5 & -6 & -5 & -1 & 0 \\ 0 & 0 & -1 & -1 & -1 & 0 & 0 \end{bmatrix}$$

(a)

$$L_h = \left[\begin{array}{ccccccccccccccc}
 & & & & & 0 & & 0 & & & & & & & \\
 & & & & -1 & & -2 & & -2 & & -1 & & 0 & & \\
 & 0 & & -2 & & -6 & & -7 & & -6 & & -2 & & 0 & \\
0 & & -2 & & -7 & & 10 & & 10 & & -7 & & -2 & & \\
 & -1 & & -6 & & 10 & & 49 & & 10 & & -6 & & -1 & \\
 & & -2 & & -7 & & 10 & & 10 & & -7 & & -2 & & 0 \\
 & 0 & & -2 & & -6 & & -7 & & -6 & & -2 & & 0 & \\
 & & 0 & & -1 & & -2 & & -2 & & -2 & & & & \\
 & & & & & & & 0 & & 0 & & & & &
\end{array}\right]$$

(b)

Fig. 4.2. The different masks used in the Laplacian of Gaussian edge detector implementation (a) square (b) hexagonal.

requires convolutions with two masks, each of which require seven multiplications and six additions. Thus a λ-layer image requires $2 \times 7^{\lambda+1}$ multiplications and $12 \times 7^{\lambda}$ additions.

Laplacian of Gaussian edge detector

Instead of using the maxima in the gradients, an alternate methodology to find the edges in an image is to use the zero crossings of the second derivative. First proposed by Marr [114], the Laplacian of Gaussian (LoG) edge detector first smoothes the image with a Gaussian before performing the second derivative computation. The smoothing can be tuned to make the edge detector sensitive to different sized edges. On the square lattice, the LoG function is approximated by a $n \times n$ mask where n depends on the degree of smoothing desired. As the LoG function is isotropic, the larger this mask, the closer the approximation is to the ideal function. However, large mask sizes will result in increased computation and so mask sizes of approximately 7×7 are typically used. Edge detection using the LoG edge detector consists of the following steps: convolving the image with the mask, thresholding the result and finally detecting the zero crossings in the thresholded image.

The hexagonal lattice is very much suited for designing isotropic kernel functions such as the LoG function, due to the excellent fit it provides for the required circular base of support. Implementation of the LoG edge detector using HIP is a simple exercise. The computation of the mask first requires the conversion of the HIP address to Cartesian coordinates. Next, since a HIP image is centred around the origin, the discrete LoG function is obtained by substituting the above coordinates into the LoG expression. The equivalent of a 7×7 square LoG mask, would be a two-layer HIP mask with $7^2 = 49$ pixels. Sample masks for both square and hexagonal cases are illustrated in Figure 4.2. Edge detection in the HIP framework using a m-layer LoG operator $L(\boldsymbol{x})$ on a λ-layer image $f(\boldsymbol{x})$, requires convolving an m-long vector with a 7^λ long HIP image vector as follows:

> **for all** $\boldsymbol{x} \in \mathbb{G}^\lambda$ **do**
> $f_2(\boldsymbol{x}) = f(\boldsymbol{x}) \circledast L(\boldsymbol{x})$
> $f_3(\boldsymbol{x}) = \text{threshold}(f_2(\boldsymbol{x}, \textit{level}))$
> **end for**
> find addresses $\boldsymbol{x} \in \mathbb{G}^\lambda$ such that $f_3(\boldsymbol{x}) \approx 0$

The bulk of the computational cost of edge detection with the LoG operator is primarily due to the convolution of the image with the LoG mask. For a mask of length 49 this requires 49 multiplications and 48 additions to compute every pixel in $f_2(\boldsymbol{x})$. Thus a total of $7^{\lambda+2}$ multiplications and $48 \times 7^\lambda$ additions are required for the entire λ-layer image. This cost is significantly more than the cost for the Prewitt edge detector, which explains the former's popularity.

Canny edge detector

The Canny edge detector [115] was designed to be an optimal edge detector. The criteria used for optimality are detection, localisation, and single response. The first criterion requires maximum true positive and minimum false negative (spurious) edges in the detector output. The second criterion requires minimising the distance between the located and actual edges. The final criterion serves to minimise multiple responses to a single edge which is a common problem with basic edge detectors. The devised solution employs Gaussian smoothing and then directional derivatives to estimate the edge directions. In this way the Canny edge detector is a combination of the Prewitt and LoG edge detection algorithms. Canny also introduced a hysteresis in the thresholding stage. Thus if the candidate edge pixel is above the highest threshold then the point is definitely an edge and if it is below the lowest threshold then it is definitely not an edge. Intermediate points could possibly be an edge depending on the state of the neighbouring points. The operation of the Canny edge detector first involves derivative computations in the horizontal and vertical directions using 7×7 masks. The squared responses are combined to generate a candidate edge map which is thresholded to generate the final edge map.

The implementation of the Canny edge detector using HIP can proceed as follows. The Gaussian smoothing and derivative operations are combined to compute two masks h_2 and h_3 oriented as in the Prewitt edge detector.

The HIP address is converted to Cartesian coordinates before computing the oriented masks. The specific mask weights are found by evaluating the directional derivatives of a Gaussian function at the Cartesian coordinates. The requisite algorithm for implementing the Canny edge detector for a λ-layer HIP image is as follows:

for all $\boldsymbol{x} \in \mathbb{G}^\lambda$ **do**
  $f_2(\boldsymbol{x}) = f(\boldsymbol{x}) \circledast h_2(\boldsymbol{x})$
  $f_3(\boldsymbol{x}) = f(\boldsymbol{x}) \circledast h_3(\boldsymbol{x})$
  $M(\boldsymbol{x}) = \sqrt{f_2(\boldsymbol{x})^2 + f_3(\boldsymbol{x})^2 - f_2(\boldsymbol{x})f_3(\boldsymbol{x})}$
end for
for all $\boldsymbol{x} \in \mathbb{G}^\lambda$ **do**
  $f_4(\boldsymbol{x}) = \text{thresh}(\boldsymbol{x}, level_1, level_2)$
end for

The computational requirements of the Canny edge detector are naturally greater than the previous two methods. Once again the most expensive stage is the convolution stage. Assuming the masks employed consist of 49 points, computing an output pixel after convolution requires 49 multiplications for each of the two masks and 48 additions. Thus, for a λ-level image this results in $2 \times 7^{\lambda+2}$ multiplications and $96 \times 7^\lambda$ additions. The high computational cost is offset by much better performance than the other two detectors that have been discussed.

Comparison of edge detector performance

Edge detectors were implemented using HIP and tested on three test images. Two of these (T1 and T2) were synthetic while the third (T3) was a real image of a coin. These images, of size 256 by 256 pixels (see figures 4.3(a) to 4.3(c)), were chosen as they contain a mixture of curves and lines along with a variation in contrast. The hexagonal test images were obtained by resampling the square images into a five-layer HIP structure (see figures 4.3(d) to 4.3(f)). Both qualitative and computational examinations of the edge detector performance were performed. For fair comparison, for each of the three edge detection techniques the threshold was tuned to produce the best qualitative results and then the ratio of edge pixels to the image size was computed. Computational performance was measured by examining the complexity of the algorithms.

The results for the Prewitt edge detector are shown in figures 4.4(a) to 4.4(c). The results of processing the synthetic images illustrate that the representation of circles in a hexagonal image is very good. This is illustrated by the smoothness of the circles. Furthermore, the diagonal dividing lines are also smooth. However, the vertical lines are noticeably ragged. This is due to the nature of the hexagonal lattice which offsets the individual vertical pixels by half a pixel per row. Examination of the edge detected image T3, shows that (i) the shape of the coin is circular, and (ii) the kiwi bird, the fern, and

Fig. 4.3. The test images used in edge detection. (a) T1 (b) T2 (c) T3 are square sampled and (d) T1 (e) T2 (f) T3 are hexagonally sampled.

the number '20' all appear clearly. Overall, the appearance of the hexagonal image shows little noise. The ratios of edge pixels to image size, for the test images T1, T2 and T3 are 11.5%, 11.2% and 13.3% respectively.

The results of edge detection with the LoG operator, using the same test images are shown in figures 4.4(d) to 4.4(f). Once again, the curves are shown with good clarity. There is an improvement over the Prewitt case in that the leaves of the fern are also revealed clearly in T3. However, the edges are thicker which is typical of a second order derivative operator which produces double edges. The Gaussian smoothing does serve to reduce the distinct ragged nature of vertical lines that was found in the Prewitt edge detector output. The ratios of edge pixels to image size are 19.4%, 18.3% and 18.0% for T1, T2, and T3 respectively.

The results of Canny edge detector are illustrated in figures 4.4(g) to 4.4(i). As expected, this edge detector shows improved performance over the previous two cases. In all cases, the edges appear less noisy due to the maximal suppression step. The ratios of edge pixels are 18.7%, 19.9% and 18.2% for T1, T2 and T3 respectively. The lines are thicker than for the Prewitt case due to the smoothing effect of the Gaussian.

There are a number of conclusions that can be drawn from the above study. In all cases, the number of edge pixels was roughly the same but the

Fig. 4.4. Results of edge detection: (a) Prewitt T1 (b) Prewitt T2 (c) Prewitt T3 (d) LoG T1 (e) LoG T2 (f) LoG T3 (g) Canny T1 (h) Canny T2 (i) Canny T3.

Canny edge detector appears to perform the best. This is to be expected as it is a more sophisticated edge detection methodology. The results in all cases are especially pleasing for images containing curved features (T1 and T3). This stems from the consistent connectivity of the pixels in hexagonal images which aids edge detection of curved structures. Pixels without consistent connectivity show up as discontinuities in the contour (in the input image) and typically result in breaks in the edge image. This problem is generally noticeable in curved objects on square lattices. Vertical lines however do suffer from appearing ragged on hexagonal lattices. The Canny and LoG detectors serve to remove this difficulty via in-built smoothing.

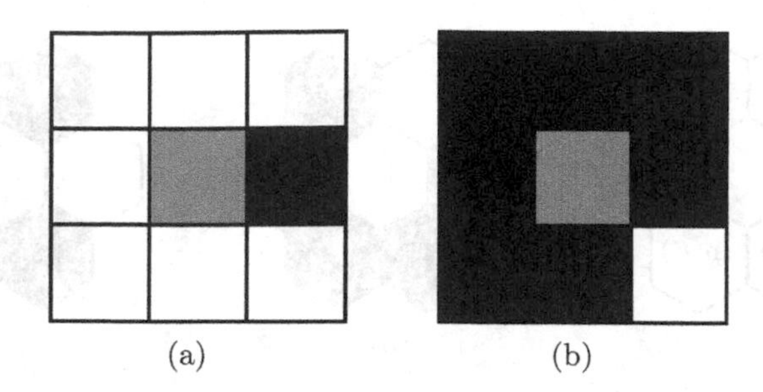

Fig. 4.5. Skeletonisation on a square image showing cases where removal of the gray pixel is bad: (a) end of stroke (b) erosion.

4.1.2 Skeletonisation

Skeletonisation plays an important role in many image preprocessing stages. Once an image has been edge-detected it is often necessary to skeletonise or thin the edges. This serves to remove all the redundant points which are contained within the edge image whilst retaining its basic structure and characteristics. The first recorded work on skeletonisation can be dated to Blum [116] who called it the medial axis transform. Over the years, many studies of this process on square [117] as well as hexagonal [67, 76] lattices have been reported.

Skeletonisation is a morphological operation which is usually implemented by iterative application of thinning. Hence, it heavily depends on a pixel's connectivity for its retention or deletion. In this operation, edge pixels, which are termed as foreground pixels, are selectively removed or deleted by applying some empirical rules. The rules are formulated based on desired characteristics of a skeleton: one-pixel thickness, mediality and homotopy. The first two properties ensure non-redundancy of the information preserved, while the last property preserves the topology of the given image.

On a square lattice, a simple algorithm for skeletonisation is based on the following rules [118]:

1. A pixel should have more than one but less than seven neighbours.
2. Do not erode a one-pixel wide line.

Having one foreground neighbour (see Figure 4.5(a)) implies that the pixel is the end of a stroke and having seven or eight neighbours (see Figure 4.5(b)) would yield an erosion upon deletion. Lines which are only one-pixel thick will definitely be eroded to produce a discontinuity.

These rules can be applied repeatedly, to find candidate pixels first in the east and south directions and then in the west and north. The candidate points are deleted after each pass. An interesting point to note in the above algorithm is the inconsistency in the type of neighbourhood considered when

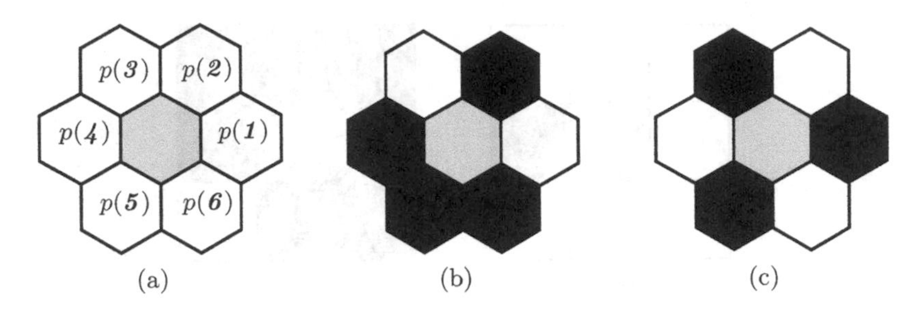

Fig. 4.6. Hexagonal thinning (a) neighbours of a candidate point (b and c) examples.

it comes to foreground and background pixels. For the foreground, an eight-neighbourhood is used whereas for the background a four-neighbourhood is used. Pixels in the latter are at unit distance whereas those in the former are at different distances from the central pixel. The reason for such usage goes back to the Jordan curve theorem for partitioning a plane $\mathbb{R}^2$ using a simple curve and the difficulty in translating it to a sampled space $\mathbb{Z}^2$. The theorem states that any simple closed curve partitions the space into two unconnected regions, namely an unbounded exterior and a bounded interior. It can be shown that with square sampling, using only four (or eight) connectivity for both the foreground and background pixels can lead to paradoxical results for partitioning. For example, we can end up with situations where a simple closed curve partitions the $\mathbb{Z}^2$ space into more than two regions or into two regions which are connected. Such problems can be avoided and the theorem can be translated for the square grid only by using different types of connectivity for the curve and its complement.

On a hexagonally sampled space however, there exists only one type of natural connectivity (six) and hence there is no such problem. As a result, the algorithm for skeletonisation on a hexagonal lattice can be greatly simplified. We will now show how to develop the skeletonisation algorithm for a hexagonal grid using the HIP framework. Figure 4.6(a) illustrates an example thinning scenario. Let a HIP structure p be indexed by $\boldsymbol{x} \in \mathbb{G}^1$. There are two computations carried out in thinning operations. The first is the number of non-zero neighbours of the origin:

$$N = p(\mathit{1}) + \cdots + p(\mathit{5}) + p(\mathit{6}) \tag{4.1}$$

The second is the the crossing number which is the number of 0 to 1 transitions in a complete transition around all the neighbourhood points:

$$S = \sum_{x \in \mathbb{G}^1} |p(\boldsymbol{x} \oplus_{\mathit{1}} \mathit{1}) - p(\boldsymbol{x})| \tag{4.2}$$

Note that the addition above is the closed addition operator described in Section 3.2.3. Two examples are illustrated in Figures 4.6(b) and in 4.6(c) where the dark pixels stand for foreground pixels and have value 1. The gray pixels are the pixels of interest. In these examples, $N = 4$, $S = 2$ and $N = 3$, $S = 3$, respectively. It is possible to generate a set of axioms that can be applied to skeletonise, which are analogous to the rules applied to square images:

1. $N > 1$ and $N < 6$
2. $S = 2$
3. $p(\mathit{1})p(\mathit{2})p(\mathit{3}) = 0$
4. $p(\mathit{1})p(\mathit{2})p(\mathit{6}) = 0$
5. $p(\mathit{1}) = 1$ and $S_{\mathit{1}} \neq 2$

Here, S_{1} refers to the crossing number centred on the point with HIP address ***1***. However, these rules will generate a skeletonisation with a bias in that they will tend to lie more towards HIP address ***1***. This is remedied by three additional rules:

6. $p(\mathit{4})p(\mathit{5})p(\mathit{6}) = 0$
7. $p(\mathit{3})p(\mathit{4})p(\mathit{5}) = 0$
8. $p(\mathit{4}) = 1$ and $S_{\mathit{4}} \neq 2$

These three rules are the same as rules 3 to 5 but with the HIP addresses negated which, as described in Section 3.2.2, is the same as a 180° rotation. The process of skeletonisation as described above is a two-pass algorithm. The first pass applies rules 1 - 5 listed above. The second pass applies rules 1, 2, and 6 - 8. The algorithm for a λ-level image, $f(\boldsymbol{x})$, proceeds as follows:

while have valid candidates for deletion **do**
  for all $\boldsymbol{x} \in \mathbb{G}^\lambda$ **do**
    if conditions 1 - 6 are true **then**
      $list \leftarrow list \cup \{\boldsymbol{x}\}$
    end if
  end for
  pop and delete all points in *list*
  for all $\boldsymbol{x} \in \mathbb{G}^\lambda$ **do**
    if conditions 1 - 2 and 6 - 8 are true **then**
      $list \leftarrow list \cup \{\boldsymbol{x}\}$
    end if
  end for
  pop and delete all points in *list*
end while

During the first pass, points which obey all five criteria are flagged for deletion. However, they are not deleted until the end of the second pass. This is to prevent the structure of the image being changed whilst the algorithm is progressing. In the second pass, valid candidates are once again flagged for deletion. After the second pass, all flagged points are deleted and the

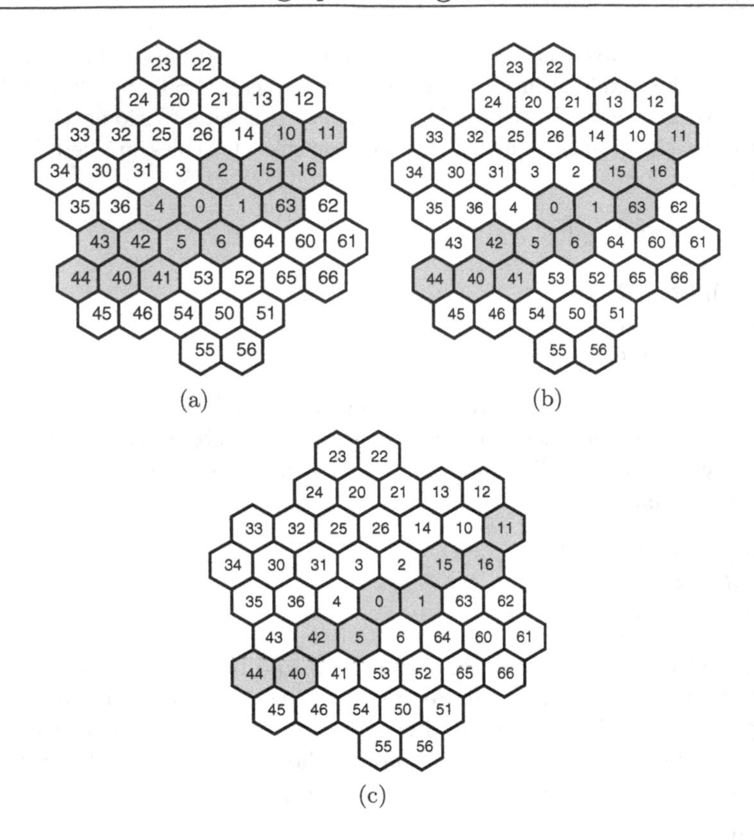

Fig. 4.7. Steps in the skeletonisation algorithm (a) original image (b) after first pass (c) after second pass of the algorithm..

process is repeated until no further points are marked for deletion, at which point the algorithm terminates. Figure 4.7 illustrates the result of applying this algorithm to a previously edge-detected two-level HIP image. In the first phase, the pixels in the northwest are flagged for removal while in the second phase pixels in the southeast are flagged.

4.2 Frequency Domain Processing

Frequency domain processing offers several computational advantages for image processing and analysis and the discrete Fourier transform(DFT) is by far the most widely used tool for this purpose. It is thus desirable that a fast, efficient method for computation of the discrete Fourier transform be developed. Several attempts have been made to develop fast algorithms for computation of DFT on the hexagonal lattice. More recently, one approach starts with a hexagonal image but computes the DFT on a square lattice and uses the Chinese remainder theorem to sort the input and output indices [78]. Yet another approach resamples the hexagonal image onto a larger (double in one direction) square lattice and computes the DFT [79]. Thus both these approaches transform the problem on to a square sampled space to solve the problem using established techniques. Hexagonal aggregates and their symmetry properties have been exploited in [80] to develop a fast algorithm for the DFT. The addressing scheme used here follows the GBT scheme which is similar to but different from that used in the HIP framework. All of the reported algorithms are based on the decimation in time technique proposed in the well known Cooley-Tukey algorithm [119] for fast computation. In this section, we develop a fast algorithm for computing the DFT within the HIP framework. We then showcase the use of DFT with a popular application, namely, linear filtering.

4.2.1 Fast algorithm for the discrete Fourier transform

As previously stated, direct implementation of the hexagonal discrete Fourier transform (see Section 3.4.5) is far too slow to be of any practical use for real images. For example, computing one point in a λ-layer HIP image requires 7^λ complex multiplications along with $7^\lambda - 1$ complex additions. A way to speed up the process is to use look-up tables to store the complex exponentials in the DFT. However a λ-level HIP image will require an array storing $7^{2\lambda}$ values, which means the memory requirements can be prohibitive. The methodologies to speed up computations are based on observations of the redundancy contained within the complex exponential terms. This is the basis for the Cooley-Tukey algorithm, and there is significant historical evidence that the methodology can actually be originally attributed to Gauss [120].

The specific method used here to derive the fast version of the discrete Fourier transform is the vector-radix form of the Cooley-Tukey algorithm [111]. In the following, only the forward DFT will be discussed since the inverse transform can be computed similarly, requiring only a change in the sign for the exponential term. The DFT can be formulated in a matrix form thus:

$$\mathbf{X} = \mathbf{E}\mathbf{x} \tag{4.3}$$

where, both $\mathbf{X}$ and $\mathbf{x}$ are 7^λ-long row vectors, whilst $\mathbf{E}$ is the transformation matrix. A comparison with the HDFT can be made by examining equation (3.54) which is repeated here for clarity:

$$X(\boldsymbol{k}) = \sum_{\boldsymbol{n} \in \mathbb{G}^\lambda} x(\boldsymbol{n}) \exp\left[-2\pi \mathrm{j} H(\boldsymbol{k})^T \mathbf{N}_\lambda^{-1} h(\boldsymbol{n})\right]$$

The solution to computational speed-up can be found by examining the matrix $\mathbf{E}$. This is defined as follows:

$$\mathbf{E} = \left[\exp\left(-2\pi \mathrm{j} H(\boldsymbol{k})^T \mathbf{N}_\lambda^{-1} h(\boldsymbol{n})\right)\right] \tag{4.4}$$

This is defined for all possible combinations of HIP addresses $\boldsymbol{k}$ and $\boldsymbol{n}$ which belong to $\mathbb{G}^\lambda$. $\mathbf{E}$ is a large matrix of size $7^\lambda \times 7^\lambda$. As previously stated, there is a large degree of redundancy in this matrix. For example, a trivial redundancy can be observed whenever $\boldsymbol{k}$ or $\boldsymbol{n}$ is $\boldsymbol{0}$ which results in the corresponding element of $\mathbf{E}$ becoming 1. A second redundancy can be found by noting that the exponential term in equation (4.4) represents the n roots of unity. The only difference between the different columns of the matrix is the order of the roots. This means that effectively $\mathbf{E}$ has only 7^λ unique values. As an example, Table 4.1 shows the various redundancies involved in computation of a discrete Fourier transform of a one-layer HIP image. Note that the result of the inner product for each of these combinations is a constant number. Also note, that the denominator in the inner product is the determinant of $\mathbf{N}$ and the numerator is one of the seven permissible values when using radix-7 notation. These redundancies also occur when computing the HDFT of an arbitrary sized HIP image and are exploited in the derivation of the vector-radix Cooley-Tukey algorithm.

The preceding discussion serves to illustrate the fact that the redundancy of the matrix is directly related to the specific periodicity matrix. The periodicity matrix can be factored into a product of two integer matrices as follows:

$$\mathbf{N}_\lambda = \mathbf{N}_\alpha \mathbf{N}_\beta \tag{4.5}$$

Next, we will show that any sample point in the HIP image can be written in terms of these matrices by using the notion of congruency. Two integer vectors, $\mathbf{a}$ and $\mathbf{b}$, are said to be congruent with respect to a matrix $\mathbf{B}$ if

$$\mathbf{a} = \mathbf{b} + \mathbf{B}\mathbf{r}, \qquad \forall \mathbf{r} \in \mathbb{Z}^n \tag{4.6}$$

Thus, due to the periodic nature of the lattice, every sample in an image, $x(\mathbf{n})$, is congruent to a sample from the sample space $I_{\mathbf{N}_\lambda}$. We will denote a vector $\mathbf{m} \in I_{\mathbf{N}_\lambda}$ which is congruent to $\mathbf{n}$ as:

Table 4.1. Redundancies in the computation of a HIP discrete Fourier transform.

$(\boldsymbol{n},\boldsymbol{k})$	$(\boldsymbol{n},\boldsymbol{k})$	$(\boldsymbol{n},\boldsymbol{k})$	$(\boldsymbol{n},\boldsymbol{k})$	$(\boldsymbol{n},\boldsymbol{k})$	$(\boldsymbol{n},\boldsymbol{k})$	$H(\boldsymbol{k})^T\mathbf{N}^{-1}h(\boldsymbol{n})$
$(1,1)$	$(2,6)$	$(3,5)$	$(4,4)$	$(5,3)$	$(6,2)$	$-\frac{2}{7}$
$(1,2)$	$(2,1)$	$(3,6)$	$(4,5)$	$(5,4)$	$(6,3)$	$\frac{1}{7}$
$(1,3)$	$(2,2)$	$(3,1)$	$(4,6)$	$(5,5)$	$(6,4)$	$\frac{3}{7}$
$(1,4)$	$(2,3)$	$(3,2)$	$(4,1)$	$(5,6)$	$(6,5)$	$\frac{2}{7}$
$(1,5)$	$(2,4)$	$(3,3)$	$(4,2)$	$(5,1)$	$(6,6)$	$-\frac{1}{7}$
$(1,6)$	$(2,5)$	$(3,4)$	$(4,3)$	$(5,2)$	$(6,1)$	$-\frac{3}{7}$

$$\mathbf{m} = \langle \mathbf{n} \rangle_{\mathbf{N}}$$

Using equation (4.5), any sample $\mathbf{n} \in I_{\mathbf{N}_\lambda}$ can then be expressed as:

$$\mathbf{n} = \langle \mathbf{p} + \mathbf{N}_\alpha \mathbf{q} \rangle_{\mathbf{N}} \tag{4.7}$$

where, $\mathbf{p} \in I_{\mathbf{N}_\alpha}$ and $\mathbf{q} \in I_{\mathbf{N}_\beta}$. Cardinalities of $I_{\mathbf{N}_\alpha}$ and $I_{\mathbf{N}_\beta}$ are $|\det \mathbf{N}_\alpha|$ and $|\det \mathbf{N}_\beta|$ respectively. Now, $\mathbf{p}$ and $\mathbf{q}$ serve to partition the sample space into two subsets so that any pair of vectors, one from $\mathbf{p}$ and one from $\mathbf{q}$, will yield a unique vector, $\mathbf{n}$, in $I_{\mathbf{N}_\lambda}$. Similarly, the frequency domain samples can be partitioned as follows:

$$\mathbf{k}^T = \left\langle \mathbf{l}^T + \mathbf{m}^T \mathbf{N}_\beta \right\rangle_{\mathbf{N}^T} \tag{4.8}$$

where, $\mathbf{l} \in J_{\mathbf{N}_\beta}$ and $\mathbf{m} \in J_{\mathbf{N}_\alpha}$. Once again, the cardinalities of $J_{\mathbf{N}_\alpha}$ and $J_{\mathbf{N}_\beta}$ are $|\det \mathbf{N}_\alpha|$ vectors and $|\det \mathbf{N}_\beta|$ respectively. The definition of the DFT as given in equation (3.47) can be rewritten using this partitioning as follows:

$$\begin{aligned} X(\mathbf{l} + \mathbf{N}_\beta^T \mathbf{m}) = & \sum_{\mathbf{p} \in I_{\mathbf{N}_\alpha}} \sum_{\mathbf{q} \in I_{\mathbf{N}_\beta}} x(\langle \mathbf{p} + \mathbf{N}_\alpha \mathbf{q} \rangle_{\mathbf{N}}) \cdot \\ & \exp\left[-2\pi \mathrm{j} (\mathbf{l}^T + \mathbf{m}^T \mathbf{N}_\beta) \mathbf{N}_\lambda^{-1} (\mathbf{p} + \mathbf{N}_\alpha \mathbf{q})\right] \end{aligned} \tag{4.9}$$

Expanding the exponential term and using $\mathbf{N}_\lambda = \mathbf{N}_\alpha \mathbf{N}_\beta$ we can simplify the above equation as follows:

$$X(\mathbf{l} + \mathbf{N}_\beta^T \mathbf{m}) = \sum_{\mathbf{p} \in I_{\mathbf{N}_\alpha}} \left(C(\mathbf{p}, \mathbf{l}) \exp\left[-2\pi \mathrm{j} \mathbf{l}^T \mathbf{N}_\lambda^{-1} \mathbf{p}\right]\right) \exp\left[-2\pi \mathrm{j} \mathbf{m}^T \mathbf{N}_\alpha^{-1} \mathbf{p}\right] \tag{4.10a}$$

where

$$C(\mathbf{p}, \mathbf{l}) = \sum_{\mathbf{q} \in I_{\mathbf{N}_\beta}} x(\langle \mathbf{p} + \mathbf{N}_\alpha \mathbf{q} \rangle_{\mathbf{N}}) \exp\left[-2\pi \mathrm{j} \mathbf{l}^T \mathbf{N}_\beta^{-1} \mathbf{q}\right] \tag{4.10b}$$

These two relations are the first level of decomposition of a decimation in space Cooley-Tukey FFT algorithm. Equation (4.10b) represents a two dimensional DFT of the image $x(\langle \mathbf{p} + \mathbf{N}_\alpha \mathbf{q} \rangle_{\mathbf{N}})$ taken with respect to the periodicity matrix $\mathbf{N}_\beta$. The region of support for this is $I_{\mathbf{N}_\beta}$ which is exactly one period of $x(\langle \mathbf{p} + \mathbf{N}_\alpha \mathbf{q} \rangle_{\mathbf{N}})$ over the vector variable $\mathbf{q}$. A different matrix-$\mathbf{N}_\beta$ DFT needs to be evaluated for each vector $\mathbf{p}$. Thus, the total number of transforms needed to compute the entire DFT is $|\det \mathbf{N}_\alpha|$. The other equation (4.10a) prescribes how to combine the outputs of the different matrix-$\mathbf{N}_\beta$ DFTs to produce the final matrix-$\mathbf{N}_\lambda$ DFT. The exponential terms $\exp\left[-2\pi \mathrm{j} \mathbf{l}^T \mathbf{N}_\lambda^{-1} \mathbf{p}\right]$ that multiply $C(\mathbf{p}, \mathbf{l})$ are commonly known as twiddle factors. The results of these multiplications are then combined using a number of matrix-$\mathbf{N}_\alpha$ DFTs.

Up to this point, the derivation has been in terms of vectors. However, the HDFT defined in Section 3.4.5 was purely in terms of HIP addresses. Thus the previous equation needs to be redefined in terms of HIP addresses. This can be achieved via the mapping functions which were previously defined in equations (3.52) and (3.53). The desired mappings are:

$$\mathbf{p} = h(\boldsymbol{p}) \qquad \mathbf{q} = h(\boldsymbol{q})$$
$$\mathbf{m} = H(\boldsymbol{m}) \qquad \mathbf{l} = H(\boldsymbol{l})$$

where $\boldsymbol{p}, \boldsymbol{m} \in \mathbb{G}^{\alpha}$ and $\boldsymbol{q}, \boldsymbol{l} \in \mathbb{G}^{\beta}$. In other words $\boldsymbol{p}$ and $\boldsymbol{m}$ are HIP addresses with α digits. Also, $\boldsymbol{q}$ and $\boldsymbol{l}$ are HIP addresses with β digits. Thus we have for the congruency relations:

$$\mathbf{p} + \mathbf{N}_{\alpha}\mathbf{q} = h(\boldsymbol{qp}) \tag{4.11a}$$
$$\mathbf{l} + \mathbf{N}_{\beta}^{T}\mathbf{m} = H(\boldsymbol{ml}) \tag{4.11b}$$

For example if $\boldsymbol{q} = \mathit{4}$ and $\boldsymbol{p} = \mathit{2}$ then vector $\mathbf{n}$ associated with $\mathbf{p} + \mathbf{N}_{\alpha}\mathbf{q}$ has the HIP address of $\mathit{42}$. The decimation in space equations can thus be rewritten in terms of the HIP addresses as:

$$X(\boldsymbol{ml}) = \sum_{\boldsymbol{p} \in \mathbb{G}^{\alpha}} C(\boldsymbol{p}, \boldsymbol{l}) \exp\left[-2\pi \mathrm{j} H(\boldsymbol{l})^{T} \mathbf{N}_{\lambda}^{-1} h(\boldsymbol{p})\right] \exp\left[-2\pi \mathrm{j} H(\boldsymbol{m})^{T} \mathbf{N}_{\alpha}^{-1} h(\boldsymbol{p})\right] \tag{4.12a}$$

where

$$C(\boldsymbol{p}, \boldsymbol{l}) = \sum_{\boldsymbol{q} \in \mathbb{G}^{\beta}} x(\boldsymbol{qp}) \exp\left[-2\pi \mathrm{j} H(\boldsymbol{l})^{T} \mathbf{N}_{\beta}^{-1} h(\boldsymbol{q})\right] \tag{4.12b}$$

Since the twiddle factor is a function of $\boldsymbol{p}$ and $\boldsymbol{l}$ it is convenient to denote it by a weight function used commonly for DFTs as follows:

$$W_{\alpha}(\boldsymbol{p}, \boldsymbol{l}) = \exp\left[-2\pi \mathrm{j} H(\boldsymbol{l})^{T} \mathbf{N}_{\alpha}^{-1} h(\boldsymbol{p})\right] \tag{4.13}$$

For convenience, we also define a new function, D, to denote the product of C and W:

$$D_{\lambda}(\boldsymbol{p}, \boldsymbol{l}) = C(\boldsymbol{p}, \boldsymbol{l}) W_{\lambda}(\boldsymbol{p}, \boldsymbol{l})$$

We now illustrate the details involved in the above decomposition using an example for a two-level image. In this case, the periodicity matrix $\mathbf{N}_2$ can be decomposed into two other, identical, periodicity matrices each of which is $\mathbf{N}_1$. Equations (4.12) can thus be written as:

$$C(\boldsymbol{p},\boldsymbol{l}) = \sum_{\boldsymbol{q}\in\mathbb{G}^1} x(\boldsymbol{qp}) W_1(\boldsymbol{q},\boldsymbol{l})$$

$$X(\boldsymbol{ml}) = \sum_{\boldsymbol{p}\in\mathbb{G}^1} C(\boldsymbol{p},\boldsymbol{l}) W_2(\boldsymbol{p},\boldsymbol{l}) W_1(\boldsymbol{p},\boldsymbol{m})$$

In these equations, $\boldsymbol{p}$, $\boldsymbol{l}$, and $\boldsymbol{m}$ all belong to $\mathbb{G}^1$. Using matrix notation, $C(p,l)$ is written as:

$$\begin{bmatrix} C(\boldsymbol{p},\boldsymbol{0}) \\ C(\boldsymbol{p},\boldsymbol{1}) \\ C(\boldsymbol{p},\boldsymbol{2}) \\ C(\boldsymbol{p},\boldsymbol{3}) \\ C(\boldsymbol{p},\boldsymbol{4}) \\ C(\boldsymbol{p},\boldsymbol{5}) \\ C(\boldsymbol{p},\boldsymbol{6}) \end{bmatrix} = \begin{bmatrix} 1 & 1 & 1 & 1 & 1 & 1 & 1 \\ 1 & b^* & a & c & b & a^* & c^* \\ 1 & a & c & b & a^* & c^* & b^* \\ 1 & c & b & a^* & c^* & b^* & a \\ 1 & b & a^* & c^* & b^* & a & c \\ 1 & a^* & c^* & b^* & a & c & b \\ 1 & c^* & b^* & a & c & b & a^* \end{bmatrix} \begin{bmatrix} x(\boldsymbol{0p}) \\ x(\boldsymbol{1p}) \\ x(\boldsymbol{2p}) \\ x(\boldsymbol{3p}) \\ x(\boldsymbol{4p}) \\ x(\boldsymbol{5p}) \\ x(\boldsymbol{6p}) \end{bmatrix} \tag{4.14}$$

where $a = \exp[\frac{-2\pi j}{7}]$, $b = \exp[\frac{-4\pi j}{7}]$, and $c = \exp[\frac{-6\pi j}{7}]$; a^* denotes the complex conjugate of a and $\boldsymbol{p} \in \mathbb{G}^1$. Hence, the term $W_1(\boldsymbol{q},\boldsymbol{l})$ represents the seven roots of unity. In a similar fashion, it is possible to write a matrix form of equation (4.12):

$$\begin{bmatrix} X(\boldsymbol{l0}) \\ X(\boldsymbol{l1}) \\ X(\boldsymbol{l2}) \\ X(\boldsymbol{l3}) \\ X(\boldsymbol{l4}) \\ X(\boldsymbol{l5}) \\ X(\boldsymbol{l6}) \end{bmatrix} = \begin{bmatrix} 1 & 1 & 1 & 1 & 1 & 1 & 1 \\ 1 & b^* & a & c & b & a^* & c^* \\ 1 & a & c & b & a^* & c^* & b^* \\ 1 & c & b & a^* & c^* & b^* & a \\ 1 & b & a^* & c^* & b^* & a & c \\ 1 & a^* & c^* & b^* & a & c & b \\ 1 & c^* & b^* & a & c & b & a^* \end{bmatrix} \begin{bmatrix} C(\boldsymbol{0},\boldsymbol{l})W_2(\boldsymbol{0},\boldsymbol{l}) \\ C(\boldsymbol{1},\boldsymbol{l})W_2(\boldsymbol{1},\boldsymbol{l}) \\ C(\boldsymbol{2},\boldsymbol{l})W_2(\boldsymbol{2},\boldsymbol{l}) \\ C(\boldsymbol{3},\boldsymbol{l})W_2(\boldsymbol{3},\boldsymbol{l}) \\ C(\boldsymbol{4},\boldsymbol{l})W_2(\boldsymbol{4},\boldsymbol{l}) \\ C(\boldsymbol{5},\boldsymbol{l})W_2(\boldsymbol{5},\boldsymbol{l}) \\ C(\boldsymbol{6},\boldsymbol{l})W_2(\boldsymbol{6},\boldsymbol{l}) \end{bmatrix} \tag{4.15}$$

The matrix is identical to that shown in equation (4.14). Hence, both stages require only a radix-7 computation. Finally, a matrix for all the associated twiddle factors can be computed:

$$\mathbf{W}_2(\boldsymbol{p},\boldsymbol{l}) = \begin{bmatrix} 1 & 1 & 1 & 1 & 1 & 1 & 1 \\ 1 & f^* & d^* & e & f & d & e^* \\ 1 & d^* & e & f & d & e^* & f^* \\ 1 & e & f & d & e^* & f^* & d^* \\ 1 & f & d & e^* & f^* & d^* & e \\ 1 & d & e^* & f^* & d^* & e & f \\ 1 & e^* & f^* & d^* & e & f & d \end{bmatrix} \tag{4.16}$$

Here $d = \exp[\frac{-6\pi j}{49}]$, $e = \exp[\frac{-10\pi j}{49}]$, and $f = \exp[\frac{-16\pi j}{49}]$. The matrix $W_2(\boldsymbol{p},\boldsymbol{l})$ and equations (4.14) and (4.15) are sufficient to describe the entire reduced DFT for a second-order HIP image. For the example under consideration, these equations are also the complete DFT. The processing steps involved are illustrated in Figure 4.8. The figure is divided into two sides. Each side illustrates an individual stage in the HFFT process. Each of the blocks labelled DFT is a radix-7 DFT, such as the one used to compute $C(\boldsymbol{p}, \boldsymbol{l})$. After the results of the first set of DFTs are computed, the results are multiplied by the twiddle factors, $W_2(\boldsymbol{p}, \boldsymbol{l})$ to form $D(\boldsymbol{p}, \boldsymbol{l})$. These serve as inputs to the DFTs on the right hand side of Figure 4.8. Each of the individual radix-7 DFT boxes is further expanded in Figure 4.9. The solid lines represent multiplication by the individual exponential terms and the dashed lines represent multiplication by their complex conjugates.

The reduction in computations in the entire process should now be apparent. For example, to compute the DFT of a two-layer HIP image directly, will require 2352 complex additions and 2401 complex multiplications. By contrast, the reduced form, as shown in the figure, requires only 588 complex additions and 735 complex multiplications.

The example given above naturally leads to the complete Cooley-Tukey decimation in space decomposition decomposition for the HIP framework. In the example, a single decomposition was performed to reduce the order of the DFT by one. It is possible to continue this process an arbitrary number of times to produce a complete decomposition and the FFT. This can be achieved by starting with equation (4.12) and using the weight function in equation (4.13). This allows the DFT decomposition to be written in a single equation as:

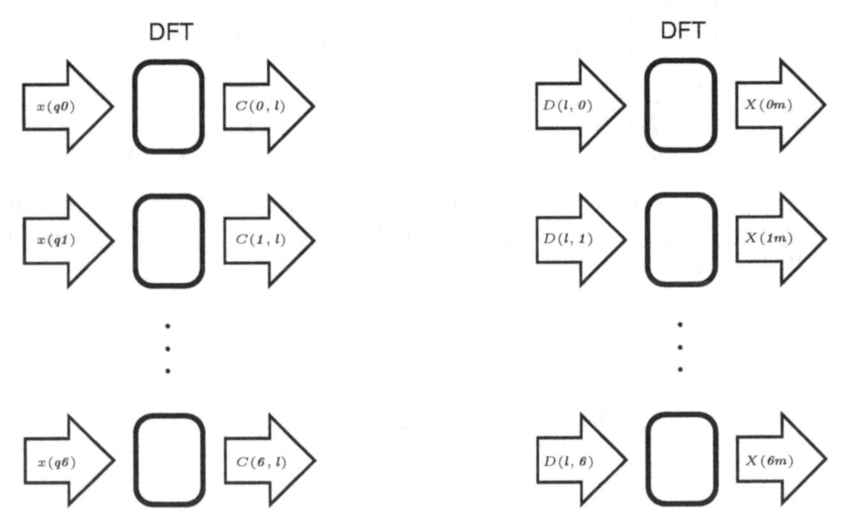

Fig. 4.8. FFT processing stages for a two-layer HIP image.

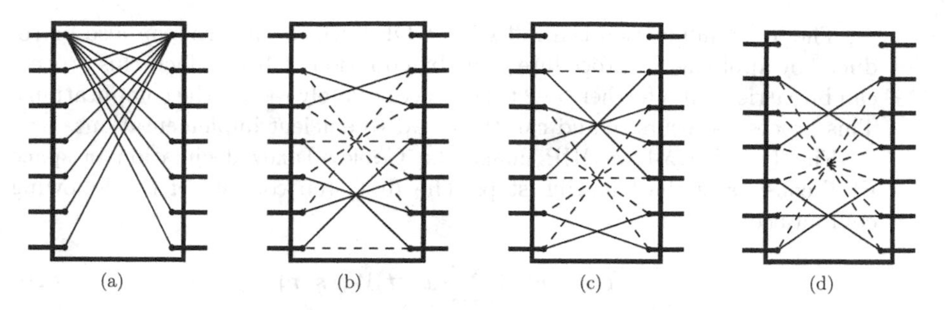

Fig. 4.9. An expansion of the DFT box in Figure 4.8: exponential term equals (a) 1, (b) $\exp[\frac{-2\pi \jmath}{7}]$, (c) $\exp[\frac{-4\pi \jmath}{7}]$, and (d) $\exp[\frac{-6\pi \jmath}{7}]$.

$$X(\boldsymbol{ml}) = \sum_{\boldsymbol{p} \in \mathbb{G}^{\alpha}} \left(\left(\sum_{\boldsymbol{q} \in \mathbb{G}^{\beta}} W_{\beta}(\boldsymbol{q}, \boldsymbol{l}) x(\boldsymbol{qp}) \right) W_{\lambda}(\boldsymbol{p}, \boldsymbol{l}) \right) W_{\alpha}(\boldsymbol{p}, \boldsymbol{m}) \tag{4.17}$$

The order of this can be reduced, as in the example, using a single periodicity matrix $\mathbf{N}_1$, which yields the following:

$$X(\boldsymbol{ml}) = \sum_{\boldsymbol{p} \in \mathbb{G}^{1}} \left(\left(\sum_{\boldsymbol{q} \in \mathbb{G}^{\beta}} W_{\beta}(\boldsymbol{q}, \boldsymbol{l}) x(\boldsymbol{qp}) \right) W_{\beta+1}(\boldsymbol{p}, \boldsymbol{l}) \right) W_{1}(\boldsymbol{p}, \boldsymbol{m}) \tag{4.18}$$

This equation shows two distinct DFT computations: a radix-7 DFT characterised by the outer summation and a radix-7^{β} DFT as seen in the inner equation. The inner summation is, of course, multiplied by an appropriate twiddle factor. The inner DFT can also be reduced in a similar fashion. Before doing this some definitions are required:

$$\begin{aligned} \boldsymbol{r} &= \boldsymbol{r}_{\beta-1}\overline{\boldsymbol{r}} = \boldsymbol{r}_{\beta-1} \cdots \boldsymbol{r}_1 \boldsymbol{r}_0 \\ \boldsymbol{q} &= \overline{\boldsymbol{q}}\boldsymbol{q}_0 = \boldsymbol{q}_{\beta-1} \cdots \boldsymbol{q}_1 \boldsymbol{q}_0 \end{aligned}$$

Here $\overline{\boldsymbol{r}}$ are the digits of $\boldsymbol{r}$ with the highest order digit, $\boldsymbol{r}_{\beta-1}$, removed. Similarly, $\overline{\boldsymbol{q}}$ is $\boldsymbol{q}$ without the first digit, $\boldsymbol{q}_0$. Hence we rewrite equation (4.12) as:

$$\begin{aligned} C(\boldsymbol{p}, \boldsymbol{r}) &= \sum_{\boldsymbol{q} \in \mathbb{G}^{\beta}} W_{\beta}(\boldsymbol{q}, \boldsymbol{r}) x(\boldsymbol{qp}) \\ &= \sum_{\boldsymbol{q}_0 \in \mathbb{G}^{1}} \left(\left(\sum_{\overline{\boldsymbol{q}} \in \mathbb{G}^{\beta-1}} W_{\beta-1}(\overline{\boldsymbol{q}}, \overline{\boldsymbol{r}}) x(\boldsymbol{qp}) \right) W_{\beta}(\boldsymbol{q}_0, \overline{\boldsymbol{r}}) \right) W_1(\boldsymbol{q}_0, \overline{\boldsymbol{r}}) \end{aligned} \tag{4.19}$$

The inner summation is a radix-$7^{\beta-1}$ DFT which, in turn, can also be reduced by applying the procedure given by equation (4.18). In fact, this process can be carried out a further $\beta-2$ times to completely reduce the computations. This process is recursive and can thus lead to efficient implementations.

Finally, for a λ-layer HIP image the Cooley-Tukey decimation in space FFT consists of the following steps The first step consists of the following computation:

$$C_1(\boldsymbol{t},\boldsymbol{r}) = \sum_{\boldsymbol{s}\in\mathbb{G}^1} x(\boldsymbol{st})W_1(\boldsymbol{s},\boldsymbol{r}) \tag{4.20}$$

where $\boldsymbol{t} \in \mathbb{G}^{\lambda-1}$ and $\boldsymbol{r} \in \mathbb{G}^1$. The subsequent steps can be written as:

$$C_\gamma(\boldsymbol{t},\boldsymbol{rn}) = \sum_{\boldsymbol{s}\in\mathbb{G}^1} C_{\gamma-1}(\boldsymbol{st},\boldsymbol{n})W_1(\boldsymbol{s},\boldsymbol{r})W_\gamma(\boldsymbol{s},\boldsymbol{n}) \tag{4.21}$$

where $\boldsymbol{t} \in \mathbb{G}^{\lambda-\gamma}$, $\boldsymbol{r} \in \mathbb{G}^1$, and $\boldsymbol{n} \in \mathbb{G}^{\gamma-1}$. The final step which gives the DFT is:

$$X(\boldsymbol{rn}) = \sum_{\boldsymbol{s}\in\mathbb{G}^1} C_{\lambda-1}(\boldsymbol{s},\boldsymbol{n})W_1(\boldsymbol{s},\boldsymbol{r})W_\lambda(\boldsymbol{s},\boldsymbol{n}) \tag{4.22}$$

where $\boldsymbol{r} \in \mathbb{G}^1$ and $\boldsymbol{n} \in \mathbb{G}^{\lambda-1}$. In equations (4.21) and (4.22), the final term $W_\gamma(\boldsymbol{s},\boldsymbol{n})$ and $W_\lambda(\boldsymbol{s},\boldsymbol{n})$ respectively, are the twiddle factors for the current stage of the FFT algorithm. This scales the inputs in a particular stage of the FFT algorithm, as stated previously. Thus at each stage of the algorithm, a radix-7 DFT is performed.

The above procedure can be extended to compute the inverse hexagonal FFT in an iterative manner. It requires the use of the complex conjugate of the previously defined twiddle factor, $W_\alpha^*(\boldsymbol{p},\boldsymbol{l})$. Thus, the inverse FFT for a λ-layer HIP frequency image can be written as:

$$K_1(\boldsymbol{t},\boldsymbol{r}) = \frac{1}{7}\sum_{\boldsymbol{s}\in\mathbb{G}^1} X(\boldsymbol{st})W_1^*(\boldsymbol{s},\boldsymbol{r}) \tag{4.23}$$

where $\boldsymbol{t} \in \mathbb{G}^{\lambda-1}$ and $\boldsymbol{r} \in \mathbb{G}^1$. The subsequent steps can be written:

$$K_\gamma(\boldsymbol{t},\boldsymbol{rn}) = \frac{1}{7}\sum_{\boldsymbol{s}\in\mathbb{G}^1} K_{\gamma-1}(\boldsymbol{st},\boldsymbol{n})W_1^*(\boldsymbol{s},\boldsymbol{r})W_\gamma^*(\boldsymbol{s},\boldsymbol{n}) \tag{4.24}$$

where $\boldsymbol{t} \in \mathbb{G}^{\lambda-\gamma}$, $\boldsymbol{r} \in \mathbb{G}^1$, and $\boldsymbol{n} \in \mathbb{G}^{\gamma-1}$. The final step which gives the resulting IDFT is:

$$x(\boldsymbol{rn}) = \sum_{\boldsymbol{s}\in\mathbb{G}^1} K_{\lambda-1}(\boldsymbol{s},\boldsymbol{n})W_1^*(\boldsymbol{s},\boldsymbol{r})W_\lambda^*(\boldsymbol{s},\boldsymbol{n}) \tag{4.25}$$

where $\boldsymbol{r} \in \mathbb{G}^1$, and $\boldsymbol{n} \in \mathbb{G}^{\lambda-1}$. The twiddle factors for the IHFFT are the complex conjugates of the HFFT results. An example of these computations

Table 4.2. Computations required to compute the HFFT and IHFFT for different sized HIP images.

λ	HFFT	IHFFT
1	$X(\boldsymbol{r}) = \sum x(\boldsymbol{s})W_1(\boldsymbol{s},\boldsymbol{r})$	$x(\boldsymbol{n}) = \frac{1}{7}\sum X(\boldsymbol{s})W_1^*(\boldsymbol{s},\boldsymbol{r})$
2	$C_1(\boldsymbol{t},\boldsymbol{r}) = \sum x(\boldsymbol{st})W_1(\boldsymbol{s},\boldsymbol{r})$ $X(\boldsymbol{rn}) = \sum C_1(\boldsymbol{s},\boldsymbol{n})W_1(\boldsymbol{s},\boldsymbol{r})W_2(\boldsymbol{s},\boldsymbol{n})$	$K_1(\boldsymbol{t},\boldsymbol{n}) = \frac{1}{7}\sum X(\boldsymbol{st})W_1^*(\boldsymbol{s},\boldsymbol{r})$ $x(\boldsymbol{rn}) = \frac{1}{7}\sum K_1(\boldsymbol{s},\boldsymbol{n})W_1^*(\boldsymbol{s},\boldsymbol{r})W_2^*(\boldsymbol{s},\boldsymbol{n})$
3	$C_1(\boldsymbol{t},\boldsymbol{r}) = \sum x(\boldsymbol{st})W_1(\boldsymbol{s},\boldsymbol{r})$ $C_2(\boldsymbol{t},\boldsymbol{rn}) = \sum C_1(\boldsymbol{st},\boldsymbol{n})W_1(\boldsymbol{s},\boldsymbol{r})W_2(\boldsymbol{s},\boldsymbol{n})$ $X(\boldsymbol{rn}) = \sum C_2(\boldsymbol{s},\boldsymbol{n})W_1(\boldsymbol{s},\boldsymbol{r})W_3(\boldsymbol{s},\boldsymbol{n})$	$K_1(\boldsymbol{t},\boldsymbol{r}) = \frac{1}{7}\sum X(\boldsymbol{st})W_1^*(\boldsymbol{s},\boldsymbol{r})$ $K_2(\boldsymbol{t},\boldsymbol{rn}) = \frac{1}{7}\sum K_1(\boldsymbol{st},\boldsymbol{n})W_1^*(\boldsymbol{s},\boldsymbol{r})W_2^*(\boldsymbol{s},\boldsymbol{n})$ $x(\boldsymbol{rn}) = \frac{1}{7}\sum K_2(\boldsymbol{s},\boldsymbol{n})W_1^*(\boldsymbol{s},\boldsymbol{r})W_3^*(\boldsymbol{s},\boldsymbol{n})$

for both the HFFT and the IHFFT is shown in Table 4.2. The first row is the basic level computation which is the trivial case. Examination of the formulae given yields useful information about the computational aspects of the HFFT algorithm that has been presented. A given λ-layer HIP image will require λ stages to completely compute the HDFT. Each stage requires a maximum of $7^{\lambda+1}$ complex multiplications. At each of these stages, there is an HFFT requiring $(\lambda - 1)$ stages and 7^{λ} multiplications. Thus, the overall number of complex multiplications is $\lambda 7^{\lambda+1} + (\lambda - 1)7^{\lambda}$. Due to redundancies as illustrated in Table 4.1, the number of multiplications can be reduced to $\lambda 7^{\lambda}$ actual multiplications. For an image of size N $(=7^{\lambda})$ then this leads finally to $N \log_7 N$ complex multiplications or $O(N \log_7 N)$.

4.2.2 Linear Filtering

Linear filtering is a commonly employed tool in both spatial and frequency domain image processing. Filtering, as a technique, serves to modify the input image to emphasise desirable features and eliminate unwanted features. The resulting image can then be used for subsequent processing. This is analogous to the processing that is performed in the visual cortex in the human brain (see Section 2.1.2). Linear filtering is of importance due to a wide range of applications. When the filter kernels are large, it is more efficient to implement the filtering operation in the frequency domain since the HFFT algorithm will help in reducing the computational expense.

Given a spatial domain HIP image $f(\boldsymbol{n})$ and a linear, shift-invariant operator $h(\boldsymbol{n})$, the result of filtering the former with the latter is the image, $g(\boldsymbol{n})$, which is found as:

$$g(\boldsymbol{n}) = h(\boldsymbol{n}) \circledast f(\boldsymbol{n})$$

Taking Fourier transform of both sides we have:

$$G(\boldsymbol{k}) = H(\boldsymbol{k})F(\boldsymbol{k})$$

where, G, H, and F are the Fourier transforms of the original HIP images g, h, and f, respectively. In other words, given the Fourier transform operator defined as $\mathscr{F}$, then:

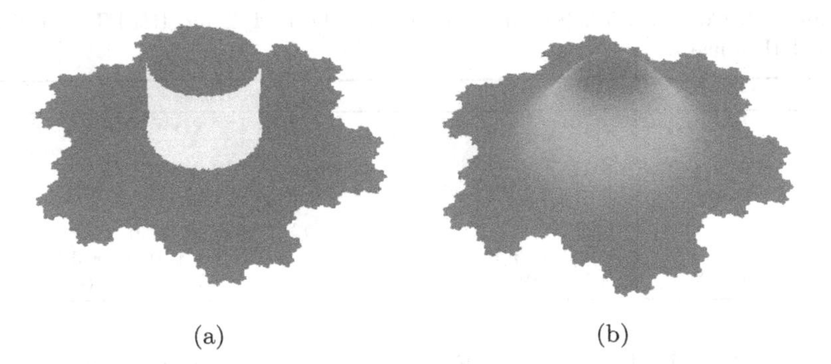

(a) (b)

Fig. 4.10. The lowpass filter functions: (a) ideal (b) Butterworth.

$$g(\boldsymbol{n}) = \mathscr{F}^{-1}[H(\boldsymbol{k})F(\boldsymbol{k})] \tag{4.26}$$

Typically, for a given image f, the problem is to select H so that the resulting image, g, exhibits some desired characteristics. This linear filtering case study will focus on lowpass and highpass filters. In each category, two examples of filtering were examined: ideal and non-ideal. A Butterworth kernel function was chosen for the non-ideal filter study. The chosen kernels are as below. For the ideal filter:

$$H_{IL}(\boldsymbol{g}) = \begin{cases} 1, & |\boldsymbol{g}| \leq R \\ 0, & |\boldsymbol{g}| > R \end{cases} \tag{4.27}$$

and for the non-ideal filter

$$H_{BL}(\boldsymbol{g}) = \frac{1}{1 + C(\frac{|\boldsymbol{g}|}{R})^{2n}} \tag{4.28}$$

where $|\boldsymbol{g}|$ is the radius from the image centre. It is measured using Euclidean distance (measure d_9, equation (3.33)); R is the cutoff frequency; C is the magnitude at the point where $|\boldsymbol{g}| = R$; and n is the filter order. These filter functions are shown in Figure 4.10 in the frequency domain. The highpass filter kernels H_{IH} and H_{BH} respectively, are obtained by reversing the equalities in equation (4.27) and negating the exponential in equation (4.28). Thus we have

$$H_{IH}(\boldsymbol{g}) = \begin{cases} 1, & |\boldsymbol{g}| > R \\ 0, & |\boldsymbol{g}| \leq R \end{cases} \tag{4.29}$$

and

Fig. 4.11. The test images for filtering: (a) synthetic image (S1) (b) real image (R1) (c) magnitude spectrum of S1 (d) magnitude spectrum of R1.

$$H_{BH}(\boldsymbol{g}) = \frac{1}{1 + C(\frac{R}{|\boldsymbol{g}|})^{2n}} \tag{4.30}$$

Here, the parameter C is set to be $(\sqrt{2} - 1)$ which causes the filter's amplitude to drop by approximately 3 dB at the cutoff frequency.

Two test images, as illustrated in Figure 4.11, were chosen for the filtering experiment. The illustrated images employ a five-layer HIP structure. The synthetic image, Figure 4.11(a), was chosen as it has rings of increasing frequency which should be affected when the image is filtered. The Fourier magnitude spectra of S1 and R1 are illustrated in figures 4.11(c) and 4.11(d) respectively. For S1, the rings are clearly depicted in the frequency domain though they are copied along the six axes of symmetry of the HIP image. For R1, a peak corresponding to the DC component of the image can be observed. Apart from this peak, R1 has no other distinguishing features in the frequency spectrum. Comparison between the ideal and non-ideal filters can thus be performed by requiring that the power loss incurred during the filtering operation is constant. Thus, the differences in visual quality of the filtered image are solely due to the filter's characteristics. For a HIP image, F, with λ layers, the power ratio is defined as:

$$\beta = \frac{\sum_{|\boldsymbol{g}| \leq R} |F(\boldsymbol{g})|^2}{\sum_{\boldsymbol{g} \in \mathbb{G}^\lambda} |F(\boldsymbol{g})|^2} \tag{4.31}$$

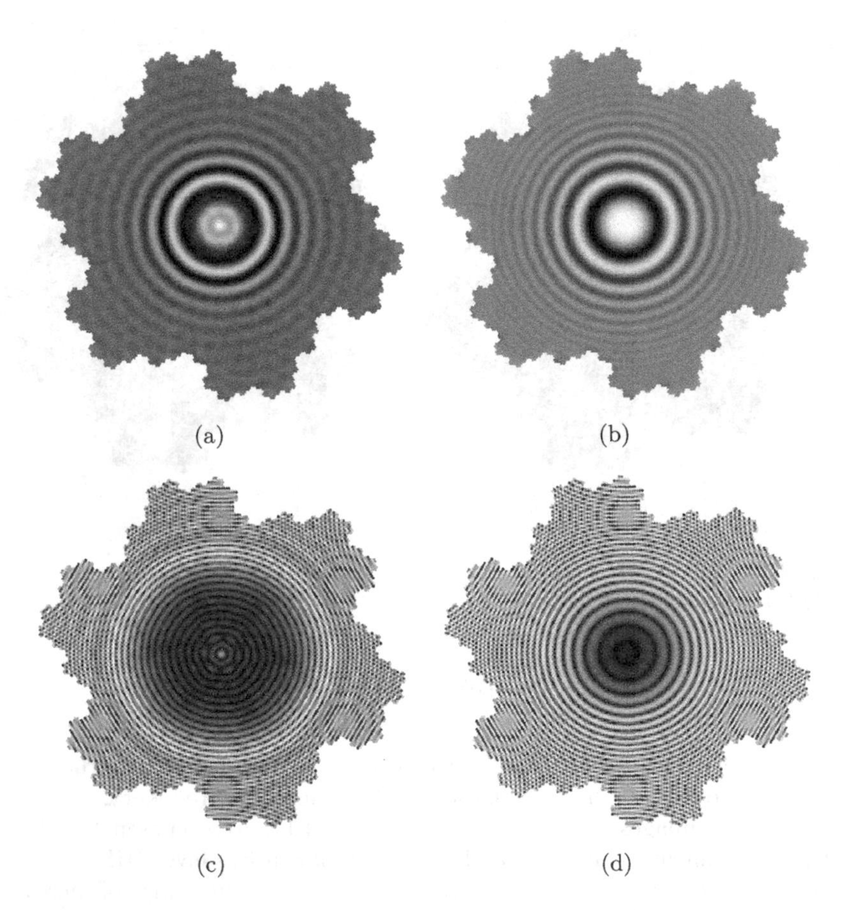

Fig. 4.12. The results for S1 after linear filtering: (a) ideal LPF (b) non-ideal LPF (c) ideal HPF (d) non-ideal HPF.

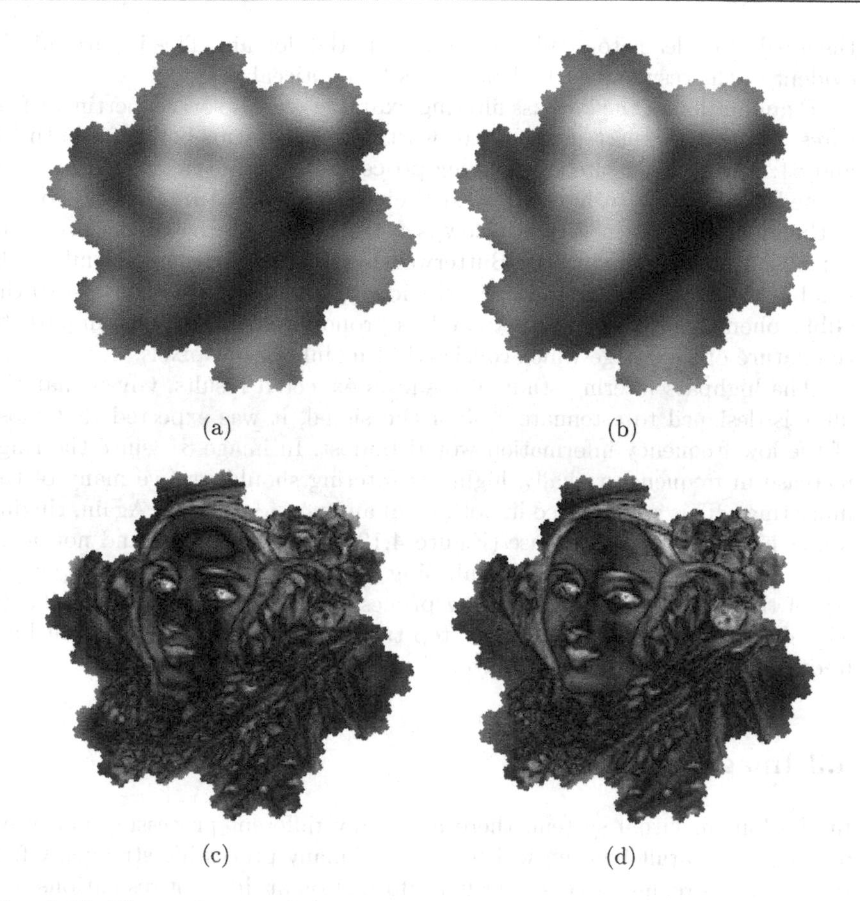

Fig. 4.13. The results for R1 after linear filtering: (a) ideal LPF (b) non-ideal LPF (c) ideal HPF (d) non-ideal HPF.

Thus β is the ratio of power contained within the passband and the total power in the given image. For the case study β was set to be 0.2. For the synthetic image, Figure 4.10(a), this yielded the cutoff frequency R to be 330 and for the real image, Figure 4.10(b), this gave a cutoff frequency of 25 points. A second order Butterworth filter was implemented for the comparison ($n = 2$). Keeping β the same, the cutoff frequency was 145 for the synthetic image and 14 for the real image. The lowpass and highpass filtered images are shown in Figure 4.12 for the synthetic image and in Figure 4.13 for the real image .

There is more ringing in the results of the ideal filter compared with the Butterworth filter, which is to be expected. The ringing is due to the Gibbs phenomenon wherein a sharp transition in the frequency domain, as present in the ideal filter, leads to oscillations in the spatial domain. This is particularly evident in the results for S1, though it is less noticeable for R1.

Examination of the lowpass filtering example reveals several pertinent features. Firstly, as expected, the Butterworth filter performs better for both R1 and S1. Secondly, despite the blurring process, the visible rings in S1 are well formed and appear to have retained their shape well. An exception to this is the central circle in the ideal lowpass filter case which has an obviously hexagonal shape. Finally, the Butterworth filter has produced significantly less blurring on R1 as compared to the ideal filter. The distortion due to the Gibbs phenomena also appears to be less pronounced. This is due, in part, to the nature of the image which consisted of mainly bright pixels.

The highpass filtering study also shows expected results. Given that the filter is designed to attenuate 20% of the signal, it was expected that most of the low frequency information would be lost. In image S1, since the rings increase in frequency radially, highpass filtering should remove many of the inner rings. This is confirmed in both ideal and non-ideal cases. Again, ringing was exhibited in the ideal case (Figure 4.12(c)). In both ideal and non-ideal filtering, there was an additional aliasing effect observed around the periphery of the image due to the filtering process. For R1, the effect is similar to edge detection since edges, being a step transition in intensity, represent high frequency information in the image.

4.3 Image pyramids

In the human visual system, there are many different processing pathways operating in parallel within which there are many processing streams. Often these streams require access to visual information at different resolutions. For instance, in the analysis of form, it is advantageous to process a low resolution version of the object and use this to define the shape without the confusion caused by excessive detail. The simple and complex cells in the primary visual cortex have a range of tuning parameters such as radial frequency, orientation and bandwidth. These imply that the scale at which an image is being analysed is variable [3, 121]. Multiresolution image analysis is also an important strategy employed in computer vision. Image pyramids have been used in many applications such as content-based image retrieval and image compression. The topic of hexagonal image pyramids has been explored in depth by various researchers [15, 52]. In this section we examine the task of pyramidal decomposition of a given HIP image. The hierarchical nature of the HIP addressing scheme should facilitate the computations needed for this task. Given an image, the raw image intensity data can be manipulated to produce a set of lower resolution image representations. This is done using two different methodologies: subsampling and averaging. We will show how both these methods can be implemented in the HIP framework.

4.3.1 Subsampling

For a given L-layer image, the image at the lower resolution level $L-1$ is found by retaining one of a set of pixels at level L. In the HIP framework, seven is the natural choice for the reduction factor as successive layers of the image are order seven apart. Since the HIP image is stored as a vector, the subsampling operation requires retaining pixels which are spaced seven apart, with the first pixel taken to be anywhere among the first seven samples. In terms of implementation, subsampling of a HIP image can be shown to be equivalent to fixing the last digit of an address to any value x which ranges between 1 and 6. A λ-level HIP image contains 7^λ points. It is indexed using HIP addresses which are λ digits in length. At the next level, the image has $7^{\lambda-1}$ pixels with addresses which are of length $\lambda - 1$ digits. The HIP address structure is as follows:

$$\boldsymbol{g} = g_{\lambda-1} \cdots g_2 g_1 g_0, \qquad \boldsymbol{g} \in \mathbb{G}^\lambda \tag{4.32}$$

By letting the least significant digit in the address be $\boldsymbol{x}$ we get:

$$g_{\lambda-1} \cdots g_2 g_1 g_0 \rightarrow g_{\lambda-1} \cdots g_2 g_1 \boldsymbol{x}$$

The last digit is $\boldsymbol{x} \in \mathbb{G}^1$, and this can have any value from $\boldsymbol{0}$ to $\boldsymbol{6}$. This operation implicitly divides the address by seven as seen in the following example. As an example, consider a two-layer image with addresses $\boldsymbol{g} \in \mathbb{G}^2$. There are a total of 49 pixels in this image. By replacing g_0 with $\boldsymbol{x}$ however, we get only seven possible addresses all ending in $\boldsymbol{x}$. The results for $\boldsymbol{x} = \boldsymbol{3}$ is shown below:

$$\{\boldsymbol{0}, \boldsymbol{1}, \boldsymbol{2}, \boldsymbol{3}, \boldsymbol{4}, \boldsymbol{5}, \boldsymbol{6}, \boldsymbol{10}, \cdots, \boldsymbol{66}\} \rightarrow \{\boldsymbol{3}, \boldsymbol{13}, \boldsymbol{23}, \boldsymbol{33}, \boldsymbol{43}, \boldsymbol{53}, \boldsymbol{63}\}$$

This operation also introduces a rotation due to the spiral shift of the pixel with lowest order address in each layer. Two further illustrative examples are shown in Figure 4.14 for $\boldsymbol{x} = \boldsymbol{0}$ and $\boldsymbol{x} = \boldsymbol{5}$. The original five-layer HIP image in Figure 4.14(a) is reduced to a four-layer HIP image in both figures 4.14(b) and 4.14(c). However, the two examples give slightly different results as they use a different subset of the original data. The reduced images are also rotated as expected. Reduction of the order by seven results in a rotation of the image by $-40.9°$. For display purposes, it is easy to compensate for this rotation, however the rotation is left here.

Closed multiplication can also be used for reduction in resolution as shown in Section 3.2.3 since it is equivalent to down-sampling of the image with unequal offsets. For comparison, closed multiplication by $\boldsymbol{10}$ is shown in Figure 4.14(d). The central portion of this image is identical to the reduction of order using $\boldsymbol{x} = \boldsymbol{0}$ (Figure 4.14(b)). The portion immediately below and to the right of the image's centre is identical to Figure 4.14(c). Thus, by using closed multiplication and selecting an appropriate portion of the image, a reduction

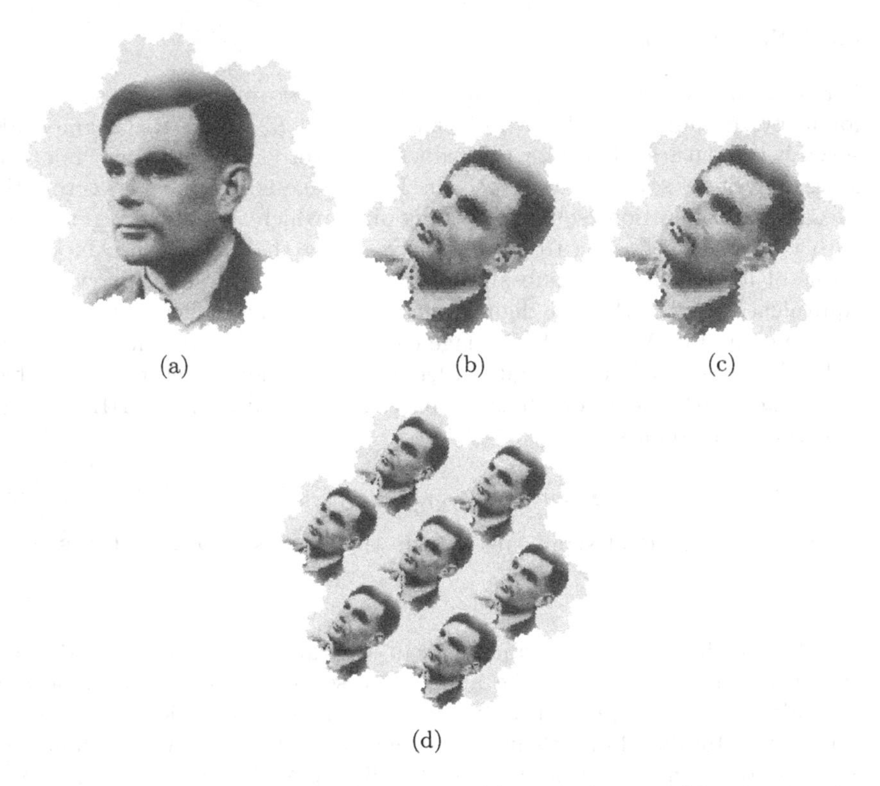

Fig. 4.14. Reducing the order of an image (a) original image (b) using ***0*** point from each group of seven (c) using ***5*** point from each group of seven (d) using closed multiplication ($\circledast_\lambda$) by ***10***.

in resolution can be achieved. Furthermore, the closed multiplication method will allow access to all versions of reductions by seven, simultaneously. This approach can be performed on the original data without the need for creation of a new image as it requires just a simple manipulation of the HIP addresses.

4.3.2 Averaging

The subsampling methodology for generating multiresolution representations can lead to potential problems in terms of loss of information. This can be a particular problem when the image data varies rapidly and the reduction order is high. For example, in an edge-detected image some of the edge information may be lost in the reduction process, potentially resulting in breaks in the edge. An averaging operation preceding the subsampling step is the standard approach to address such problems. Typically local averaging techniques are used with straightforward or weighted (as in Gaussian) forms of averaging. For a λ-level image, this can be implemented as follows:

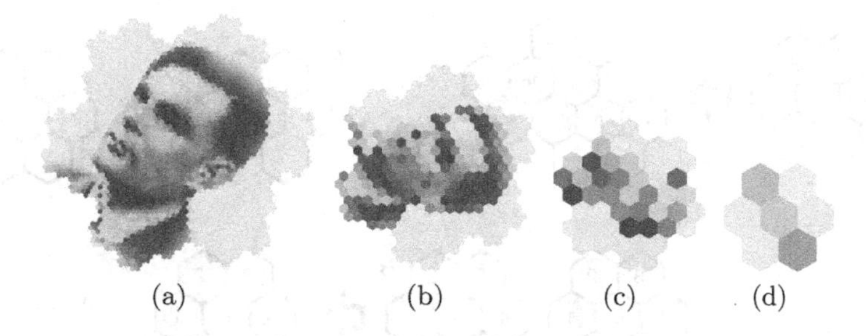

Fig. 4.15. Reducing the order of an image using averaging (a) reduced by seven (b) reduced by 49 (c) reduced by 343 (d) reduced by 2401.

$$f(\boldsymbol{g}) = \sum_{\boldsymbol{i} \in R} a(\boldsymbol{i}) o(\boldsymbol{i} \oplus \boldsymbol{g} \otimes \boldsymbol{10}) \qquad \boldsymbol{g} \in \mathbb{G}^{\lambda-1}, f : \mathbb{G}^{\lambda-1} \to \mathbb{Z}, o : \mathbb{G}^{\lambda} \to \mathbb{Z} \quad (4.33)$$

The $\sum$ operation uses HIP arithmetic to increment the index, $\boldsymbol{i}$. f is a new image indexed by a $\lambda-1$-layer HIP addressing scheme, o is the original λ-layer image, and a is an averaging function that governs how each of the neighbouring points are weighted. The averaging function a is uniform or a Gaussian. The set R has to be chosen such that it tiles the space. The cardinality of R gives the order of reduction. For instance, choosing $R = \mathbb{G}^1$ will produce a reduction by seven while $R = \mathbb{G}^2$ will reduce the image by 49. Alternatively, a reduction by four, as proposed by Burt [52] (see Figure 2.10(a)), can also be achieved by setting $R = \{\boldsymbol{0}, \boldsymbol{2}, \boldsymbol{4}, \boldsymbol{6}\}$. For the input image illustrated in Figure 4.14(a) an image pyramid is illustrated in Figure 4.15. The top layer of the pyramid is the same as Figure 4.14(a). A simple averaging function was chosen for $a(\boldsymbol{i})$. As in the subsampling case, the rotation for subsequent layers of the pyramid is visible. It can be remedied by suitable rotation of the images.

Implementation of this method is simple. A given λ-layer HIP image can be considered to be a row vector with 7^{λ} elements (see Section 3.2). Starting at the first element of the vector, every seventh element corresponds to a multiple of $\boldsymbol{10}$ in the HIP addressing scheme. Furthermore, every 49th element corresponds to a multiple of $\boldsymbol{100}$. This process can be continued for all permissible multiples of seven less than 7^{λ}. By considering smaller vectors from the total HIP image and vectors containing the appropriate averaging function, a reduction of seven can be written:

$$\begin{aligned}
\mathbf{f}_{\frac{i}{7}} &= \mathbf{a}^T \mathbf{o}_{i:i+6} \\
&= \begin{bmatrix} a_0 \cdots a_6 \end{bmatrix} \begin{bmatrix} o_i \\ \vdots \\ o_{i+6} \end{bmatrix}
\end{aligned}$$

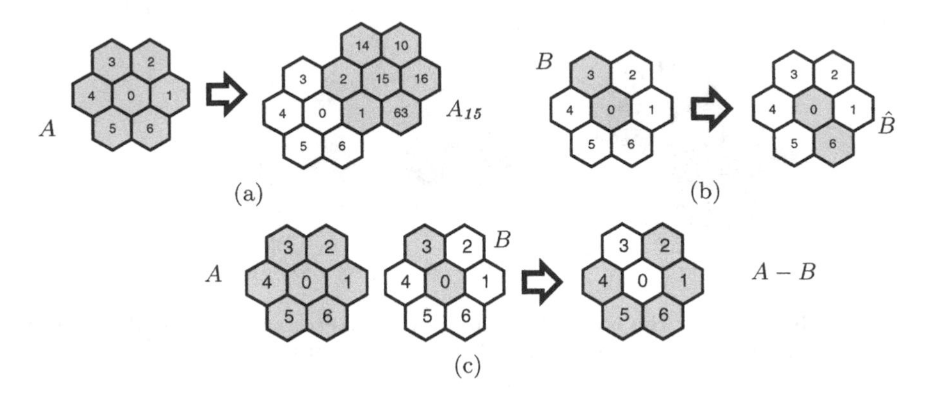

Fig. 4.16. Primitive set operations: (a) translation (b) reflection (c) difference.

Here i is a multiple of seven. Of course in this example $\boldsymbol{i}$ must be less than $(7^\lambda) - 7$. For pure averaging $a^T = \left[1 \cdots 1\right]$. Thus, in order to reduce a λ-layer image by a factor of seven would require $7^{\lambda-1}$ such vector multiplications. Other size reductions, such as illustrated in Figures 4.15(b) to 4.15(d) can be performed in the same fashion but using different sized vectors for $\mathbf{a}$ and $\mathbf{o}$.

4.4 Morphological processing

Morphological operations are concerned with the extraction of form and structure of a given image and use a set-theoretic approach to image processing. Generally, mathematical morphology is considered to be a unified approach to numerous image processing problems [48]. The set theoretic approach that was used to develop HIP addressing and image representation facilitates the development of morphological processing techniques within the HIP framework. This section will give an overview of simple morphological operations for binary images and provide examples of their operation.

First, some definitions regarding basic set manipulations are in order. Let $A = \{\boldsymbol{a_1}, \cdots, \boldsymbol{a_n}\}$ and $B = \{\boldsymbol{b_1}, \cdots, \boldsymbol{b_n}\}$ be sets of pixel addresses in $\mathbb{G}$. Each of these sets consists of n distinct HIP addresses. The translation of A by $\boldsymbol{x} \in \mathbb{G}$, denoted by $A_{\boldsymbol{x}}$ is:

$$A_{\boldsymbol{x}} = \{\boldsymbol{c_i} | \boldsymbol{c_i} = \boldsymbol{a_i} \oplus \boldsymbol{x}, \forall \boldsymbol{a_i} \in A\}$$

Figure 4.16(a) illustrates the translation $A_{\boldsymbol{15}}$. The gray pixels illustrate the members of the sets. The reflection of B, denoted by $\hat{B}$, is defined as:

$$\hat{B} = \{\boldsymbol{c_i} | \boldsymbol{c_i} = \ominus \boldsymbol{b_i}, \forall \boldsymbol{b_i} \in B\}$$

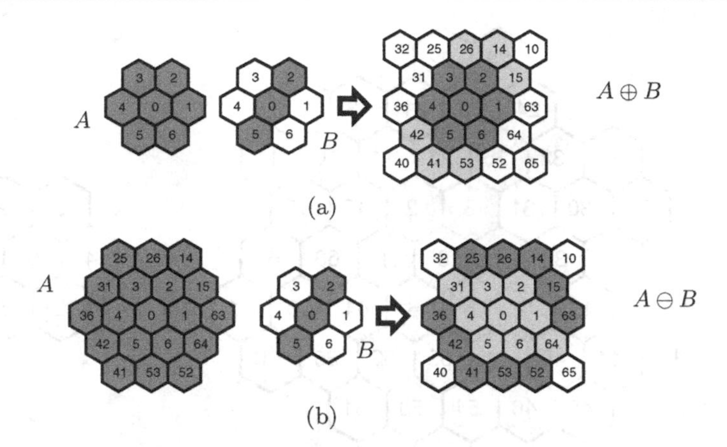

Fig. 4.17. Primitive morphological operations: (a) dilation (b) erosion.

Note that, due to the properties of HIP arithmetic, the reflection could also be achieved by multiplying all of the addresses in set B by 4. This operation is illustrated in Figure 4.16(b). The complement of set A is:

$$A^c = \{\boldsymbol{c}_i | \boldsymbol{c}_i \notin A\}$$

It is possible that A^c could be a large set. This is due to the fact that the set A is chosen from $\mathbb{G}$ which is an infinite set. In practice however, we deal with bounded images in which case A^c will also be bounded. The final simple operation is the difference between the two sets, denoted by $A - B$ and illustrated in Figure 4.16(c). It is defined as:

$$A - B = \{\boldsymbol{c}_i | \boldsymbol{c}_i \in A, \boldsymbol{c}_i \notin B\} = A \cap B^c$$

The first morphological operator of interest is *dilation*. For two sets, A and B as previously defined, dilation is defined as:

$$A \oplus B = \left\{ \boldsymbol{x} | \hat{B}_{\boldsymbol{x}} \cap A \neq \varnothing \right\}$$

where $\varnothing$ is the null set. Hence, the result of a dilation process is a set of all displacements $\boldsymbol{x}$ such that $\hat{B}$ and A overlap by at least one non-zero element. A second fundamental morphological operator is *erosion* which is defined as:

$$A \ominus B = \{\boldsymbol{x} | B_{\boldsymbol{x}} \subseteq A\}$$

The result of erosion is the set of all points $\boldsymbol{x}$ such that B when translated by $\boldsymbol{x}$ is contained in A. These two operators are illustrated in figures 4.17(a)and 4.17(b). In the figure, the light gray points illustrate the new pixels that have been added to or deleted from the original set through dilation or erosion. Combining these two operations leads to the generation of

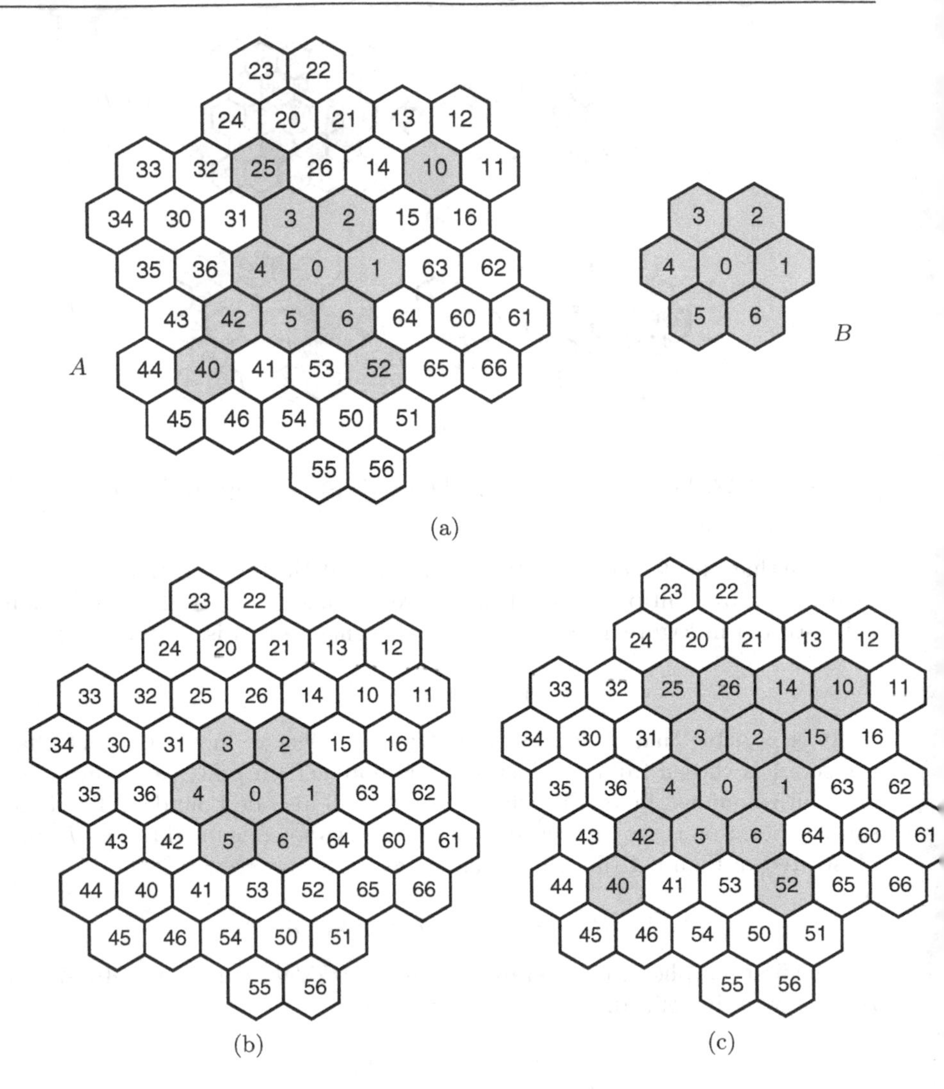

Fig. 4.18. Primitive morphological operations: (a) input sets A and B (b) opening (c) closing.

two further morphological operators. The first of these is *opening* and it is defined as:

$$A \circ B = (A \ominus B) \oplus B$$

This is just an erosion of set A by B and then a dilation by set B. It serves to smooth the contour in an image by breaking narrow bridges and eliminating thin protrusions. The second morphological operator is *closing* which is defined as:

$$A \bullet B = (A \oplus B) \ominus B$$

This operation first dilates set A by B and erodes the result by set B. As opposed to the opening operator, closing smoothes the external contour by fusing narrow breaks to eliminate small holes and filling the gaps in the contour. Figure 4.18 illustrates both of these operations. Using the above four operators many more complex morphological operators can be defined. A good review can be found in Serra [48].

In practical morphological processing, there is a notion of an object which is shaped or probed with a structural element. Thus for dilation of A by B, A is viewed as an object which is being dilated using B which is the structural element (SE). The shape of the SE and the neighbourhood used have a great deal of influence on the result of the operation. For instance, a 3×3 SE is the closest fit for an isotropic (circular) SE in a square image. However, using such an element to erode curved objects such as finely spaced curves on a fingerprint can result in the introduction of unwanted discontinuities. This is because the elements in the SE are at different distances from the centre. The consistent connectivity of pixels in hexagonal images is an attractive feature which allows for better isotropic SE definition. For example, a small circular SE can be constructed using only seven pixels, which is equivalent to a one-layer HIP image. Additionally, the HIP structure also offers an advantage in neighbourhood computations. As previously stated in Section 3.4.3, computation of neighbourhoods is particularly simple in HIP addressing due to its hierarchical nature. This advantage is directly carried over to morphological processing. Complexities that may arise due to image boundary can be handled by using a closed form of the HIP arithmetic (see Section 3.2.3). These features make morphological processing within the HIP framework computationally efficient.

4.5 Concluding remarks

This chapter showcased the utility of the HIP framework by demonstrating how several simple image processing techniques (spatial and frequency domain) could be implemented within the framework. A significant amount of theory and examples were provided covering a variety of aspects of image processing. Application-specific details were included to aid the understanding of the practical aspects of implementation. This chapter in conjunction with the previous chapter should provide enough information to enable developing real applications using the HIP framework. Examples of how this can be performed can be found in Chapter 5.

5

Applications of the HIP framework

Several applications that employ the HIP framework are presented in this chapter. The central aim is to demonstrate that the HIP framework can be used interchangeably with conventional square image processing. Three main examples are given in this chapter: critical point extraction, shape extraction, and shape discrimination. Finding critical points within an image is similar to the human visual system's saccadic mechanism. Such algorithms are commonly used in active vision systems. In shape extraction, a chain code representation is extracted for a given object. The chain uses the HIP indexing scheme. For shape discrimination, we examine the task of discriminating between different logos. The three applications combine various processing techniques that have been discussed in Chapters 3 and 4. The first uses the hierarchical nature of HIP to find points of interest in an image. The second uses edge detection and skeletonisation to follow the boundary of an object while the third employs frequency domain techniques to analyse the energy signatures of different shapes.

5.1 Saccadic search

As the application domain of computer vision expands, the characterisation of the image increasingly depends on gathering information about the structures within an image. These structures are often made up of irregular curves. It is important to find efficient methodologies to represent and analyse these structures.

The human visual system (see Section 2.1) (HVS) seems to perform this task with little problem. It is known [122] to be organised in a hierarchical fashion. The lower levels of the system directly process visual input whereas the higher levels of the system process increasingly abstract information, using prior knowledge, in two distinct, yet major pathways. These are the *what* and *where* pathways [123]. The *what* pathway deals with processes pertaining to object features and the *where* pathway deals with processing involved in the representation of spatial information. It has been suggested that an "attention" mechanism serves to couple the lower level visual processes and the higher level visual pathways. Consequently, the higher level of the HVS can control what the lower level of the HVS is examining. Thus, the processing in the HVS can function in a goal directed manner. This section attempts to partially model the attentional mechanism.

5.1.1 Saccadic exploration

When humans explore a visual scene they do so via a set of movements of the eyes. Each such movement serves to image a part of the scene which is usually a region of interest, on to the central part of the retina called the fovea where the spatial resolution is maximum. Hence, this process is also called foveation or fixation. The resulting large jumps in the successive foveations are known as saccades and they are mediated by higher-level cognitive processes [124, 125]. In addition to the biological studies, it has been successfully applied to computer vision [126, 127]. The search process mediated by saccades, dramatically reduces the amount of information required for a detailed analysis of the scene as only a part of the scene is "attended to" at any given time. In this section, a scheme for attention-driven search based upon a model of saccadic eye movements is presented. The HIP framework is an ideal candidate for a model of the saccadic system. Firstly, the hexagonal sampling of visual information is analogous to that carried out by photoreceptors in the fovea [9]. Additionally, the aggregate-based data structure facilitates processing to be performed at many resolutions simultaneously as in the HVS.

A simple illustration of the operation of the saccadic system is illustrated in Figure 5.1. The principle behind the operation is to analyse a given image via a series of attention windows centred at successive foveation points. The

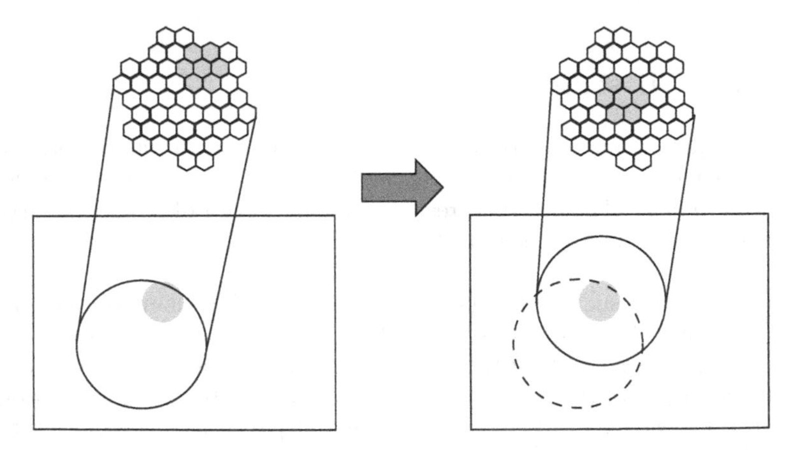

Fig. 5.1. Illustrative example of the proposed saccadic system.

foveation points are points where the image is more *interesting*. Thus, the large circle in the figure indicates the current field of view or the attention window, while the darker circle indicates the next centre of focus deemed to be a subregion of interest. The discrimination over whether one portion of an image is more interesting than another is made on the basis of featural information. The lower levels of the HVS are known [114, 122, 128] to involve the extraction of primitive features about the image (see section 2.1.2). Some example features, and the ones employed here, are local orientation and relative intensity . More details about these features will be given later.

A block diagram of the proposed algorithm is illustrated in Figure 5.2. At each instance, a subset of a given image is selected and sampled into the hexagonal array using the resampling scheme discussed in Section 6.1. The selected subset of the image is centred upon a particular point within the image. This is the basis of the attention window. The data contained within the attention window is then analysed. Specifically, features such as the weight and orientation of the data are extracted and analysed.

The weight is computed as the first moment of the data and is just a sum of the individual pixel values. For a particular image point, $\boldsymbol{x}$, this can be computed as:

$$w(\boldsymbol{x}) = \sum_{\boldsymbol{i} \in \mathbb{G}^{\lambda}} f(\boldsymbol{x} \oplus \boldsymbol{i})$$

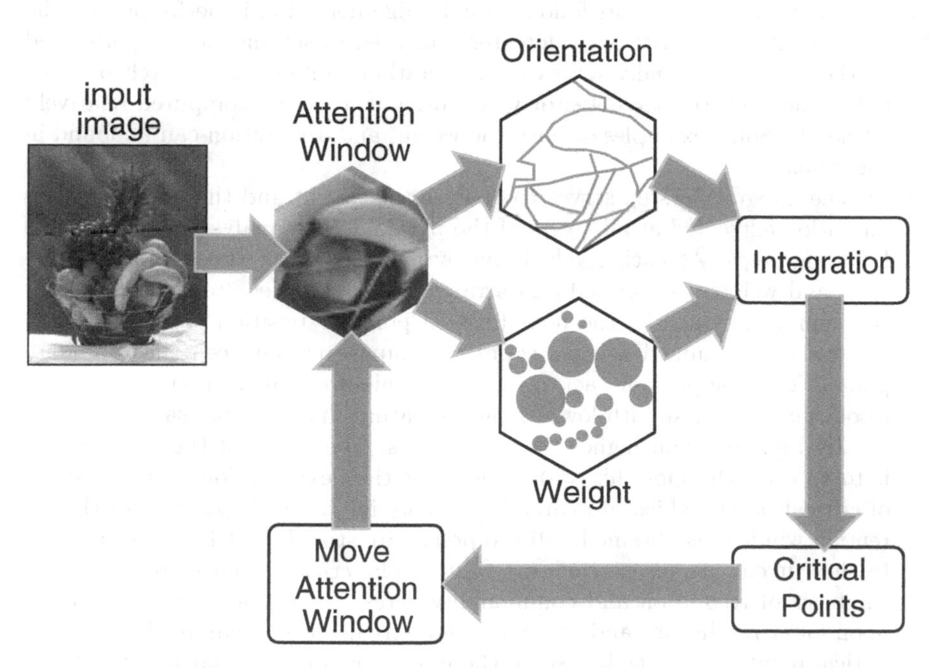

Fig. 5.2. The proposed saccadic search algorithm.

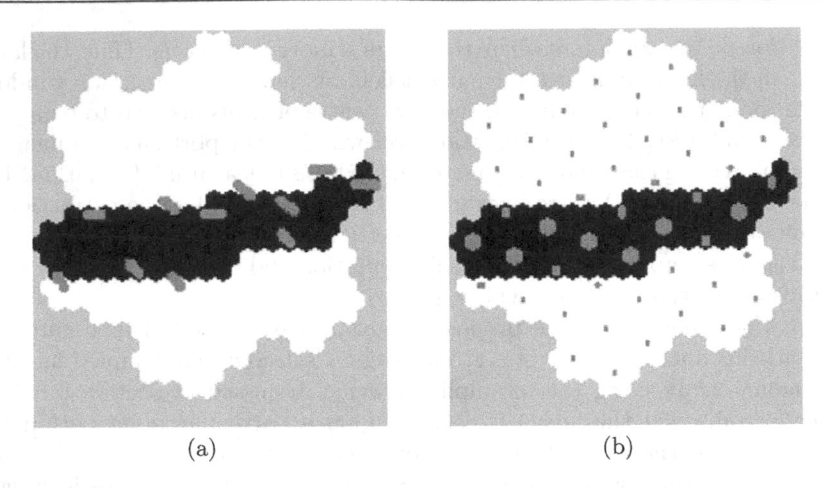

(a) (b)

Fig. 5.3. Example of (a) orientations and (b) weights.

where the f is the given λ-layer HIP image. The resulting weights are normalised by dividing by the maximum weight. This analysis is carried out in a hierarchical fashion at all layers of the pyramid implicit within the HIP image. The weight thus obtained is easily comparable between different layers of the HIP image. The other information computed is orientation, which is computed by a multi-step process as follows. First, edge detection is performed on the data using the Prewitt edge detector and then skeletonisation is performed on the edge map. Finally, a series of orientation templates are matched to the data. The Prewitt edge detector was chosen as it can be computed relatively efficiently. Some examples of both the weight and orientation can be found in Figure 5.3.

The integration step serves to analyse the weight and the orientation information extracted at all layers of the pyramid. The analysis consists of the following steps. At each level of the pyramid, a list of the features (directions and weights) is derived and arranged in rank order, from most to least common. Subsequently, the next fixation point or position of the attention window is determined using a series of comparisons between the candidate points. The comparisons act in a strictly hierarchical fashion from lowest resolution to highest with low resolution features having more say in the candidate direction than higher resolution ones. The result of the comparisons is to vote for the most likely directions for the next fixation point. A series of critical points which are candidate points for the next position of the attention window, is obtained. All the points are stored as HIP indices relative to the current attention window's centre. The critical points are ranked on the basis of resolution and commonality across resolutions. The ones based upon lower resolutions and which are more frequent are examined first. Each critical point is examined to see if the movement would result in a valid new

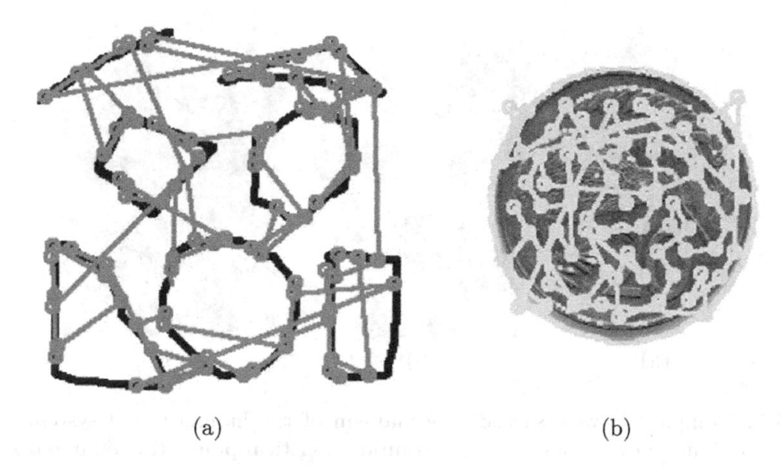

Fig. 5.4. The saccadic search illustrated on two images: (a) synthetic (b) real.

fixation point. A point is considered to be valid if it does not involve visiting a previously examined point. Occasionally, when no effective candidates can be found, a random movement is performed. The specific movement is stored as a HIP address and employed as an additive offset to the current attention window.

5.1.2 Discussion

The saccadic search system was evaluated using a series of synthetic and real images. The resolution of the images used was 256 by 256. The attention window was of size 7^3 (343) points. This provided three layers (resolutions) for analysis by the algorithm. Both the weight and orientation were analysed using blocks of seven points. This resulted in a maximum of 114 (57 for each feature) critical points. However, eliminating previously visited points and repetitions (movements to the same place) can reduce the number of critical points to a great extent.

The results for two test images are illustrated in Figure 5.4. In each subfigure, the original image is shown along with the fixation points which are superimposed on top. The fixation points are illustrated in the images by small circles. The linking lines in the image show a traversal between two successive fixation points. There are two important things to notice about the system. Firstly, it spends relatively more time around the regions containing objects. Secondly, it shows a preference for following contours. These trends are especially noticeable in the synthetic image but are also evident in the real image. For the real image, 72 fixation points were found and for the synthetic image the corresponding number of fixation points was 62. In fact, experimentation indicates that more complicated images yield no dramatic

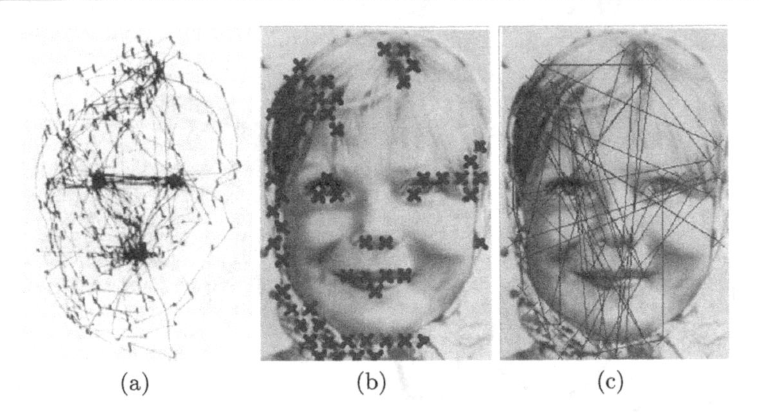

Fig. 5.5. Comparison with saccadic mechanism of the human visual system. (a) Saccades in human eye (from [125]) (b) computed critical points (c) computed saccade.

increase in fixation points. This is perhaps due to the fact that there is a finite amount of interesting information in an image. Moreover, this implies that the methodology should scale well with different resolutions and complexity of images.

As a second example of the proposed system, a comparison is provided with the work of Yarbus [125] where eye movements were tracked while human subjects performed visual inspections. Figure 5.5(a) shows the saccades generated by a human when looking at the face in figure 5.5(b). Here a line indicates the path of eye movement and a dot indicates fixation point. The critical points extracted by the proposed system are found in figure 5.5(b). It is seen that the critical points are mainly concentrated around the eyes, mouth, and head periphery, where minimas in the image intensity or maximas in the gradient (signalling strong edges) are present. These are also the regions of interest to human subjects as indicated by the clusters of fixation points in Figure 5.5(a). By drawing lines connecting the critical points in the order in which they were generated, we have the results shown in Figure 5.5(c). Comparing this with the eye movements of humans, we find a weaker similarity because the algorithm used was not tuned to replicate the human saccadic system.

To recap, this section presented an attention-driven image exploration mechanism that, in a limited domain, mimics the human saccadic system. The main features of the proposed system are the use of the HIP framework and the inherent hierarchical decomposition contained within it. The system performed well at the task of exploration both by following contours and by showing a distinct preference for regions of high image content. The advantages of the system are that it inherits the hexagonal lattice's utility at representing curves and that it is computationally efficient. It has potential application in general recognition problems as it provides a methodology to decompose an image into a series of contextual features. Additionally, it provides an alternative approach to problems such as image inspection and raster to vector conversion, by performing the scanning process for analysing the images. It could also be extended to target tracking applications by including motion as a feature of interest.

5.2 Shape extraction

Feature extraction is of interest in many computer vision applications, of which content-based image retrieval is an area which has been prominent in the last few years [129–132]. A typical content-based image retrieval system performs a query via a number of generic classes or features which have been extracted from the image. These include but are not limited to texture, colour, and shape.

The work presented here focuses on the problem of shape extraction. The shape of an object is generally defined by its dominant edges. The approach we have taken to shape extraction is based on the HVS which starts with a saccadic search described in the previous section. The next step is contour following which involves following the dominant edges from these saccadic fixation points [133]. The proposed algorithm is built around a hexagonal lattice with successive saccades extracting critical points. By linking these points together, a simple object-centred description of the underlying shape is generated.

5.2.1 Shape extraction system

The proposed shape extraction system is illustrated in Figure 5.6. The first stage of processing is saccadic search where a square image is preprocessed to find the critical points in the entire image. The first critical point is chosen as a starting point and information about the surrounding region in the image is extracted using a hexagonal attention window. Two sorts of features are extracted from the attention window, namely, the data weight and orientation. The integration process takes this feature information along with the information on existing critical points and decides the next location of the attention window. The current information is also added to the overall description of the image. Each of these stages will now be discussed in turn.

5.2.2 Critical point extraction

The purpose of this subsystem is to generate potential fixation points. These are points that contain sufficiently interesting features to warrant further investigation. Fixation points can be employed to aid in contour following or

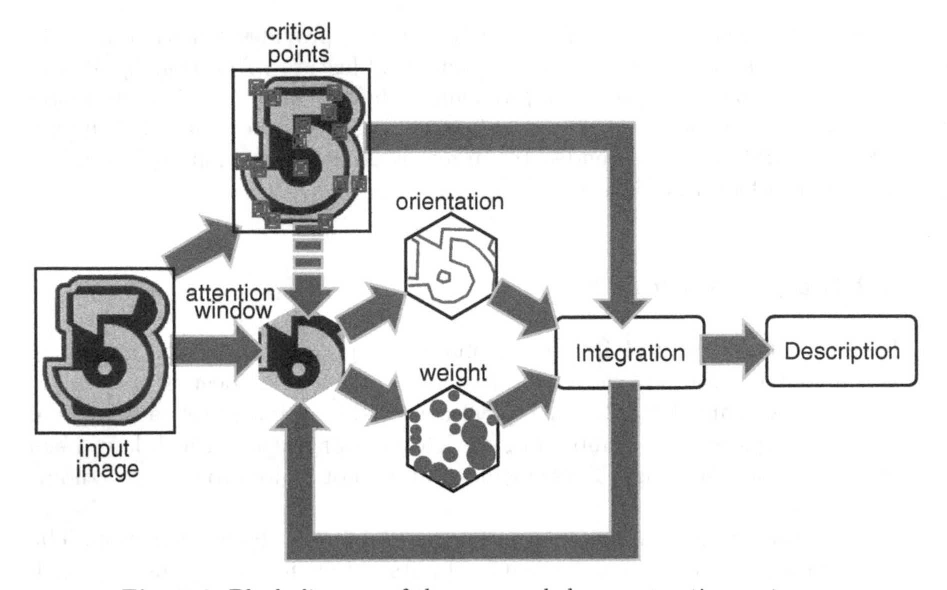

Fig. 5.6. Block diagram of the proposed shape extraction system.

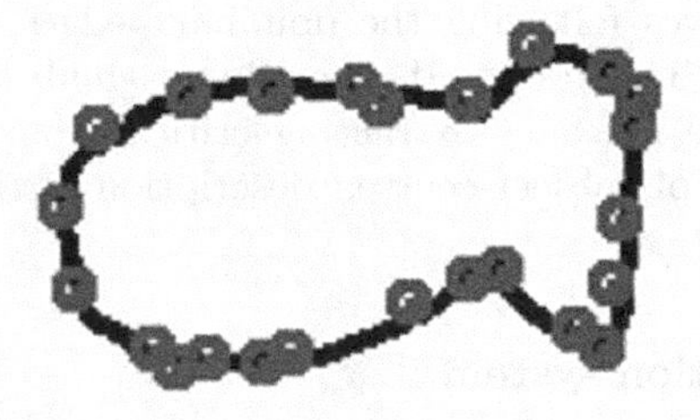

Fig. 5.7. The critical points extracted for a test image.

to resolve potential conflicts. To find candidate points, every point in the image is examined through an attention window (as described in Section 5.2.3). The critical points are defined as points where the weight in a given region exceeds a particular threshold. An example of a simple candidate image and the discovered critical points are illustrated in Figure 5.7. The coordinates of the critical points are stored using HIP indexing relative to the centre of the image.

5.2.3 Attention window

In the HVS, the attention window reduces the amount of processing by examining small subsets of the visual field at higher resolutions than the remainder of the visual field. These small subsets have been found to contain features of interest. Thus, in the shape extraction system outlined here, the attention window performs a similar role. The attention window is created by hexagonally sampling about a current point of interest using the resampling scheme described in Section 6.1. The attention window itself is stored as a HIP image.

5.2.4 Feature extraction

There are two sorts of features extracted from the image data, namely, weight and dominant orientation. They are both computed in a hierarchical fashion exploiting the HIP data structure used in the attention window. The methods are similar to those given in Section 5.1 for the saccadic search.

The weight of the image refers to the sum of the gray levels which comprise it. For the hexagonal structure that is used for the attention window, the weight can be computed for several resolutions easily. For instance, for a two-layer system, the weight can be computed by summing the cells with the same first digit. This process can be repeated to produce the weight at the lowest possible resolution (1 pixel). Thus, for an attention window with 7^L pixels there will be $\frac{7^L-1}{6}$ distinct weight values. This is the sum of the number of weights for each layer in the image. For example, the two-layer system has eight weights. This involves one weight for the first layer (the entire image) and seven for the second layer. Once found, the weights were normalised relative to the maximum weight value.

Extraction of the orientation image from the attention window is a multi-step process. First, the image is filtered using the Prewitt edge operators (Section 4.1.1). Second, the result is thinned using the skeletonisation process detailed in Section 4.1.2. Finally, the dominant orientation is found using a bank of oriented filters. For every group of seven pixels, the dominant direction is assigned to be the direction at the central pixel (as given using the two distinct Prewitt masks). By reducing the order of the image (as described for weight extraction) this process can be repeated again. There are $\frac{7^L-1}{6}$ distinct orientation features for an attention window of size 7^L. As HIP addressing has vector-like properties, the orientation is also stored as a HIP index. For practical reasons these are scaled so as to point to the nearest group of seven pixels. The dominant orientation of the entire attention window is taken to be the orientation of the image at the lowest resolution of the pyramid.

5.2.5 Integration

The information from the feature extraction stage is integrated in this stage to determine the next location of the attention window. Information is examined hierarchically from low to high resolution. Generally, this entails a comparison of the window movements given by the orientations and the weights. If the movements are complementary (in roughly the same direction) then this is a good candidate for the direction of movement. From this top-down process, a list of candidates for possible movement is generated. These are ordered from low to high resolution. Once a candidate for the movement is found, it is then examined for three things. Firstly, it should not cause a reversal in the direction of movement (i.e. travel back to where it came from in the last iteration) as shown in Figure 5.8(a), as it is redundant. Hence, in such a case, the candidate is removed and the next in the list is considered. Secondly, if the movement passes sufficiently close to a critical point then that becomes the next location for the window as shown in Figure 5.8(b). The movement is adjusted to the correct value by subtracting the coordinates. Finally, a candidate is examined to see if it will result in a movement that closes a loop as shown in Figure 5.8(c). A closed loop is considered to indicate closure of the shape of the object and hence a signal for the algorithm to end.

Whilst the algorithm is progressing, a new point is produced at every iteration that is added to the list that represents the object's shape. Each successive point is stored as a HIP index, relative to the image's centre. Upon completion of a contour the addresses can then be normalised to be relative the object's centre. This shape representation is thus very compact. Furthermore, since movements are also stored using HIP indices, the new fixation point or attention window location is found through a simple addition of the two indices corresponding to the movement and the current position.

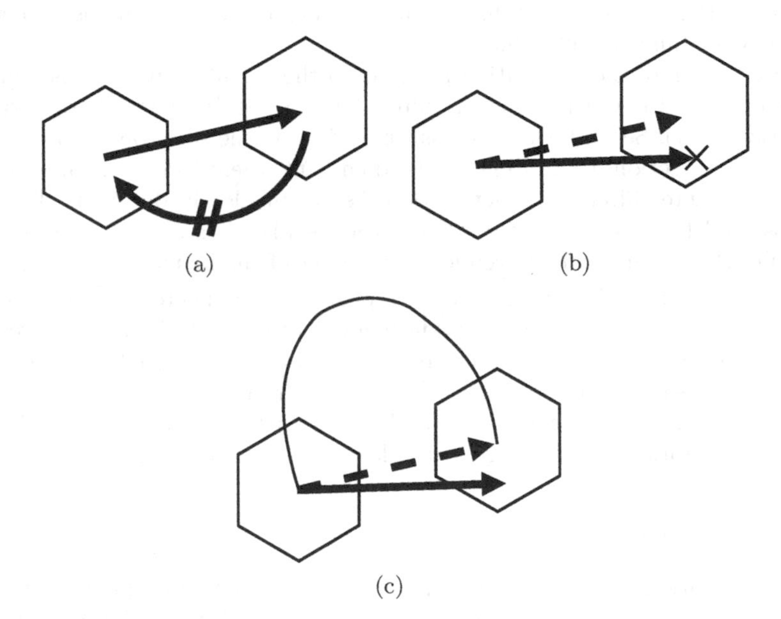

Fig. 5.8. Rules in the integration stage: (a) redundancy (b) capture (c) closure.

Table 5.1. Comparison of the number of incorrect points.

Shape	n	$\overline{\Delta}$	σ^2
square	35	0.29	0.46
fish	34	0.29	0.35

5.2.6 Discussion

The shape extraction algorithm was evaluated using simple shapes. Binary images, sampled on a square lattice, of size 256×256 pixels were used for testing. Real images can also be handled after passing through an edge detector. The attention window employed had a three-level hexagonal structure (343 hexagonal pixels). Some representative results are illustrated in Figures 5.9(a) and 5.9(b). In each image, the original image is shown (in gray) along with the shape extracted (in black). Note that the original image is artificially lightened to aid in illustrating the performance of the shape extraction algorithm. The results illustrate performance with two distinctly different shapes, namely a square and a fish shape. The square is made predominantly of straight lines and the fish is made of a mixture of straight and curved lines.

Examination of both the images and the derived shape show that the algorithm performs well. By and large, the derived shape lies within the original shape. For the square, the top left corner is poorly represented but the result is still good enough to distinguish the shape. The fish shape is much better represented with most significant aspects of its shape being correctly identified. There are a few points which are outside the original image in both shapes. An analysis of the derived representation is given in Table 5.1. In both cases the number of points, n, is practically the same, as is the mean difference, $\overline{\Delta}$. The mean squared error, σ^2, is different between the two shapes. This is consistent with the visually evident discrepancies between the real and extracted results for the square shape. However, in both cases the error is within acceptable limits.

Viewpoint invariance is a desired feature for any generalised object recognition scheme. This is possible by using object-centred representations which are invariant to affine transformations. The extracted shape in the above scheme can be easily manipulated by exploiting the positional indexing scheme of HIP. An image-centred shape representation can then be derived by a simple translation, i.e., by adding this value to all the points in the shape. Since (see Chapter 3) addition results in translation and multiplication results in a rotation and scaling of the HIP addresses, further manipulations of the representation are simple. In Figures 5.9(c) and 5.9(d), transformations on the shape are illustrated. In Figure 5.9(c), the data has a constant offset added to it and then is multiplied by 4 (the same as rotation by 180°). Figure 5.9(d), illustrates how the data can be easily scaled by a simple multiplication. In this case the data was multiplied by 63 (a scale factor of 2).

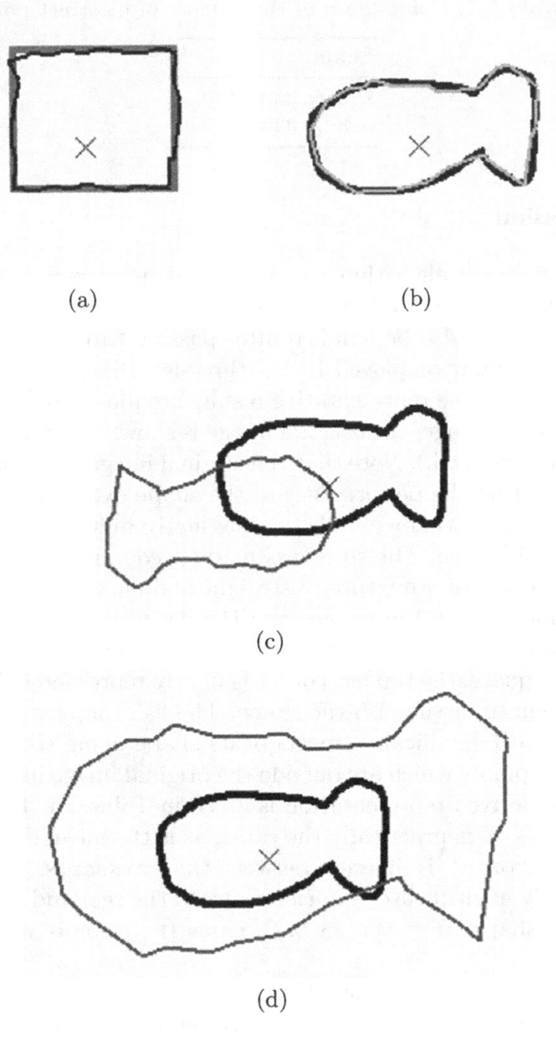

Fig. 5.9. Results of the shape extraction algorithm for (a) a square and (b) fish shaped image. X indicates the centre of a shape. Results of rotation and scaling of the extracted fish shape are in (c) and (d) respectively.

This section presented an attention-driven exploration strategy that has been applied to the problem of shape extraction on monochrome images. This task was accomplished using the HIP framework and the hierarchical structure contained within it. As a result of using HIP addressing, simple geometric transformations can be very easily performed upon the extracted shape. The strength of the approach lies in the use of the novel object-centred representation scheme for the shape. This representation scheme could be applied to a wider class of recognition problems. The manipulation aspect gives it a significant advantage over representations based on primitives (such as geons [134] or generalised cones [135]) which can be computationally expensive. Extensions to the existing scheme could be to make it robust to the existence of multiple objects in the image. This would require segmentation of the original image to distinguish between the objects. Also, erroneous points in the representation could be eliminated by interpolating between the critical point found and the nearest point on the object's edge.

5.3 Logo shape discrimination

The logo associated with a given company is an important, uniquely identifying visual feature. A logo is thus as individual as a person's fingerprints. However logos are not innate to the organisation and they are easy to copy either deliberately or unintentionally. The prevention of this typically falls to an intellectual property office of the country where the company is registered. This office holds all existing logos and makes recommendations on whether to accept new logos. In most countries, the current methodology for performing this is labour-intensive and occasionally results in erroneous decisions. It is desirable to speed up this search process by exploiting the distinctiveness of individual logos. There are many unique features about logos that allow people to delineate them. These include, but are not limited to, colour, textural content, and shape. Of these features, shape is important as it can allow a preliminary, and quick, categorisation of the logo. Furthermore, the shape of an object is relatively invariant to transformations.

Several researchers have applied shape-based retrieval to trademark images. The ARTISAN [136] project is an extension to an existing system used by the UK Patent office. First the boundary is extracted and this is used to group logos into a family of similar shapes. Another approach is the use of invariant moments to index and retrieve trademarks. The approach examines the entire logo and extracts features from it. IBM has developed a commercial system called QBIC [136] which can be used to search on a variety of features including shape. A good review of these and other approaches to logo shape discrimination is available in [132].

The rest of this section will present details of the proposed approach to logo shape discrimination. These will be discussed individually along with implementation details. Finally, the proposed scheme will be evaluated by examining its performance on a real database of logos.

5.3.1 Shape extraction system

This section presents a frequency domain approach to extracting shape information and then broadly classifying logos from an existing logo database.

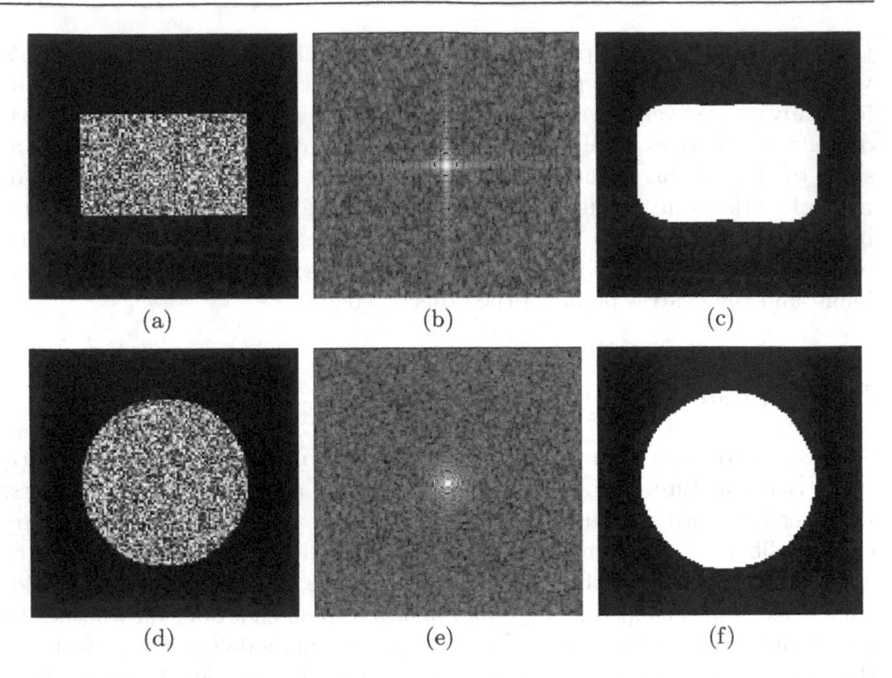

Fig. 5.10. Magnitude spectra of different shapes: (a) and (d) are input images; (b) and (e) the corresponding magnitude spectra; (c) and (f) are results of lowpass filtering.

Only square and circular shapes are examined, though the methodology could be refined for other shapes. The shape of an object can be considered to be a coarse feature. Consequently, only low frequency information is required to define it. This idea is illustrated in Figure 5.10.

Consider the images of two shapes, namely, a rectangle and a circle in Figures 5.10(a) and 5.10(d) respectively. These have random noise added to them to simulate some sort of texture. The Fourier magnitude spectra are presented in Figures 5.10(b) and 5.10(e). The rectangle has most of its energy aligned along the major and minor axes of the rectangle. The circle has the energy arranged in rings around the centre where the spatial frequency is 0. In both cases, however, most of the energy is concentrated at low frequencies.

The results of lowpass filtering and thresholding these images are shown in Figures 5.10(c) and 5.10(f). The cutoff frequency of the lowpass filtering operation is set to be a small region about the image centre. Notice that in both cases the noise has been rejected while the shape of the object, whether rectangular or circular, remains. The filtering procedure can be performed for other shapes with similar results.

This simple example serves to show that the low-frequency magnitude spectra of images are rich sources of information about the shape of an object.

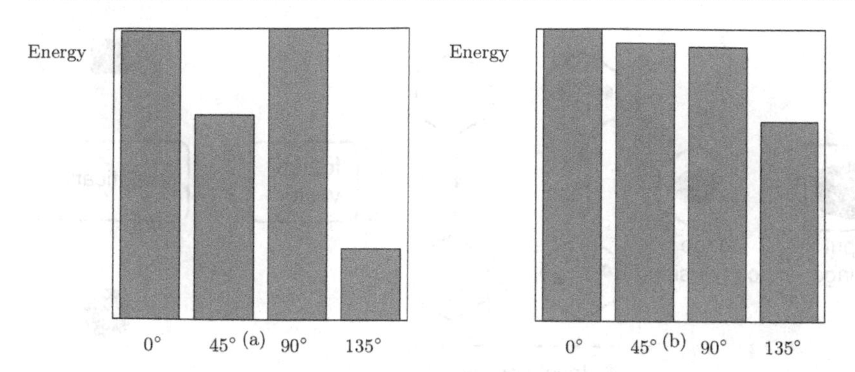

Fig. 5.11. Energy in different oriented energy bands for (a) rectangle (b) circle.

Energy signatures have been investigated previously for pattern discrimination in the local energy model based approaches, where the local energy at a point in an image refers to the sum of the squared responses of a conjugate symmetric pair of filters [137]. It has also been successively applied to discriminate linear and curved features [138]. Since, rectangles and circles are objects composed of either linear (rectangles) or curved (circles) features, this approach can be employed for the shape discrimination problem.

The Fourier energy distribution profile for Figures 5.10(a) and 5.10(d) are illustrated in the two graphs in Figure 5.11. Only four orientation channels were used in this example. The individual channels themselves were 45° wide and were centred on the orientations 0° (leftmost bar), 45°, 90°, and 135° (rightmost bar). The main feature to notice here is that the rectangular image generated two distinct peaks, namely, one at 0° and one at 90°. The circle on the other hand shows a spread of energy across multiple energy channels. This implies that an image such as a rectangle will have only two peaks in the energy whilst an image containing a curve will show a spread of energy. Thus, the discrimination between squares and circles can be found by investigation of the energy signature at low frequencies.

The proposed scheme for shape discrimination is illustrated in Figure 5.12. The scheme consists of three main stages: image mapping, feature extraction, and classification. The purpose of the image mapping stage is first to locate the logo within the image and then to hexagonally resample the image using the HIP framework. The feature extraction stage filters the resulting hexagonal image using a bank of filters which vary in both orientation and frequency range. The local energy is then computed from the filtered images to construct a feature vector. The feature vectors are then classified using LVQ. Each of these three processes are now discussed in detail.

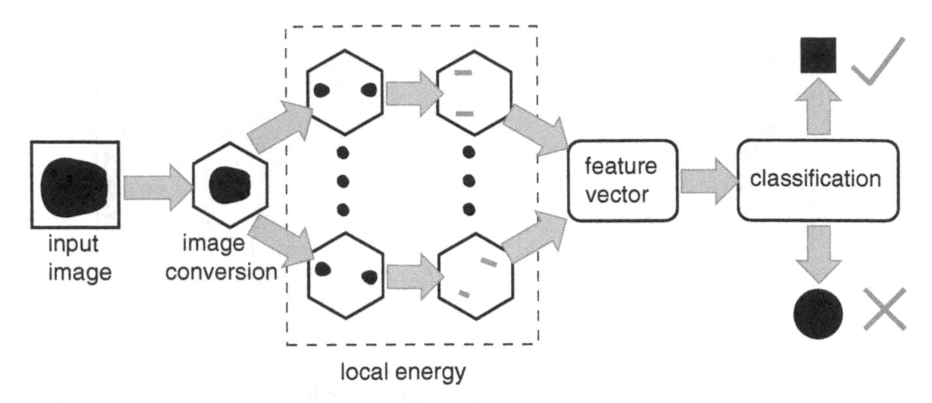

Fig. 5.12. Block diagram of the proposed system for shape discrimination.

5.3.2 Image conversion

The image conversion step serves two main purposes. The first is to segment the logo within the image, and the second is to re-sample the logo on to a hexagonal lattice. Segmenting the logo is possible simply by using a bounding box approach and assuming that the logo is clearly identifiable from the background. Two arrays are generated containing the horizontal and vertical intensity profiles for the image (see Figure 5.13(a)). These are then analysed to find the maximum and minimum values. The logo background will either correspond to the maximum (in the case of a white background) or the minimum (in the case of a black background). For a white background, the boundary of the shape is found by stepping inwards from the edge until the value in the array deviates from the maximum significantly. For a black background, the deviation is measured with respect to the minimum. In reality, the amount of deviation is not critical and can be tuned empirically to make sure that the resulting bounding box is bigger, though not significantly, than the logo. An example of a bounded image is illustrated in Figure 5.13(b). In this figure the centre point of the image is also highlighted.

Next, the content of the bounding box in Figure 5.13(b) is resampled to a HIP image using a process described in Section 6.1. The process starts at the image's centre as highlighted by the cross (Figure 5.13(b)). This along with a suitable scale factor (or the spacing between the individual hexagonal pixels) can thus be used to sample the bounded region of the original image to exactly fill the hexagonal image. An example of the output of this process is illustrated in Figure 5.13(c).

5.3.3 Local energy computation

The local energy model [137] was originally proposed as a model for generalised feature detection in the HVS. Here, the given image is filtered by a set of

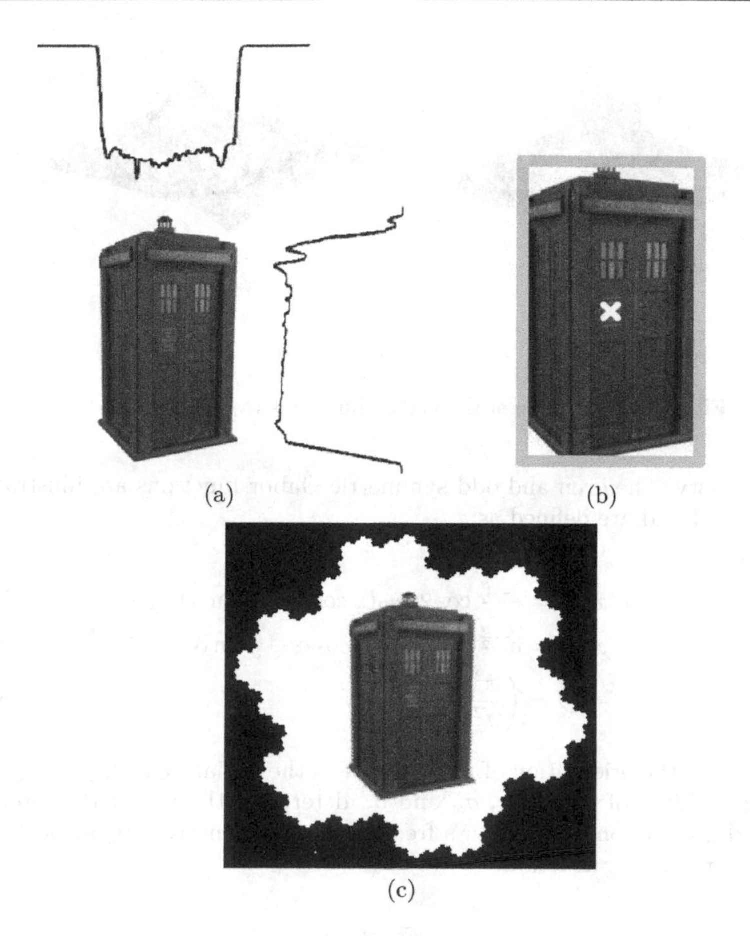

Fig. 5.13. Sample image (a) horizontal and vertical intensity profiles (b) image bounding box (c) HIP image.

filters which have orthogonal phase spectra but identical magnitude spectra. The local energy at every point of an image I can be found by computing the square root of the sum of squared magnitude responses of these filters:

$$E_L(I) = \sqrt{e^2(I) + o^2(I)} \tag{5.1}$$

Here $e(I)$ is the response of an even-symmetric filter and $o(I)$ is the response of an odd-symmetric filter. The maximal response will correspond to either an edge or line feature in the original.

To implement the filters, Gabor functions were chosen though any pair of functions which have orthogonal phase and identical magnitude spectra could be chosen. Gabor functions were chosen as it is easy to tune their orientation

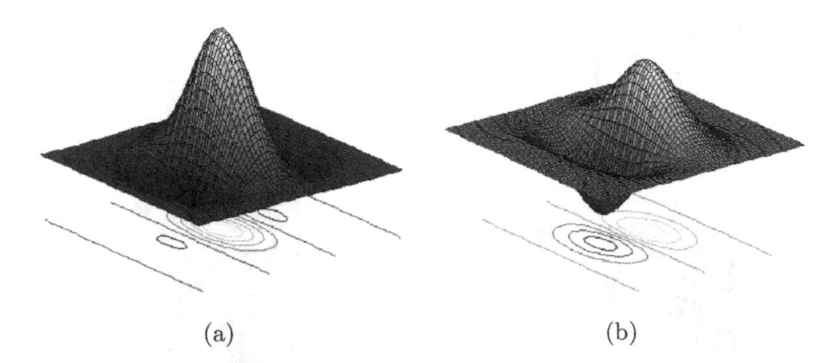

(a) (b)

Fig. 5.14. Example of the Gabor functions (a) even (b) odd.

and sensitivity. The even and odd symmetric Gabor functions are illustrated in figure 5.14 and are defined as:

$$e(x, y) = e^{-\frac{\beta}{2}} \cos 2\pi u_0 (x \cos\alpha - y \sin\alpha) \tag{5.2}$$

$$o(x, y) = e^{-\frac{\beta}{2}} \sin 2\pi u_0 (x \cos\alpha - y \sin\alpha) \tag{5.3}$$

$$\beta = \left(\frac{x^2}{\sigma_x^2} + \frac{y^2}{\sigma_y^2}\right) \tag{5.4}$$

where α is the orientation of the filter, u_0 is the radial frequency (in cycles per image width) of the filter. σ_x and σ_y determine the fall of the filter in the x and y directions. The filter's frequency and orientation bandwidth are related to σ_x and σ_y as follows:

$$\sigma_x = \frac{\sqrt{2}}{2\pi u_0} \frac{2^{B_r} + 1}{2^{B_r} - 1} \tag{5.5}$$

$$\sigma_y = \frac{\sqrt{2}}{2\pi u_0 \tan \frac{B_\theta}{2}} \tag{5.6}$$

Here B_r (in octaves) and B_θ (in degrees) are the half-peak radial and frequency bandwidths respectively. All these relationships are illustrated in Figure 5.15. Different orientations can be selected by varying α in the Euclidean plane.

Implementing the filtering in the spatial domain is expensive as the image needs to be convolved with a *bank* of filters. Furthermore, the size of the filters employed here is large. The computation can be made significantly faster by implementing the filtering in the frequency domain using the HFFT (Section 4.2.1).

The specific values for bandwidth and orientation for the filters depend on the application. The image conversion stage (Section 5.3.2) serves to convert

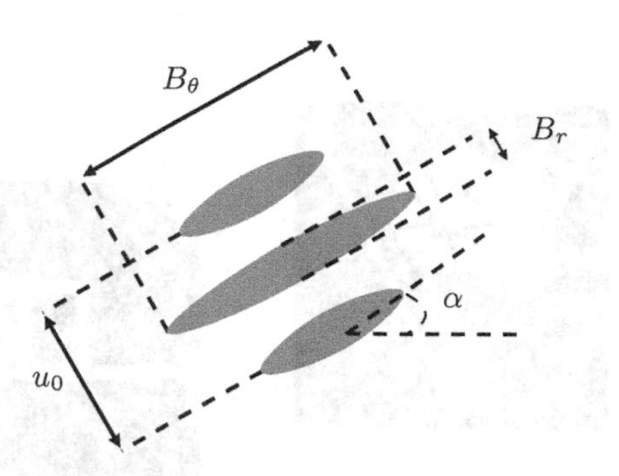

Fig. 5.15. Parameters to tune the Gabor filters.

the original input image to a HIP image of a roughly fixed size irrespective of the original size of the image. Hence, precise ranges for filter parameters can be chosen for the Gabor filters for the shape extraction application. Since the shapes of logos are described by low frequency information, the filters are tuned to low frequencies. Several distinct filters of differing bandwidths are chosen for this purpose as they individually examine different portions of the frequency spectra. Additionally, the filters are chosen to have a relatively narrow bandwidth (B_θ) to make them highly selective.

5.3.4 Feature vectors

After filtering and thresholding, the local energy is then computed, as per equation (5.1), for each orientation and at each resolution. The total energy (across all orientations) for a given resolution is then used as a normalisation factor for the resulting data. The final feature vector is thus:

$$v = [f_{11} \cdots f_{1n} \cdots f_{n1} \cdots f_{nn}]$$

where $f_{11} \cdots f_{1n}$ are the orientation features computed at the lowest resolution and $f_{n1} \cdots f_{nn}$ are orientation features computed in the n-th (highest) resolution. The value of individual feature vectors are computed using:

$$f_{ij} = \frac{e_{ij}}{\sum_{j=1}^{n} e_{ij}}$$

where e_{ij} is the energy computed in the i-th resolution and j-th orientation channel.

5.3.5 LVQ classifier

Linear vector quantisation is a well known classifier [140] for pattern recognition tasks. It splits the output space into a distinct number of codebook

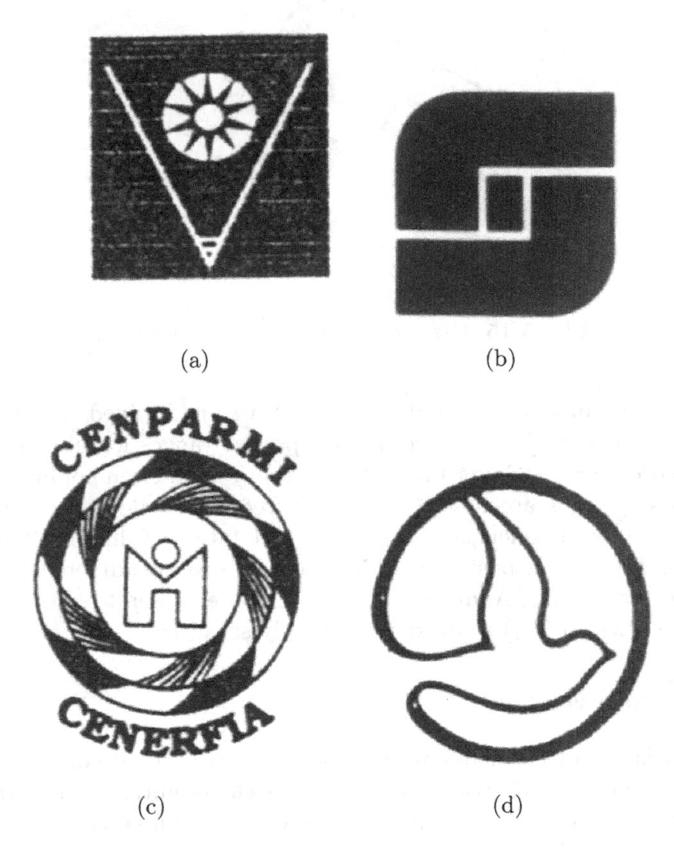

Fig. 5.16. Example images in each shape class: (a and b) square, (c and d) circular.

vectors. These individual codebook vectors are bases representing each of the classes in the problem. Learning proceeds by shifting these initial codebook vectors to more closely represent the classes in the data. It does this iteratively by successive presentation of the input/output pairs. How the codebook vectors are shifted depends on the specific variation of the algorithm (Kohonen provides three [141]) as does the penalty for codebook vectors that are not sufficiently close to the input.

5.3.6 Discussion

The performance of the proposed logo shape extraction system was tested on a real database of images in [142]. The database included 134 grayscale images (27 circular, 34 square, and 73 other). Some example images in each class are shown in Figure 5.16. For training purposes, roughly 10% of the

Table 5.2. Results of LVQ classifier.

class	entries	% correct
square	34	97.0
circle	27	92.6
total	61	95.1

images from the database in the circular and square category were used along with several *ideal* images. The ideal images were ideal in shape but with non-uniform textures mapped onto them. As a byproduct of sampling onto a hexagonal lattice the logos are rescaled to be roughly the same size. The tuning process for the filters can thus be simplified as the only variability is now in the individual features. Feature vectors from two different resolutions which were a factor of seven apart, were used for this application. Each of these resolutions was split into 12 frequency channels which were 15° apart.

The results for the trained system when evaluated on the test set are summarised in Table 5.2. This performance was achieved with 10 codebook vectors which were split with five in each class. Of the 61 images in the two classes, three of them (two circular and one square) were classified incorrectly. These cases are illustrated in Figure 5.17. One incorrectly classified circular image consisted of two circles linked together. The overall appearance of this shape is more rectangular than circular and so the misclassification is to be expected. The other shape which failed the classification process was circular but contained many thin lines of different orientations and much fine detail. This is a problem for the proposed system as it has been designed to have a very narrow (frequency) bandwidth. The incorrectly classified square image was similar to the second incorrectly classified circular image in that whilst it was square in shape it consisted of many strongly oriented internal lines.

The overall results are promising. The system achieved an average 95% success rate for discrimination on the test logo database. There however, was misclassification of some of the logos caused by complex interior textures. This needs to be remedied by further experimentation and fine tuning of the filter banks used in the local energy model.

5.4 Concluding remarks

This chapter provided some examples of developing applications using the HIP framework. The examples were chosen for their employment of different image processing operations discussed in the previous chapter.

The saccadic search scheme illustrated how by exploiting the hierarchical nature of the HIP addressing scheme, an efficient methodology can be devised to extract features from the image. The scheme performed well in following image contours and capturing regions of high image content. This scheme can

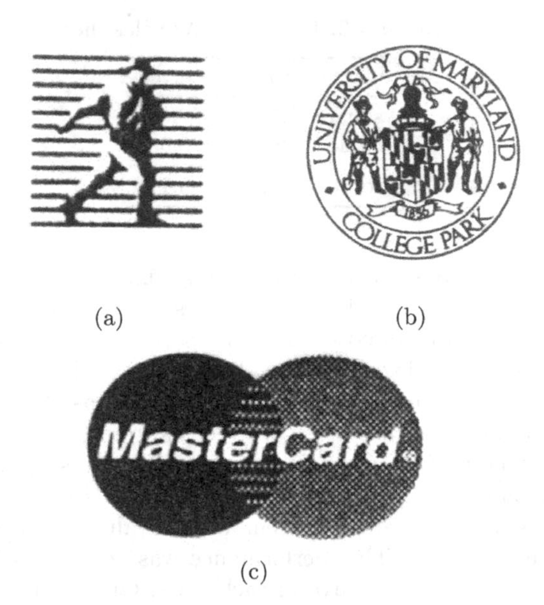

Fig. 5.17. Some misclassified logos (a) square (b and c) circular.

be used in general pattern recognition problems, image analysis, and raster to vector conversion.

The shape extraction methodology extended the saccadic search capability. Basically, the method links neighbouring points of interest together and follows contours to find the shape. The derived shape representation was a modified chain code using HIP addresses. The use of HIP addresses also permits the extraction of higher-order information about the shape such as the rate of change of direction or curvature.

The logo discrimination application was based on the use of shape information as well. However, the shape extraction was performed here in the frequency domain since linear and curved shapes are easily distinguishable in frequency space. Thus, this application provided an example of developing frequency domain solutions based on the HFFT.

The applications presented in this chapter could be alternatively implemented in a square image processing framework. However, there were two observed advantages of using the HIP framework. First was the superior efficiency of the HFFT which permits computational savings over the FFT algorithms for the square lattice. This is attractive for frequency domain solutions to applications such as content-based image retrieval, where the volume of data to be processed is generally high. Second is the ease with which multiresolution information can be extracted using HIP due to the aggregate-based data structure used for image representation.

6

Practical aspects of hexagonal image processing

Ideally, to perform hexagonal image processing (within or outside the HIP framework), one requires an imaging device which produces hexagonally sampled images and a display device that can support hexagonal pixels on a hexagonal lattice. However, the current standards for defining images use square pixels on a square lattice, notwithstanding the fact that CRT displays employ a hexagonal arrangement of RGB phosphors. Specifically, all devices for acquisition and visualisation (display) use square pixels on a square lattice. Hence, foremost among the questions that arise in the context of studying hexagonal image processing is the question of practicality. In other words, it is pertinent to ask how hexagonal image processing fits in the current scenario. The aim of this chapter is to address this question.

There are two ways in which hexagonal image processing can fit within the current scenario. One is to use appropriate solutions for image acquisition and visualisation and develop a complete hexagonal image processing system. The other is a mixed system approach wherein part of the image processing in a large system is done using hexagonal images in order to utilise its benefits, while the remaining part of processing is done with square images. This is

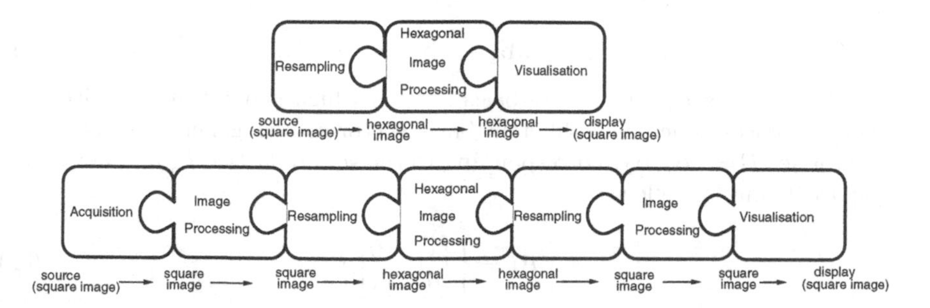

Fig. 6.1. Two ways of implementing hexagonal image processing: Complete system(top), mixed system (bottom).

possible by adopting solutions for converting between square and hexagonal images. Both these possibilities are addressed in this chapter and are illustrated in Figure 6.1. Note that the mixed approach requires more conversion steps than the complete approach.

We begin with the resampling problem which helps move from one lattice to another and then address the problem of visualisation of hexagonal images on conventional display devices.

6.1 Resampling

The process of converting an image sampled on one lattice to another is termed resampling [139]. There are two types of resampling of interest here. They are resampling an image from a square to a hexagonal lattice and, from a hexagonal to a square lattice. The first type of resampling serves as a solution for acquisition of hexagonal images given a square image source such as a camera or a processed image. This is of interest in implementing either a complete hexagonal image processing system or a mixed system. The second type of resampling is useful when further processing is desired to be done using square images. This is relevant only in mixed systems.

In devising solutions for hexagonal image acquisition through resampling, one can use an exact or an approximate solution. In the exact solution, resampling can be done such that the output samples are on a true (regular) hexagonal grid. Consequently, the acquired image will have regular hexagon shaped pixels. Alternately, one can take the samples to lie on an irregular hexagonal grid which means that the pixels in the resulting hexagonal image will not be regular hexagons. These two solutions may have different degrees of computational complexity depending on the implementation.

6.1.1 True hexagonal lattice

A discrete image, $f(m, n)$, has an associated lattice L defined as:

$$L = \{m\mathbf{b}_1 + n\mathbf{b}_2 : m, n \in \mathbb{Z}\} \tag{6.1}$$

The vectors $\mathbf{b}_1$ and $\mathbf{b}_2$ are basis vectors which generate the lattice. Different lattices hence have different basis vector sets as generators. The basis vector set $B = \{\mathbf{b}_1, \mathbf{b}_2\}$ corresponding to the square lattice L_s and hexagonal lattice L_h are as follows:

$$B_s = \left\{ \begin{bmatrix} 1 \\ 0 \end{bmatrix}, \begin{bmatrix} 0 \\ 1 \end{bmatrix} \right\} \tag{6.2}$$

and

$$B_h = \left\{ \begin{bmatrix} 1 \\ 0 \end{bmatrix}, \frac{1}{2} \begin{bmatrix} -1 \\ \sqrt{3} \end{bmatrix} \right\} \tag{6.3}$$

These lattices together with the generating basis vectors are shown in Figure 6.2.

In the process of resampling, the change in the lattice (original to desired) is effected by first reconstructing a continuous image from the original image samples via interpolation. This reconstructed image is then resampled with the desired lattice. These steps are illustrated in Figure 6.3.

Since, only samples at points on the desired lattice are of interest, the reconstruction and resampling with the desired lattice can be combined. Hence, given an image $f_s(m,n)$ defined on a square lattice with $M \times N$ points, the desired hexagonal image, $f_h(x,y)$ is found as:

$$f_h(x,y) = \sum_{m=0}^{M-1} \sum_{n=0}^{N-1} f_s(m,n) h(x-m, y-n) \tag{6.4}$$

where, h is the interpolating kernel for the reconstruction scheme and (m,n) and (x,y) are sample points in the original square and the desired hexagonal images, respectively. There are two major issues in this resampling procedure and these are the choice of the hexagonal lattice and the interpolating kernel.

Two different hexagonal sampling lattices (illustrated in Figure 6.2) can be used for resampling. The lattice in Figure 6.2(c) is obtained by rotating the one in Figure 6.2(b) clockwise, by 30°. These two basis vectors can be labelled as B_{h1} for Figure 6.2(b) and B_{h2} for Figure 6.2(c). Note that, B_{h1} and B_{h2} are also rotated versions of each other.

Both B_{h1} and B_{h2} differ from B_s only by a single basis vector as a result of which B_{h1} and B_s have the same horizontal spacing and B_{h2} and B_s have the same vertical spacing. Consequently, the samples are more densely packed in a hexagonal lattice in the direction in which its basis vector differs from

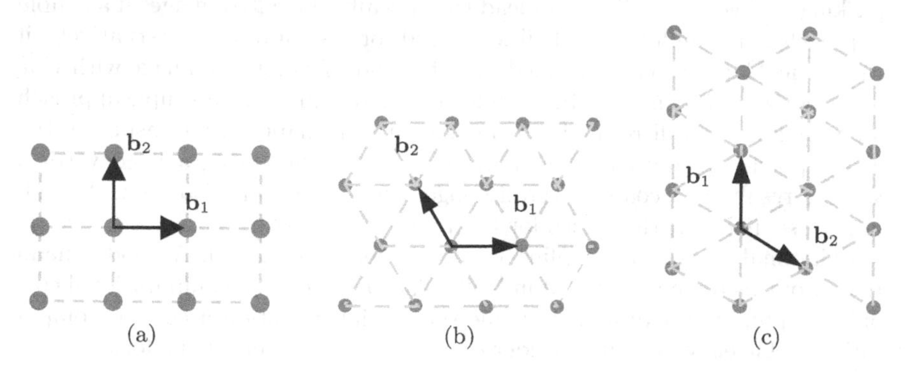

Fig. 6.2. Basis vectors for (a) square (b) horizontally aligned hexagonal lattice and (c) vertically aligned hexagonal lattice.

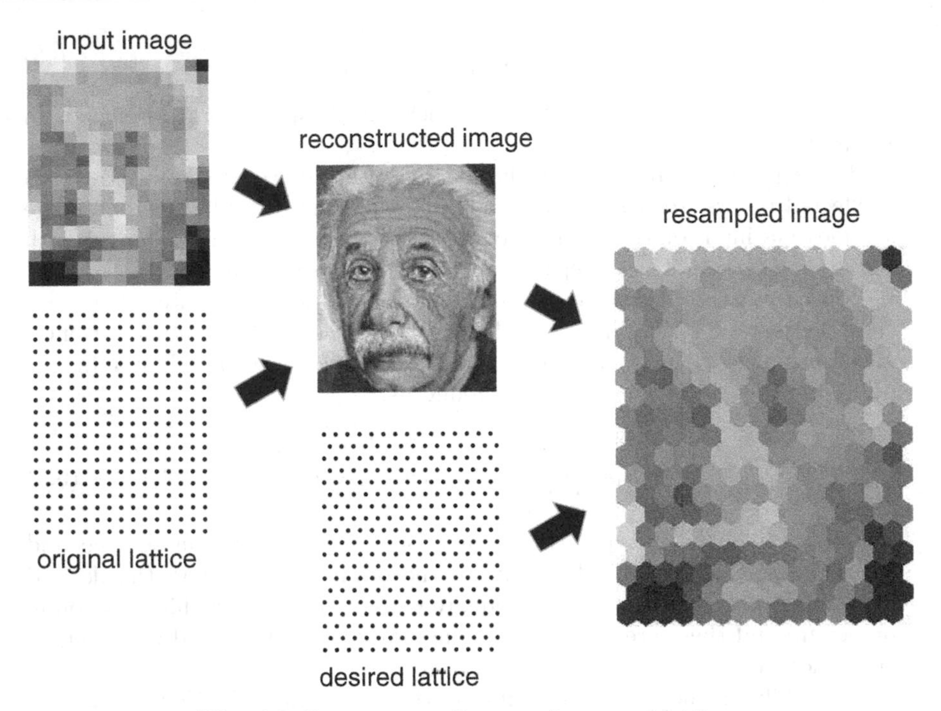

Fig. 6.3. Image resampling on a hexagonal lattice.

the square lattice. It is possible to pick either of these lattices to provide a uniform fit in either the horizontal or vertical directions. Now, if an image is to be re-sampled from an $M \times M$ square image then the choice of B_{h1} will result in one of two possibilities: horizontal lattice spacing is fixed or vertical lattice spacing is fixed. If the horizontal lattice spacing is fixed to give M points, this will result in 15% extra vertical points due to its closer packing ($\frac{\sqrt{3}}{2}$ versus 1). This can lead to vertically elongated images if a simple approach (with uniform sized tiles) is used for visualisation. Alternatively, if the vertical lattice spacing is fixed so as to yield M points, an image with 15% fewer horizontal points will be produced. Once again, using a simple approach for visualisation will result in an image with an inappropriate aspect ratio. These points are illustrated in Figure 6.4. Note that using a fixed vertical spacing results in a coarser overall image. These problems also occur, though in reverse, if B_{h2} is the choice for the resampling lattice.

The final issue in resampling is the interpolating kernel. A good general review of this topic is provided in [139]. The accuracy and computational cost of resampling are determined by the type of interpolation used. For computational efficiency, we will consider only separable kernels of the form:

$$h(x, y) = h_1(x)h_2(y) \tag{6.5}$$

(a) (b) (c)

Fig. 6.4. Two options in resampling onto a hexagonal lattice: (a) Original image (b) and (c) resampled images (top row) with fixed horizontal and vertical spacing respectively, with magnified portions (bottom row).

We will examine three separable kernels for the square to hexagonal resampling process. They are compared in terms of precision, smoothness and ease of computation. The composite image (F1) used for the comparison is illustrated in Figure 6.5. The 256 × 256 image was synthetically created. One quarter of the image contains a real image of a face. The face consists of smoothly varying regions (skin), high frequency regions (hair), and sharp transitions on curved boundaries (cheek). The rest of the image contains a mix of the synthetic ring and star images seen earlier, with different frequency content as well. These were chosen as they should exhibit aliasing with incorrect sampling. It should be noted that in all the experiments, the resampled hexagonal image has been chosen to have a nearly square boundary for convenience.

The simplest form of interpolation that can be used is the nearest neighbour method. Here, each interpolated output pixel is assigned the value of the nearest sample point in the input image. The corresponding 1-D kernel is given as:

$$h(x) = \begin{cases} 1 & \text{if } 0 \leq |x| < \frac{1}{2} \\ 0 & \text{if } |x| \geq \frac{1}{2} \end{cases} \tag{6.6}$$

The kernel function and its magnitude spectrum are illustrated in Figure 6.6(a) and Figure 6.6(b) respectively. This method requires no direct

Fig. 6.5. Input image (F1) used in re-sampling experiments.

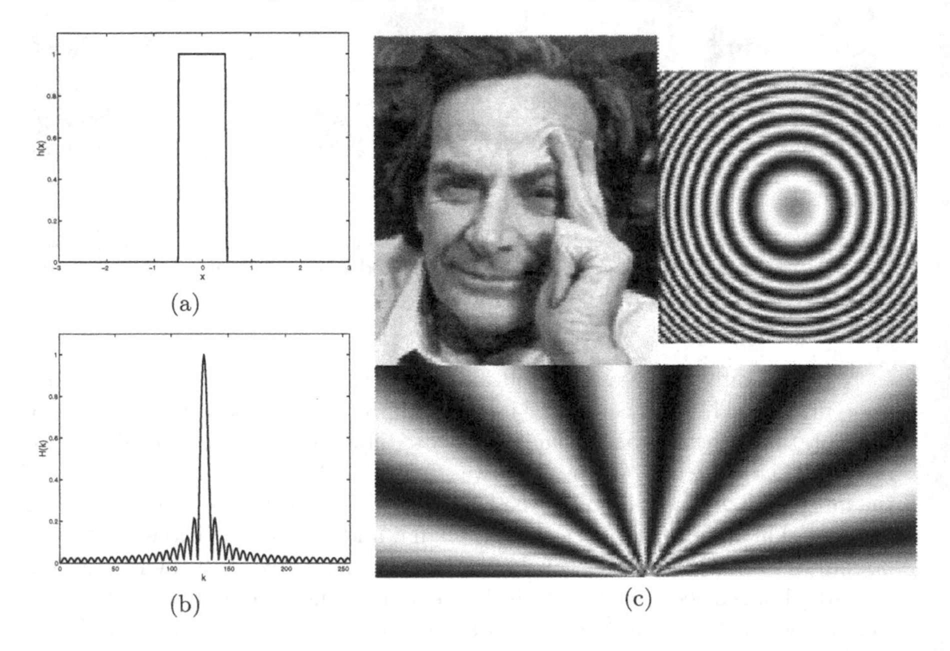

Fig. 6.6. Nearest neighbour interpolation: (a) spatial domain kernel function (b) magnitude spectrum of the kernel (c) resampled F1.

computation and only copying of values from the input to output images. The magnitude spectrum has one main lobe and many sidebands which implies that the performance of this interpolation method should be poor. This is confirmed in the resampled image shown in Figure 6.6(c). The image has many regions where it is blocky. Furthermore, there is significant aliasing in the curved portions of the image and towards the centre of the star in the star image.

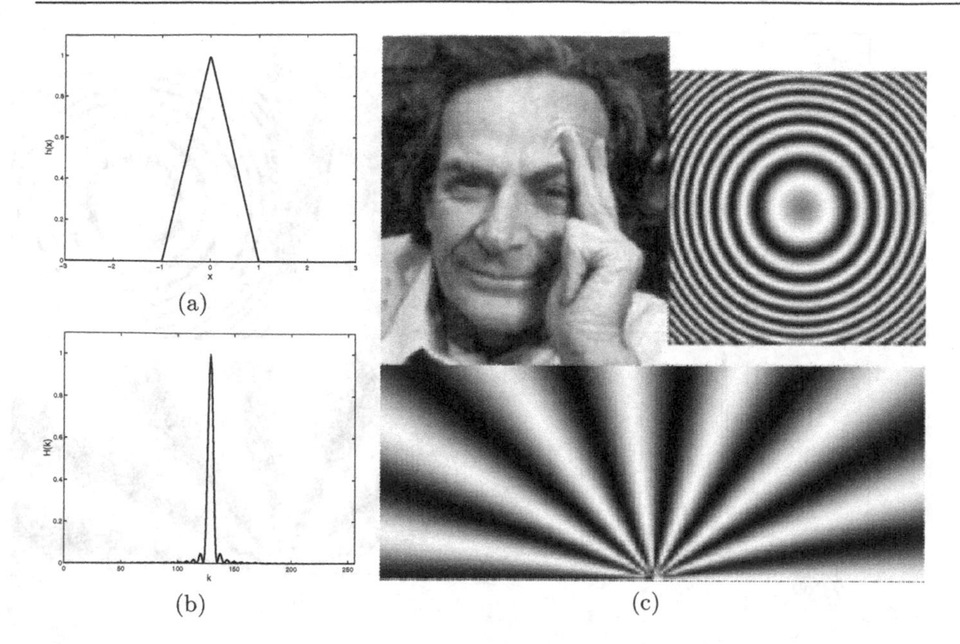

Fig. 6.7. Bi-linear interpolation: (a) spatial domain kernel function (b) magnitude spectrum of the kernel (c) resampled F1.

Bi-linear interpolation is a first-order method in which a straight line fit between pairs of points is used to estimate for an intermediate point. Bi-linear denotes that this fitting is performed in both the horizontal and vertical directions. Thus, four pixels in the input image are required to compute a single pixel in the output image. The 1-D interpolating kernel is:

$$h(x) = \begin{cases} 1 - |x| & \text{if } 0 \leq |x| < 1 \\ 0 & \text{if } |x| \geq 1 \end{cases} \tag{6.7}$$

Figure 6.7(a) shows the bi-linear kernel function and Figure 6.7(b) shows its magnitude spectrum. The first noticeable difference is the almost complete absence of side lobes in the magnitude spectrum of the kernel. This implies an improved performance over the nearest neighbour method. However, the few ripples that are present may result in some aliasing of the image. Examination of the resulting hexagonally re-sampled image in Figure 6.7(c), shows much improved performance over the previous approach. The aliasing in the curves is barely noticeable. Furthermore, the image seems to have been smoothed by the resampling process, thus helping to improve the image quality.

The next interpolation method examined is a third-order interpolation method known as bi-cubic interpolation. It is an efficient approximation of the theoretically optimal sinc interpolation function [143]. It consists of two

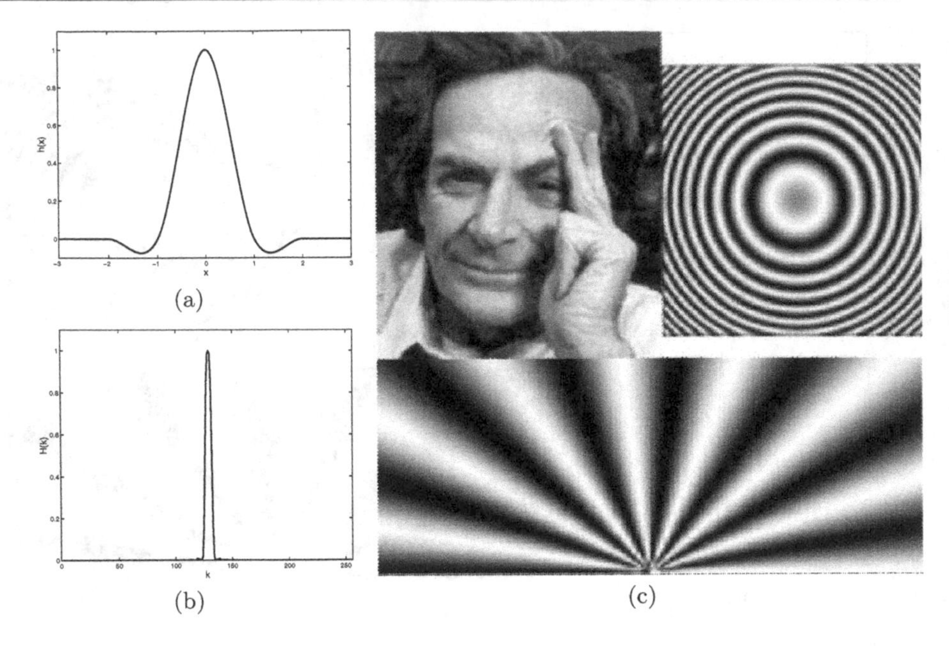

Fig. 6.8. Bi-cubic interpolation: (a) spatial domain kernel function (b) magnitude spectrum of the kernel (c) resampled F1.

piecewise cubic polynomials and requires 4×4 input pixels for every output pixel. The kernel function can be expressed as follows:

$$h(x) = \begin{cases} (a+2)\left|x\right|^3 - (a+3)\left|x\right|^2 + 1 & \text{if } 0 \leq |x| < 1 \\ a\left|x\right|^3 - 5a\left|x\right|^2 + 8a\left|x\right| - 4a & \text{if } 1 \leq |x| < 2 \\ 0 & \text{if } |x| \geq 2 \end{cases} \tag{6.8}$$

The parameter a can be used to effect a better fit to the ideal sinc function. Figures 6.8(a) and 6.8(b) show the kernel function and its magnitude spectrum. In this case a is chosen to be 0.5. It should be noted that negative intensity values will be introduced due to the profile of the interpolation function. Hence, the interpolated image data has to be normalised and rescaled. Examination of the resulting re-sampled image in Figure 6.8(c) shows that the bi-cubic kernel performs quite well. The images are relatively sharper than the bi-linear case and exhibit significantly less aliasing distortion than the nearest neighbour kernel. However, in comparison with the bi-linear approach, the results are not significantly different. Thus, given the extra computation required to achieve the result, this interpolation kernel is not considered to be as effective for the hexagonal resampling problem as the bi-linear kernel.

There exist more advanced choices for the resampling kernel, for instance, using splines [21]. However, these are more computationally intensive.

In the previous discussion, all the resampling techniques were general to hexagonal images. This has provided the necessary tools to allow resampling of square images to hexagonal images. However, much of this monograph has been concerned with the HIP framework. Hence, the remainder of this subsection will be concerned with resampling from a true hexagonal lattice to the HIP data structure and resampling directly from a square image to a hexagonal image in the HIP framework.

Assuming the hexagonal image is defined on a lattice given by B_h (equation (6.3)), a set of coordinates (m, n) for each point in the lattice can be found relative to B_h (see Figure 6.2(b). These coordinates can be converted to addresses in the HIP structure relatively easily due the vectorial nature of the data structure. To perform this conversion, we first note that the basis vectors B_h correspond to specific HIP addresses:

$$B_h = \left\{ \begin{bmatrix} 1 \\ 0 \end{bmatrix}, \frac{1}{2} \begin{bmatrix} -1 \\ \sqrt{3} \end{bmatrix} \right\} = \{\mathit{1}, \mathit{3}\}$$

Hence, an arbitrary lattice point (m, n) can be written as:

$$(m, n) \equiv \{m\mathit{1} \oplus n\mathit{3}\}$$

The scalar multiplication property of HIP addressing (see Section 3.2.2) leads to the following relation:

$$\begin{aligned} g &= m\mathit{1} \oplus n\mathit{3} \\ &= \sum_{i=0}^{m-1} \mathit{1} \oplus \sum_{j=0}^{n-1} \mathit{3} \end{aligned} \tag{6.9}$$

This expression provides an exact method to convert from an arbitrary coordinate in B_h to a HIP structure. Once this is performed the intensity or colour information can be copied over. In practice, this information can be precomputed to enable an efficient conversion using a lookup table.

The above procedure assumes that a square image is first resampled onto a hexagonal lattice using some coordinate scheme and then subsequently converts this scheme to the HIP image structure. If the goal is to generate a HIP image, the first step is not necessary as it is possible to perform the resampling directly onto the HIP structure and save on the computational cost of an extra stage. In this option, the conversion scheme for converting the HIP addresses to Cartesian coordinates (discussed in Section 3.3) is utilised to index the HIP image in terms of Cartesian coordinates which are then employed in the resampling process. The result of resampling a test image onto a HIP image is shown in Figure 6.9. The original image is the ring image which forms part of the test image in Figure 6.5. The image boundary of the HIP image is not rectangular due to the nature of the HIP data structure. Code for converting a square image to a HIP image can be found in Appendix D.

Fig. 6.9. Ring image resampled into the HIP data structure.

6.1.2 Irregular hexagonal lattice

Resampling can also be done onto an irregular hexagonal lattice. There are two candidates for the lattice as shown in Figure 6.10. The first lattice in Figure 6.10(a), is derivable from a square lattice by a lateral shift of the lattice points in alternate rows by half a unit. This is the *brick wall* lattice. It is an irregular hexagonal lattice since the distance between two closest points in successive rows is $\frac{\sqrt{5}}{2}$ and not 1, as required for a true hexagonal lattice. This is also reflected in the basis vector set for this lattice B_b shown below which is different from B_h in 6.3:

$$B_b = \left\{ \begin{bmatrix} 1 \\ 0 \end{bmatrix}, \frac{1}{2} \begin{bmatrix} -1 \\ 2 \end{bmatrix} \right\} \tag{6.10}$$

The angle between the basis vectors of the brick wall lattice is 126.9° as opposed to 120° for the true hexagonal lattice. Resampling with the brick wall lattice can be carried out in a simple manner: starting with a square image, we copy alternate rows into the desired image and generate the samples in remaining rows by averaging between two adjacent samples of the original image. This produces an image with square pixels on a brick wall lattice. Hence, this is a monohedral covering with square tiles. However, because of the increase in the distance between pixels in successive rows the resulting image is stretched in the vertical direction. This can be avoided by up-sampling first and using rectangular pixels. In one such example [18], groups of seven rows of original data are up-sampled to obtain eight rows of data followed by averaging of pixels in alternate rows.

A second type of lattice that can be used to approximate the hexagonal lattice is the well-known quincunx sampling lattice. This lattice is shown in Figure 6.10(b). The distance between the closest pixels in adjacent rows is now $\sqrt{2}$. The basis vector set of this lattice can be computed as:

$$B_q = \left\{ \begin{bmatrix} 2 \\ 0 \end{bmatrix}, \begin{bmatrix} -1 \\ 1 \end{bmatrix} \right\} \tag{6.11}$$

The angle between the basis vectors is now 135° rather than the desired 120°. Resampling a square image onto the quincunx lattice can be performed simply by setting pixels in the original image to zero if the sum of their Cartesian coordinates is non-zero and odd. This produces the familiar chessboard pattern of arrangement of pixels. Hence, this is a packing with square tiles but not a tiling. The distortions which arise due to the approximation of the hexagonal lattice with quincunx lattice can be overcome to a certain extent by up-sampling the original image (by a factor of two in the row direction and by three in the column direction) followed by quincunx sampling. An example of this approach can be found in [19], where an additional final step is also used for mapping the quincunx sampled image onto a hexagonal grid.

Of the two irregular hexagonal resampling methods discussed above, the brick wall lattice has some advantages. This lattice approximates the hexagonal lattice better than the quincunx lattice. Generating an irregular hexagonal image is also fast using this approach since it requires fewer computations. However, the disadvantage with the brick wall lattice is that the resampled image pixels lie on a non-integer grid. This is in contrast to the quincunx lat-

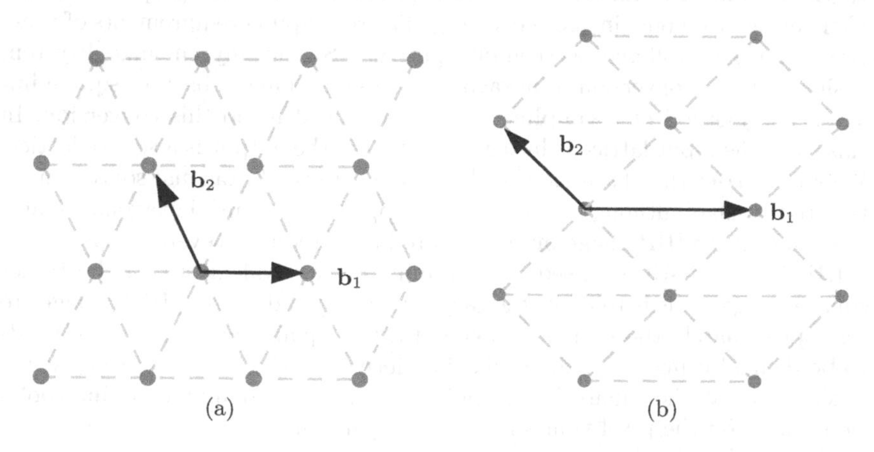

Fig. 6.10. Basis vectors for irregular hexagonal lattices: (a) brick wall (b) quincunx.

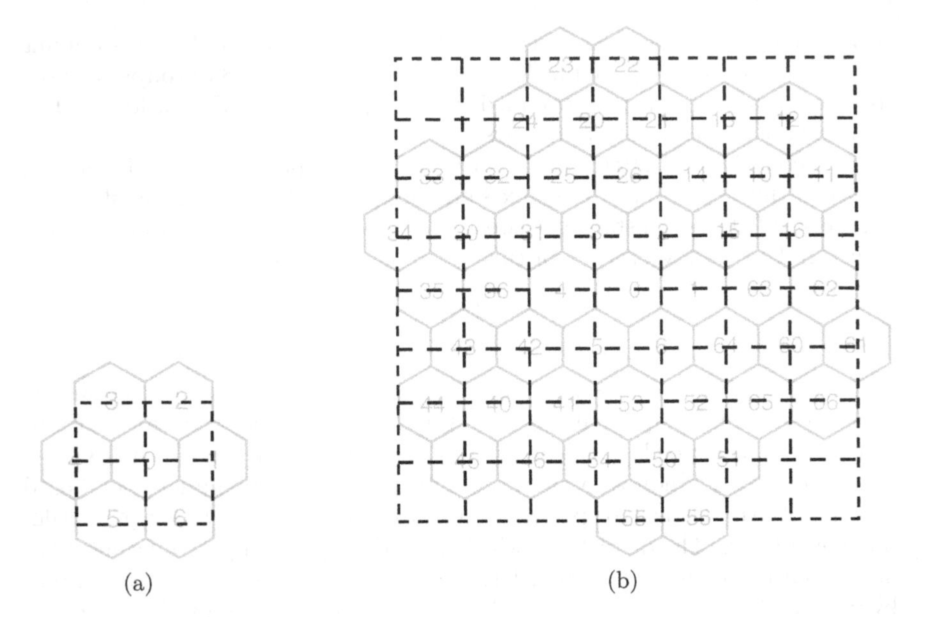

Fig. 6.11. Fitting a square lattice onto a HIP image with (a) one-layer 2) two-layers.

tice where the pixels lie on an integer grid, similar to the original image. Some image sensors developed for high speed imaging do use quincunx sampling. However, the available output is an interpolated square image [39].

6.1.3 Hexagonal to square resampling

At the beginning of this chapter, two types of systems were proposed for implementing hexagonal image processing. The resampling requirements of these two systems are different as seen in Figure 6.1. Specifically, the mixed system design calls for conversion of hexagonally sampled images back to square images. The method of resampling can be employed to do this conversion. In this case, the input lattice is hexagonal whereas the output is a square lattice. We will restrict this discussion to the case where the hexagonal source image is a HIP image, although, it is broadly applicable to any hexagonal image. Resampling the HIP image onto a square image requires several steps: Given a HIP image, firstly we need to determine the equivalent size of the target square image. This is to allow the target (square) and source (HIP) images to be approximately the same size. Secondly, the square sampling lattice needs to be defined using the size information derived in the first step. Finally, the pixel values at the square lattice points need to be computed by interpolation, based on the pixel values in the hexagonal image. These steps will now be discussed in detail.

As stated in Section 3.2, a λ-layer HIP image has 7^{λ} points which are labelled from $\boldsymbol{0}$ to $\boldsymbol{6}\cdots\boldsymbol{6}$ (which consists of λ $\boldsymbol{6}$s). Let us assume the target square image is of $N \times N$. By equating the number of points in the square and HIP images, we determine the value of N as follows:

$$N = \left\lceil e^{\frac{\lambda \log 7}{2}} \right\rceil \tag{6.12}$$

Here, $\lceil . \rceil$, rounds the number up to the nearest integer. This result defines the number of points in the required square lattice. The lattice spacing needs to be determined next. It is desirable for the square lattice to completely cover the HIP image to establish equivalence between the source and target images. In order to achieve this fit, we can start with a standard lattice with unit spacing and apply suitable scaling. The scaling factor can be determined using the boundary information of the HIP image. Square lattices of size defined by N, superimposed on some HIP images (to fully cover them), are illustrated in Figure 6.11. The points which lie on the boundary of the HIP image are actually distinct, which is a consequence of the construction process for HIP images. These points, which we will call extrema, are the set E_{λ} for a λ-level image. For the images illustrated in Figure 6.11 these are:

$$E_1 = \{\boldsymbol{1}, \boldsymbol{2}, \boldsymbol{3}, \boldsymbol{4}, \boldsymbol{5}, \boldsymbol{6}\}$$
$$E_2 = \{\boldsymbol{61}, \boldsymbol{22}, \boldsymbol{23}, \boldsymbol{34}, \boldsymbol{55}, \boldsymbol{56}\}$$

Due to the construction of the HIP image, there will always be six such extrema, two of which define the vertical limits and four that define the horizontal limits. For example, in a two-layer HIP image, these correspond to $\boldsymbol{61}, \boldsymbol{34}$ and $\boldsymbol{22}, \boldsymbol{23}, \boldsymbol{55}, \boldsymbol{56}$, respectively. The two vertical extrema are always related by HIP negation (see Section 3.2.2). The four horizontal extrema come in pairs. The first pair of addresses are related by an addition of $\boldsymbol{1}$ and the second pair by an addition of $\boldsymbol{4}$. Like the horizontal extrema, the second pair of numbers are related by negation. It is sufficient to describe these extrema just using the first two addresses. Inspection of HIP images can lead to the conclusion that the extrema for a λ-layer image are:

$$E_{\lambda} = \{\cdots \boldsymbol{661223445661}, \cdots \boldsymbol{122344566122}, \cdots \boldsymbol{122344566123}, \cdots, \\ \cdots \boldsymbol{334556112334}, \cdots \boldsymbol{455611233455}, \cdots \boldsymbol{455611233456}\}$$

Finding the coordinates of the addresses of the extrema will give a set of coordinates for a minimum bounding rectangle for the HIP image. However, since, we are only interested in a bounding square, the scaling factor can be found by only finding the horizontal extrema. The lattice spacing is the difference between the x values for the extrema divided by the number of square lattice points, N, required. It can be ascertained by looking at the HIP image that the difference between the x coordinates of the extrema is just $2x$. For example, if the first extrema gives an x-coordinate of a then the lattice spacing, s, is:

$$s = \frac{2a}{N-1} \tag{6.13}$$

The spacing information plus the bounding square are enough to define the square resampling lattice. From equation (6.1), the square lattice is defined as $L = \{m\mathbf{b}_1 + n\mathbf{b}_2\}$. The vectors $\mathbf{b}_1$ and $\mathbf{b}_2$ are the standard square basis vectors. Hence, the desired lattice is $B_s = \left\{ \left[s\ 0\right]^T, \left[0\ s\right]^T \right\}$. A down-sampling of the image can be achieved using a larger lattice which is equivalent to multiplying the basis vectors by a positive scalar that is greater than 1. Similarly, up-sampling the image is achieved by using a smaller lattice or scaling of the basis vectors by a scalar that is between 0 and 1.

Now that the square lattice has been defined the final step in HIP to square image conversion is interpolation. An interpolation procedure similar to that of the previous methods in this section can be employed. However, as the source image is hexagonal, the interpolating kernel must be computed with respect to the axes of symmetry of the hexagonal lattice. This will result in some extra computation compared to the conversion of a square image to a hexagonal image. This can be handled via a simple optimisation using the knowledge that the source is a HIP image. For a particular lattice point represented by Cartesian coordinates (x, y), we can find the nearest HIP address $\boldsymbol{g}$ and then compute an interpolation based on a neighbourhood of this point. This reduces the computations significantly and can exploit the simplicity of neighbourhood definitions for HIP images, previously discussed in Section 3.4.3. We will now describe this optimisation procedure. The Cartesian coordinates are first multiplied by a suitable conversion matrix which is the inverse of the conversion matrix described in Section 3.3 as C_{2e}:

$$\begin{bmatrix} r_1 \\ r_2 \end{bmatrix} = \frac{1}{\sqrt{3}} \begin{bmatrix} \sqrt{3} & 1 \\ 0 & 2 \end{bmatrix} \begin{bmatrix} x \\ y \end{bmatrix}$$

The result can be split into integer and fractional parts as $(r_1, r_2) \rightarrow (i_1.f_1, i_2.f_2)$. An initial estimate of the closest HIP address, $\hat{\boldsymbol{g}}$, can be found by considering the integer parts only as follows:

$$\hat{\boldsymbol{g}} = r_1 \otimes \boldsymbol{1} \oplus r_2 \otimes \boldsymbol{3}$$

Next, the fractional parts are considered. If the value on either skewed axis is more than 0.5, then an additional amount is added, with the fixed threshold factor of 0.5 being obtained based on the geometry of the lattice. As an example, if we start with the square image coordinates $(1.45, 2.13)$, the corresponding skewed axis representation is $(2.67, 2.46)$. From the integer parts we get a HIP address of $\boldsymbol{g} = \boldsymbol{1} \oplus \boldsymbol{1} \oplus \boldsymbol{3} \oplus \boldsymbol{3} = \boldsymbol{14}$. Since the fractional part is greater than 0.5, an additional $\boldsymbol{1}$ is added, yielding the final result of $\boldsymbol{10}$ for the nearest HIP address.

The pixel value at this address needs to be found via interpolation. An example of how the interpolation process operates is illustrated in Figure 6.12.

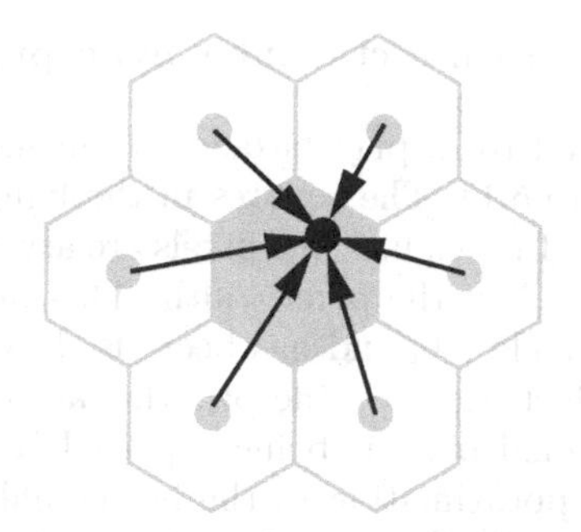

Fig. 6.12. Computation of a square lattice point using interpolation based on surrounding hexagonal pixels.

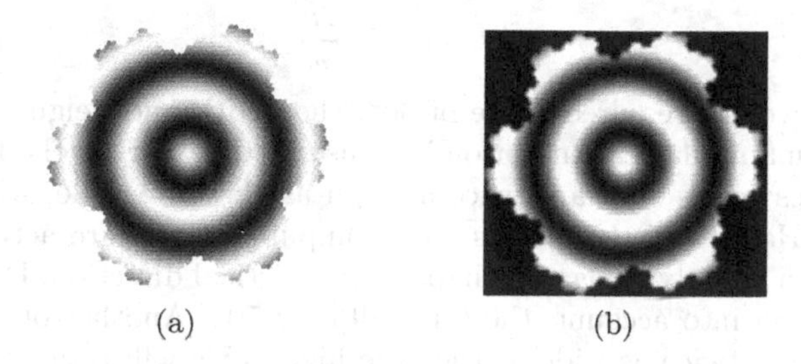

Fig. 6.13. Resampling a HIP image to a square image: (a) original (b) after resampling.

Starting with a square lattice point (illustrated as a black •), the nearest HIP address is computed (the gray hexagon) and the nearest neighbours are used to compute the intensity value of the sample point using an interpolating kernel. The size of the neighbourhood used for interpolation depends on the size of the kernel.

An example of the HIP to square image resampling is illustrated in Figure 6.13. Pixels that lie outside the original HIP image are coloured black (as a simple solution) as shown in the square image. Other means to determine the pixel values for these locations could be employed, if desired. Example code to perform the HIP to square image conversion is included in Appendix D.3.

6.2 Display of hexagonal images

Visualisation of processed data is necessary for any practical hexagonal image processing. The difficulty in displaying hexagonal images is due to the fact that available display devices use a square matrix of pixels. Thus the problem is similar to approximating the hexagonal grid using a square grid. Hence, we will see that the simple solutions that can be developed are similar to those for irregular hexagonal image acquisition.

6.2.1 Approximation with rectangular hyperpixels

A common method used to display hexagonal images is the *brick wall* approach, shown in Figure 6.14. The squares in the figure represent individual pixels within the image. In each line, the pixels are arranged so that they overlap the previous line by half the pixel width. This methodology for display is attributable to Rosenfeld [46] and has been used extensively as it requires little computational effort. However, despite the fact that the grid is pseudo-hexagonal, each hexagonal pixel is being replaced by a single square pixel. This is a very coarse approximation to the hexagonal image. In this lattice, the relationship between the inter-pixel angle θ and a pixel's width, w, and height, h, can be computed as:

$$\theta = \tan^{-1} \frac{2h}{w} \tag{6.14}$$

To achieve the required angle of 60°, the width and height of the pixels need to be manipulated. An option is to use rectangular pixels. For instance, Her [20] uses pixels that are twice as high as they are wide, which yields a θ of 76.0°. Her states that pixels on a computer screen are actually slightly oblate with a slightly elongated shape in the vertical direction. Despite taking this correction into account, the θ is still over 70°. Another option is to use pixels that are twice as wide as they are high. This will result in a θ of 45°. Neither of these situations is ideal. A better solution perhaps is to use a large accumulation of pixels (called a hyperpixel) that yield a close approximation to the ideal angle. As an example, if each hyperpixel was 17 pixels wide and 20 high then the inter-hyperpixel angle would be 59.5°. However, there is a tradeoff between the size of the hyperpixel and the maximum resolution of a complete image that can be displayed.

A caveat in designing hyperpixels is imposed by the oblique effect by virtue of which the HVS is perceptually less sensitive to edges in diagonal directions [95]. This means that the perceived results are particularly compromised when approximating hexagonal pixels with *rectangular* hyperpixels. This effect is illustrated in Figure 6.15(a). A better approach would be to use

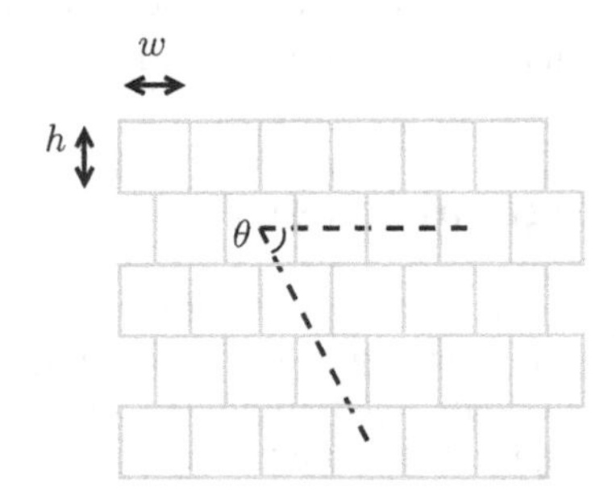

Fig. 6.14. A *brick-wall* image.

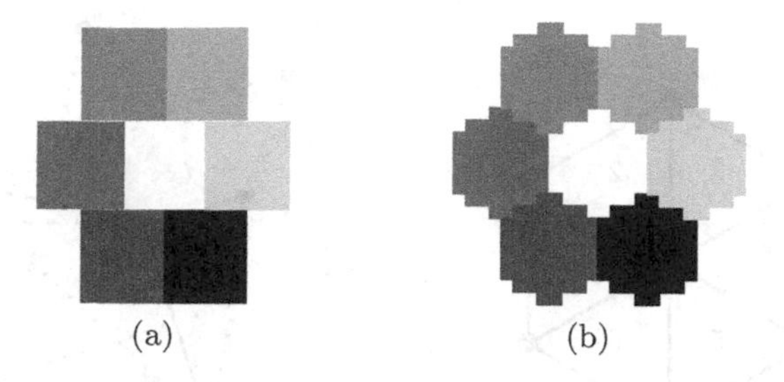

Fig. 6.15. Comparison of hyperpixels: (a) rectangular (b) hexagonal.

accumulations of pixels that represent, more closely, a hexagonal pixel. This approach, illustrated in Figure 6.15(b), is discussed in the next section.

6.2.2 Approximation with hexagonal hyperpixels

A better fit to a hexagon can be produced by constructing a hyperpixel which is roughly hexagonal in shape. This is the most general purpose method for displaying hexagonal images. The process requires no special hardware and is efficient. Figure 6.16 shows two possible hexagonal hyperpixels that can be generated in this fashion. Figure 6.16(a) consists of 30 pixels and Figure 6.16(b) consists of 56 pixels. Whilst both of these examples closely approximate a hexagon, the second case (Figure 6.16(b)) is favoured as it is slightly bigger. The Cartesian coordinates of the pixel locations in a hexagonal image are the centres of the hyperpixels in the displayed image. For a given resolution, say 1280×1024, it is possible to compute the maximum size for a

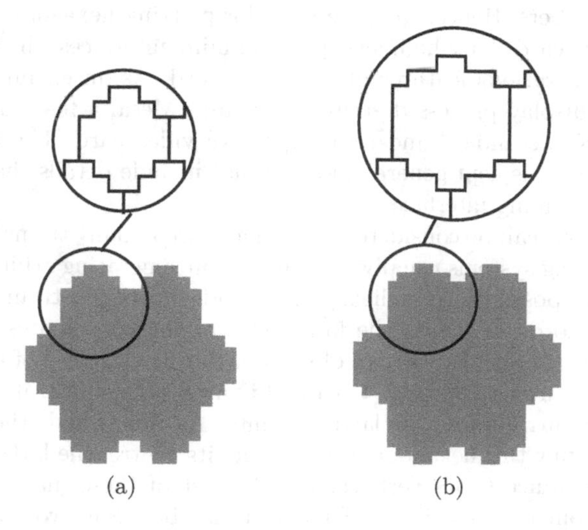

Fig. 6.16. Two possible hexagonal tiles for spatial domain.

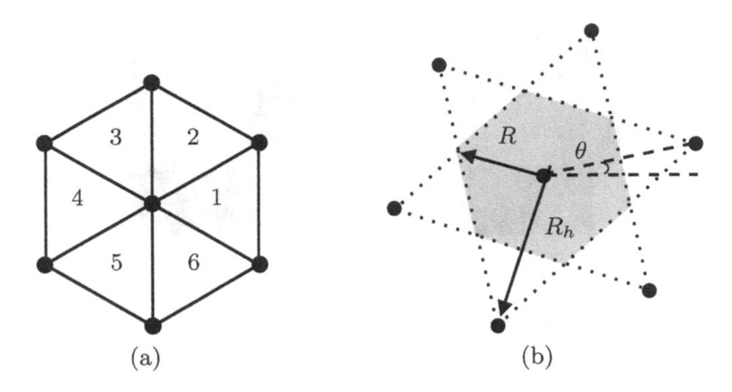

Fig. 6.17. Decomposition of a hexagon into triangles.

particular hyperpixel, for example the one in Figure 6.16(b). This hyperpixel is 8×7 in size and overlaps the next row by 2 pixels. This gives a maximum effective resolution of 160×204. Overall, this method is an improvement over the brick wall approach but is still only an approximation. Ideally it would be good if it was possible to plot accurate hexagons on the screen. This approach is covered next.

6.2.3 Approximation via polygon generation

The most accurate method for displaying hexagonal pixels is by plotting true hexagons. Most modern video cards are capable of rendering millions of triangles a second and hence are also capable of plotting hexagons. Furthermore, the video cards typically compute the vertices of the triangles using floating point numbers. Hence, they are ideal for plotting hexagons. Additionally, many modern cards have high level programming interfaces which can be used to exploit the various features of the video card. As an example of this approach, the display process described here uses Mesa, a free implementation of the OpenGL standard and an inexpensive video card. The following discussion however, is kept general and does not include details that are specific to the programming interface.

All polygons can be considered to be made up of many triangles. Therefore video processing systems usually concentrate on generating arbitrary triangles as quickly as possible. As a hexagon is made up of six triangles, the only requirement to draw one, is the knowledge of the coordinates of the centre and the six vertices. The vertices of each of the six triangles of a hexagon are labelled in Figure 6.17(a) as a '•' while Figure 6.17(b) illustrates a hexagonal tile and the surrounding six lattice points. As illustrated, the information required to draw this hexagonal tile include, its centre, the lattice orientation θ and the distance to its vertices, R. The last of these, namely, R, can be computed from the knowledge of the distance between two adjacent lattice points, R_h as:

$$R = \frac{R_h}{\sqrt{3}} \tag{6.15}$$

The i-th vertex of the hexagon can be found using the computed values of R and θ as follows:

$$\left(R\cos\left[i\frac{\pi}{3} + \frac{\pi}{2} + \theta\right], R\sin\left[i\frac{\pi}{3} + \frac{\pi}{2} + \theta\right]\right) \tag{6.16}$$

Here, $i = 1, 2, \cdots 6$ and corresponds to the vertices. Once the vertices are known, the hexagon can be generated using six triangles as illustrated in Figure 6.17(a). Furthermore, no scaling of the coordinates is required as OpenGL allows translations of the viewpoint to permit viewing at arbitrary distances. In all cases, an image generated in this fashion will be accurate, with the only limitation being the display hardware employed.

6.2.4 Displaying HIP images

As the HIP framework uses a different data structure, we discuss the visualisation of HIP images separately. Two approaches are covered in this subsection. First we examine the hexagonal hyperpixel approach and then the polygon rendering approach. Displaying a HIP image is a two step process. The first step in displaying involves coordinate conversion where the HIP address is converted into a suitable coordinate system for display such as Cartesian coordinates. The second step involves constructing a hyperpixel or plotting a hexagonally shaped tile about this coordinate.

The fundamentals of the coordinate conversion process have already been covered in Section 3.3. We now show a simple method for computing the Cartesian coordinates of a HIP address directly without going through an intermediate coordinate system. An arbitrary HIP address $\boldsymbol{g}$ can be expanded as per equation (6.17):

$$\begin{aligned} \boldsymbol{g} &= g_{\lambda-1} \cdots g_2 g_1 g_0 \\ &= \sum_i \boldsymbol{g_i} \otimes \boldsymbol{10}^i \end{aligned} \tag{6.17}$$

In the equation the summation is performed using HIP addition. We note that the set of points described by $\boldsymbol{10}^i$ lie on a spiral as shown in chapter 3. This means the radial distance of these points from the origin increases. Hence, finding the Cartesian coordinates of a HIP address just requires summing the offsets produced by each of the term in the right hand side. For $i = 0$, the digit $\boldsymbol{g_0}$ can have one of seven values ranging from $\boldsymbol{0}$ to $\boldsymbol{6}$ and this corresponds to a rotation by angle of $|\boldsymbol{g_0}| \times 60°$. In the frequency domain, the situation is a little different because the frequency sampling matrix is dependent on the size of the image in the spatial domain. Specifically, for visualisation in the

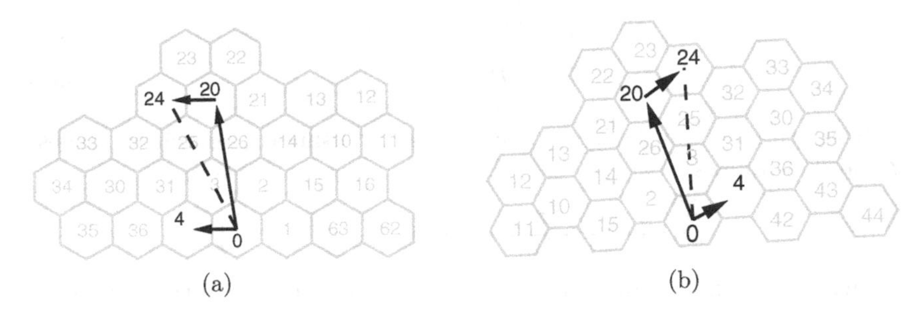

Fig. 6.18. Finding the coordinates of point *24* in the (a) spatial and (b) frequency domains.

frequency domain the rotation is by $g_0 \times -60°$ and there is an extra rotational offset induced by the rotation of the basis. This was covered in more detail in Section 3.4.5. This procedure is now illustrated with an example using the HIP address *24*. A diagram illustrating the spatial and frequency domain hyperpixels is given in Figure 6.18. In the spatial domain, the offset is:

$$\begin{array}{r} (-1, \ 0) \\ +\ (-\frac{1}{2}, \frac{9\sqrt{3}}{2}) \\ \hline (-\frac{3}{2}, \frac{9\sqrt{3}}{2}) \end{array}$$

In the frequency domain the offset is:

$$\begin{array}{r} (-\frac{18}{343}, \frac{20}{343\sqrt{3}}) \\ +\ (-\frac{21}{343}, \frac{91}{343\sqrt{3}}) \\ \hline (-\frac{3}{343}, \frac{111}{343\sqrt{3}}) \end{array}$$

Once the offsets are found, the image can be displayed on the screen using the hyperpixel approach or polygon rendering approach. These are now discussed in turn. Since the spatial and frequency domain tiles are different, these will be discussed individually, starting with the spatial domain tile generation.

On a typical screen with a resolution of 1280 × 1024 pixels, HIP images with five or fewer layers can be displayed using hexagonal hyperpixels. A five-layer HIP image is roughly equivalent to a square sampled image of 128 × 128 pixels. Larger images can be handled by displaying portions of the image in a scrollable window or scaling of the image.

Generation of tiles for the frequency domain case is a little more complicated than the spatial domain described in Section 6.2.2. This is due to the reciprocal nature of the spatial and frequency domain lattices which manifests itself as a change in scale, rotation and a reflection. For a single-layer HIP image, the rotation is 10.9° and for every subsequent layer there is an extra rotation by $\tan^{-1}\frac{\sqrt{3}}{2} = 40.9°$. However, the approach used can be identical to the method for the spatial domain tiles with different hyperpixels used for different layer images. A representative example is shown in Figure 6.19.

The tiles that are generated in this fashion are of a fixed size. Furthermore, each of these tiles has an integer size and is fixed on an integer grid. Note that in practice, however, the coordinate conversion routines will return a number which contains irrational parts which can be avoided with some scaling operation as follows. Conversion of a HIP address into a Cartesian coordinate will always yield an x-coordinate which is an integer multiple of $\frac{1}{2}$, and a y-coordinate which is an integer multiple of $\frac{\sqrt{3}}{2}$. This is because the hexagonal lattice has a fixed spacing in the horizontal and vertical directions. To convert them to integers then one just needs to multiply the x-coordinate by 2 and the y-coordinate by $\frac{2}{\sqrt{3}}$. Once a unique pair of integers corresponding to each hexagon has been found, a suitable scale factor and offset are all that are required to plot the hexagons. These depend on the hexagonal hyperpixel employed but a representative example is illustrated in Figure 6.20 for a spatial domain image. Thus, given the hyperpixel of Figure 6.16(b), to be able to display the images correctly the x-coordinate needs to be shifted by four and the y-coordinate needs to be shifted by seven. For the frequency domain, the methodology is the same but the specific offsets and scale factors are different.

The second method for displaying the HIP image is to exploit the polygon generation unit in modern video cards. The orientation of the lattice θ and the parameter R, are needed to draw the hexagonal tile as discussed in Section 6.2.3. For a HIP spatial domain image, the distance R_h between adjacent lattice points is 1 and thus R, from equation (6.15), is just $\frac{1}{\sqrt{3}}$. To draw the hexagonal tile in the frequency domain, knowledge of the frequency

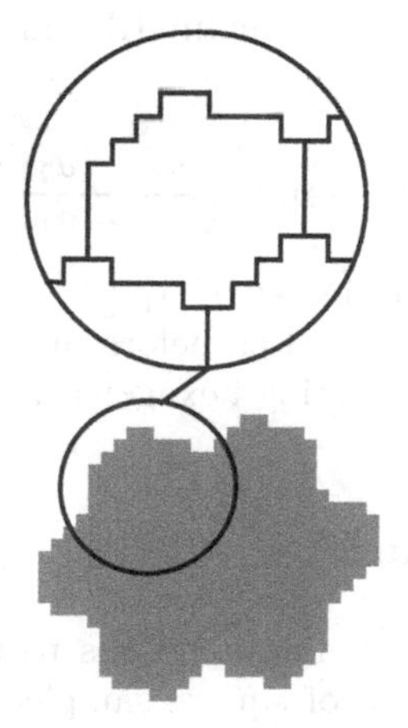

Fig. 6.19. A hexagonal tile in the frequency domain for a one-layer HIP image.

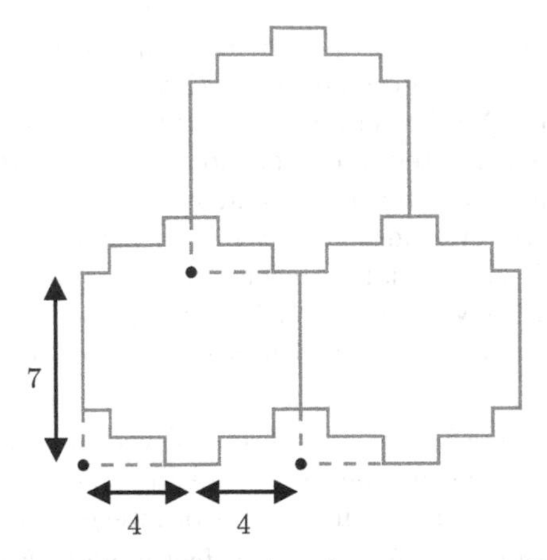

Fig. 6.20. Generating the tiles on a raster display.

domain sampling matrix, U (see equation (3.43)), is required. If the frequency sampling matrix is written as a pair of column vectors:

$$\mathbf{U} = \begin{bmatrix}\mathbf{u}_1 & \mathbf{u}_2\end{bmatrix}$$

then R_h can be computed as:

$$R_h = \sqrt{\mathbf{u}_1^T \mathbf{u}_1}$$

As the frequency sampling matrix is orthogonal to the spatial sampling matrix, the above could be computed using $\mathbf{u}_2$ instead of $\mathbf{u}_1$. The required radius of the hexagonal tile can thus be computed using equation (6.15). The orientation angle for frequency domain HIP images can also be computed using $\mathbf{U}$:

$$\theta = \tan^{-1}\left(\frac{u_{21} - u_{22}}{u_{11} - u_{12}}\right) \tag{6.18}$$

where (u_{11}, u_{12}) are the coordinates of $\mathbf{u}_1$.

Finally, the vertices are found as before, using equation (6.16). Code for displaying a HIP image by plotting hexagons can be found in Appendix D.

6.3 Concluding remarks

This chapter focused on providing solutions to using hexagonal image processing in the current regime of square sampled images, in machine vision. Two different types of usage are envisaged, namely, a complete system, which can process hexagonal images and a mixed system where both square and hexagonal images find a place. The complete system is possible in practice by adopting solutions for image acquisition and display. The literature to date in hexagonal image processing contains numerous examples of this type of system. The mixed system is being proposed to allow exploitation of the advantages of both the lattices as per an application's requirement. An example of such a system would be in processing medical images where natural structures occur frequently. Here, many of the preprocessing steps could be done using square images while structure extraction by morphological processing or frequency domain techniques can use hexagonal images as they can be more efficient. Processing images obtained from space probes can also be done using a mixed system design to efficiently preprocess and filter large volume of images.

In an effort to enable a further study of the effects of changing the sampling lattice, we have provided solutions that will permit design of both complete and mixed image processing systems. Code to help conduct such a study using the HIP framework is provided in Appendix D.

7

Processing images on square and hexagonal grids - a comparison

Up until now, the discussion has been concerned with details of how to process images sampled on a hexagonal lattice using the HIP framework. Since the widely adopted convention is to use images sampled on a square lattice, the natural question to ask is how the two compare. The aim of this chapter is to perform a detailed comparison between processing of images defined or sampled on the square and hexagonal lattices. The HIP framework will be used to do some of this comparison.

In general, as mentioned in Chapter 1, a hexagonal lattice has some attractive properties arising from the lattice geometry: isoperimetry, additional equidistant neighbours and uniform connectivity. As a consequence of these properties, it was seen from the literature survey in Chapter 2 that there are some direct advantages in using hexagonal images: *higher sampling density* [59]; *better representation* for curved structures. (This is particularly enticing since, given the preponderance of curvature in nature, there are many applications which could benefit from the use of hexagonal lattices to perform the analysis and processing); *higher computational efficiency* for some types of processing. Masks are easier to design and more efficient to implement for compact isotropic kernels and for morphological operations [25,48,62]. In addition to the above advantages, processing using the HIP framework can also be beneficial since its single index addressing scheme can further improve the efficiency of computations.

We begin with an investigation of these points first to understand, at a general level, the relative strengths and weaknesses of square and hexagonal image processing. Next, to get a deeper insight, we use specific case studies based on the HIP framework.

7.1 Sampling density

A digital image is a sampled representation of a 2-D projection of a real world scene, albeit within finite bounds. Let us consider this 2-D projection as a continuous image with a spectrum that is isotropic and bandlimited. Thus the region of support for the spectrum is a circle. The sampled image then will have a spectrum which is a periodic (in two directions) extension of the spectrum of the continuous image with the period determined by the sampling rate. We have already shown in Section 3.1 that this sampled spectrum can be related to a periodic tiling whose periodicity is determined by the prototile's size. The size restriction is dictated by the need for perfect reconstruction of the continuous image. It was shown that for perfect reconstruction, the prototile must be large enough to completely contain the baseband. Furthermore, for optimal sampling, the prototile should have a maximal fit to the baseband, with least wasted area in the prototile. This means for optimal sampling, we need a maximal packing of the baseband. Since the baseband has a circular region of support, this is equivalent to the problem of packing circles in a 2-D space. The densest packing of circles of fixed size, in a plane has been shown to be achievable only with a hexagonal lattice [144]. In terms of sampling lattices, the prototile is nothing but the corresponding Voronoi cell which is a square for a square lattice and a regular hexagon for a hexagonal lattice. It is thus clear that the size and shape of the Voronoi cell influence the reconstructability of the continuous image from the sampled image and the optimality of the sampling. We now verify that the hexagonal sampling is more efficient using the formal definitions of lattices.

Comparison of sampling on square and hexagonal lattices requires a common framework. Historically, some work in this area has been previously performed in a general periodic context by Petersen [14] and Mersereau [59]. As noted in Section 6.1, the sampling of Euclidean space can be concisely defined using a sampling lattice. A sampling lattice of a 2-D Euclidean space can be written as:

$$L_{\mathbf{V}} = \left\{ \mathbf{Vn} : \mathbf{n} \in \mathbb{Z}^2, \mathbf{V} \in \mathbb{R}^{2\times 2} \right\}$$

$\mathbf{V}$ is a sampling matrix made up of vectors $\mathbf{v}_1$ and $\mathbf{v}_2$. The vectors $\mathbf{v}_1$ and $\mathbf{v}_2$ together form a periodic basis of Euclidean space. $\mathbf{V}$ is also said to be a generating matrix for the lattice $L_{\mathbf{V}}$. For a square lattice, as illustrated in Figure 7.1(a), the generating matrix, $\mathbf{V}_s$ is:

$$\mathbf{V}_s = \begin{bmatrix} 1 & 0 \\ 0 & 1 \end{bmatrix}$$

There are many possible generating matrices for the hexagonal lattice. However, they are all related by rotations and mirroring. It is thus sufficient to illustrate one such representative lattice as shown in Figure 7.1(b). The sampling matrix, $\mathbf{V}_h$, for this is:

$$\mathbf{V}_h = \begin{bmatrix} 1 & \frac{1}{2} \\ 0 & \frac{\sqrt{3}}{2} \end{bmatrix}$$

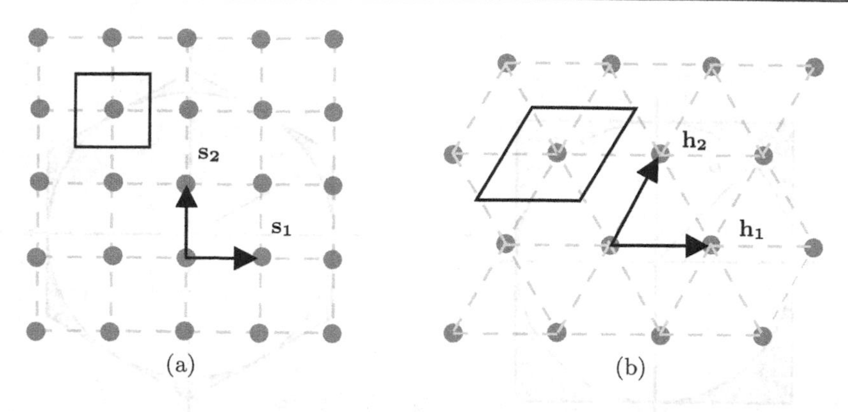

Fig. 7.1. Periodic sampling lattices and their corresponding fundamental regions: (a) square (b) hexagonal.

The columns of each of these matrices are the individual basis vectors for the sampling lattice and are labelled in a similar fashion for the general case $\mathbf{V}$. Consequently, the individual sampling lattices for square and hexagonally sampled points are $L_{\mathbf{S}}$ and $L_{\mathbf{H}}$ respectively. The only difference between $\mathbf{V}_s$ and $\mathbf{V}_h$ is the second basis vector ($\mathbf{v}_{h2}$ or $\mathbf{v}_{s2}$ respectively) which in the case of $L_{\mathbf{H}}$ depends upon $\mathbf{v}_{h1}$. This can be seen by examination of $\mathbf{V}_h$ which for a movement of 1 along $\mathbf{v}_{h2}$ will result in a movement of $\frac{1}{2}$ along $\mathbf{v}_{h1}$.

The basis vectors can be used to generate the fundamental region of a lattice which is a parallelotope. These are illustrated in black in Figure 7.1. In the case of square sampling this is a square with sides of unit length, whereas in the case of hexagonal sampling it is a parallelogram with sides of unit length. The entire plane can be tiled with these parallelotopes in such a way that the shape encloses one and only one sample point.

A circularly bandlimited function is illustrated as being fit inside a square versus hexagonal cell in Figure 7.2. Note that both the cells are of equivalent size. The bandlimited function can be defined as:

$$X(\omega_1, \omega_2) = \begin{cases} X_a(\omega_1, \omega_2), & {\omega_1}^2 + {\omega_2}^2 \leq B^2 \\ 0, & {\omega_1}^2 + {\omega_2}^2 > B^2 \end{cases}$$

where $X_a(\omega_1, \omega_2)$ is an arbitrary frequency spectra defined within the circular region with radius B. Inspection of Figure 7.2(a) shows that the square region has sides of length $2B$ and has an area $4B^2$. Similarly, Figure 7.2(b) shows the hexagon to have sides of length B and an area of $\frac{3}{2}B^2\sqrt{3}$. It can be seen that while both the regular hexagon and the square cover the circular region, the wasted space, indicated in grey, is less for a regular hexagon than for a square. In fact, the square wastes approximately a third more area.

The sampling density is generally defined as a ratio of the area enclosed in the circle to the area within the fundamental region of a sampling lattice. The

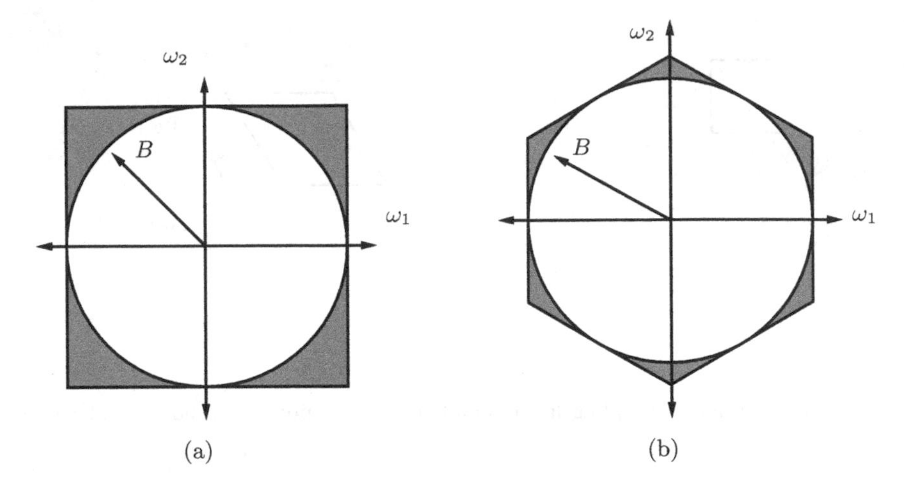

Fig. 7.2. A circularly bandlimited function inscribed in (a) a square and (b) a hexagon.

area of the fundamental region is the determinant of the generating matrix. Hence, the sampling density is:

$$\sigma = \frac{\pi B^2}{|\det \mathbf{V}|}$$

where $\mathbf{V}$ is the generating matrix. Thus for square sampling:

$$\mathbf{V}_s = \begin{bmatrix} 1 & 0 \\ 0 & 1 \end{bmatrix}$$

The sampling density for a square lattice is:

$$\sigma_s = \pi B^2 \tag{7.1}$$

The sampling density for the hexagonal case is:

$$\sigma_h = \frac{2}{\sqrt{3}} \pi B^2 \tag{7.2}$$

Thus the density of hexagonal lattice is 13.4% higher than the square lattice. As a result, the mean distance to the nearest pixel is shorter in a hexagonal image than that in a square image of similar resolution. Hence, it follows that hexagonal sampling requires 13.4% fewer sample points than square sampling to represent the same continuous bandlimited signal. This is a reason for hexagonal sampling being the most efficient sampling scheme upon a two-dimensional lattice [14]. This is an important advantage of hexagonal sampling which can be exploited in many applications.

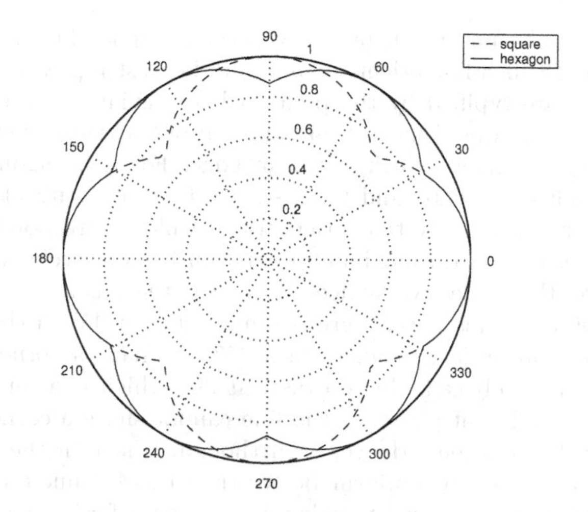

Fig. 7.3. The distance functions for a hexagon and a square.

7.2 Comparison of line and curve representation

Lines and curves are important structures in most images. In a previous chapter, a study of edge detection revealed that there could be some benefits in using hexagonal images for representing curves. The main reason for this is the fact that every pixel in the hexagonal image has six equidistant, neighbours as compared to four in the square image. This provides for a better angular resolution necessary for accurate edge following, in the hexagonal lattice. This section will investigate the relative benefits that the lattices offer for representing curves and lines.

Before investigating this however, it is informative to study the polar plot of the distance of all points on a hexagon/square to its centre. This distance r can be computed as a function of θ as follows:

$$r = \cos\theta$$

where θ is the angle subtended by the line joining a point on the perimeter and the centre, to the horizontal. This equation is only valid for a small range of θ. For a hexagon this is $\theta \in [-\frac{\pi}{6}, \frac{\pi}{6})$ and for a square this is $\theta \in [-\frac{\pi}{4}, \frac{\pi}{4})$. We can use the symmetry of the shape to find the distance for angles outside this range. The distance functions are plotted on the same graph shown in Figure 7.3. A circle is also plotted for reference, as in this case all points on the circle are equidistant from the centre.

Examining the figure one can see that overall, the distance plot for the hexagon has six lobes while that for the square has four lobes which is consistent with the number of rotational symmetry of these shapes. Furthermore, the distance plot for a hexagon is closer to the circle than that of the square.

This is due to the isoperimetric property of the hexagon. The figure also gives an indication of which line orientations would be best represented in the two lattices. These are typified by the points where deviation from the circle is a minimum since it would mean that a line drawn at this orientation would have neighbouring pixels which share an edge. For a hexagon, such orientations are multiples of 60° and for a square they are multiples of 90°. By the same token, we can say that orientations which correspond to points of minima on the plots (with maximum deviation from a circle) will be poorly represented on the respective lattices. However, there is a caveat due to the lattice geometry. For instance, there is a minimum at 45° in the plot for the square lattice. Two adjacent pixels on a 45° line have a corner in common and are actually 8-adjacent. In contrast, at 30° which is a minimum in the hexagonal plot, adjacent pixels on the line cannot share a corner or an edge as they are not 6-adjacent. Hence, even though, based on the plot alone, it appears that a 30° line will appear better than the 45° line on a square lattice, the lattice geometry must be taken into account for a correct prediction. With this in mind, we can predict that lines of orientations 0°, 60° and 120° can be represented well on a hexagonal lattice while lines of orientations 0°, 90° and 45° can be represented well on the square lattice, albeit the first two appearing smoother than the third. Representing a 30° oriented line can be expected to be erroneous on both lattices. From the figure, we note that up to 30°, the distance plot for the square and the hexagon overlap.

The figure allows some insight into curves as well. To represent curves, a lattice needs to allow depiction of rapid changes in orientations and from the distance plot we can see that such rapid changes can be perfectly depicted at points where the plots overlap with the circle. There are six such points for the hexagonal plot compared to only four for the square. Additionally, if we take the lattice geometry into consideration, the distance to the nearest neighbour is always unity for a hexagonal lattice unlike in the square lattice. These two observations together mean that the representation of curves should be better on a hexagonal lattice than on a square.

These predictions are next verified by carrying out some experiments on the practical aspects of representing lines and curves on the two lattices. The rest of this section is divided into several subsections. The first examines the Bresenhams line/circle drawing algorithm on the two lattices. The second examines the effect of image resolution on line/curve representation. Finally, the last subsection summarises the results and makes some general recommendations.

7.2.1 Algorithmic comparison

Drawing lines and curves on a particular lattice is a common task in computer graphics. Possibly, the most well known algorithms for the same are due to Bresenham [145]. These line and curve drawing algorithms work by choosing a next point which would minimise the error, in the least squares sense, for the desired function. This section will examine how to develop such algorithms for a hexagonal lattice. Then, the algorithms will be evaluated on hexagonal and square lattices.

Previous work on line drawing algorithms for hexagonal lattices has been done by Wüthrich and Stucki [50]. However, the emphasis of that work was limited to the efficiency of the proposed hexagonal algorithm compared to the traditional square equivalent. In this discussion the aim is to study the effect of representation of continuous functions, such as lines and curves, in a discrete sampled space.

A piecewise linear approximation to a curve $c(x, y)$ is as follows:

$$c(x, y) = \sum_{i=0}^{N} l_i(x, y) \tag{7.3}$$

Here, the curve can be seen to be the sum of N separate line segments $l_i(x, y)$, with different orientations. Obviously, the accuracy of the resulting representation is improved by increasing the number of distinct line segments. Ideally, N should be large enough to represent the curve with little error. Large N implies high angular resolution for the line segments as well. Equation (7.3) is expressed for real curves in Euclidean space. Thus, the line representation problem on an integer grid has to be examined first.

There has been a large amount of study devoted to the drawing of lines on integer grids [146] with Cartesian axes. This discussion will be concerned with the most well known one, namely, the Bresenham line drawing algorithm [145]. The general equation for a line using Cartesian coordinates is:

$$y = mx + c$$

where m is the slope of the line and c is the y-intercept. If the line to be drawn starts from (x_1, y_1) and ends at (x_2, y_2) then the slope of the line is:

$$m = \frac{y_2 - y_1}{x_2 - x_1} = \frac{\Delta y}{\Delta x}$$

A simple approach is to iterate the value of the x-coordinate from x_1 to x_2 using integer steps and at each point round the resulting y-coordinate to the nearest integer point. However, this requires repeated and slow real number computations. Bresenham solved this problem by examining the nature of the fraction m and what happens to it in each iteration. This allows only integer arithmetic to be performed.

A Bresenham-style line drawing algorithm for a hexagonal lattice can be developed as follows. The problem for a hexagonal lattice using a skewed axis coordinate system is illustrated in Figure 7.4. The filled hexagon illustrates the current point in the rasterisation of the gray line. The two open hexagons illustrate the two possible choices for the next point. These are the points $(x + 1, y)$ and $(x, y + 1)$. This is a point of difference from the traditional Bresenham line algorithm but, in the hexagonal case, the point $(x + 1, y + 1)$

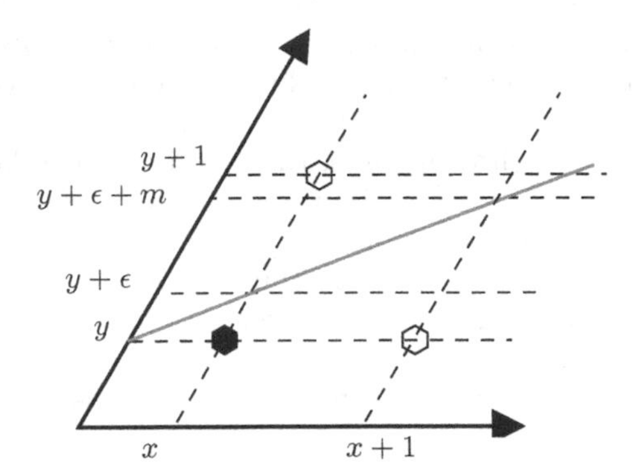

Fig. 7.4. Current point and candidates for the next point in Bresenhams line drawing algorithm on a hexagonal lattice.

results in a disjointed line. The next point in the line is decided using a simple criterion based on the error ϵ. The error, ϵ, is defined as the difference between the discrete coordinate and the actual coordinate of the point in Euclidean space. If $\epsilon + m$ is greater than $\frac{1}{2}$ (the actual point falls less than halfway between the two points) then plot point $(x+1, y)$ otherwise plot point $(x, y+1)$. This algorithm can be written concisely in terms of HIP addresses. If the original point, (x, y), is $\boldsymbol{g}$ then $(x+1, y)$ becomes $\boldsymbol{g} \oplus \boldsymbol{1}$ and $(x, y+1)$ becomes $\boldsymbol{g} \oplus \boldsymbol{2}$. For more detail refer to Appendix C.1.

Based on the above description of the algorithm, we can note the following. Starting with a current pixel, the decision about the next point is between two adjacent pixels that are 45° apart, in a square lattice, or 60° apart for a hexagonal lattice. The difference to note between these two cases is that the distance between the current and candidate pixels is not uniform for the square lattice as it is for the hexagonal lattice. Hence, there will be a bias in the case of square lattice in favour of the pixel to the right more often than in the hexagonal lattice, which can potentially lead to more errors.

We have studied the performance of the straight line algorithm for the two, for comparison. Only a subset of all possible lines were chosen for comparison, as they were thought to be representative. Also, the lines were chosen to intercept the origin as this makes the comparison simpler. The lines are at angles (with the x-axis) of 0°, 30°, 45°, 60°, 90°, and 120°. Based on the discussion in the beginning of Section 7.2, we have a set of three lines that can be depicted perfectly on each of the two lattices and one line (at 30°) which will be less perfect in both lattices. Comparison is carried out visually and quantitatively by computing the average deviation from the ideal line as:

$$e = \frac{\sum_x |y_a - y|}{n} \tag{7.4}$$

Table 7.1. Errors in depicting lines of different slopes.

angle	$\overline{e_s}$	$\max(e_s)$	σ_s	$\overline{e_h}$	$\max(e_h)$	σ_h
0°	0.00	0.00	0.00	0.00	0.00	0.00
30°	0.25	0.50	0.15	0.50	1.00	0.50
45°	0.00	0.00	0.00	1.00	2.35	0.64
60°	0.43	0.86	0.25	0.00	0.00	0.00
90°	0.00	0.00	0.00	0.50	0.50	0.50
120°	0.43	0.86	0.25	0.00	0.00	0.00

where y is the theoretical coordinate for the real line, y_a is the rasterised point and n is the number of points in the line which serves to normalise the error. Here the summation is carried out for all the x coordinates in the line.

Figure 7.5 illustrates the lines drawn using the algorithm. In each case, there are two lines representing the hexagonal and square cases which are individually labelled, with the lines artificially offset to aid comparison. To provide a fair comparison between the two cases, the square pixels have been scaled to give the same area as the hexagonal pixels. This gives the square pixels of length/width $\sqrt{\frac{3\sqrt{3}}{2}}$. The summary of the errors for each line on the two grids are given in Table 7.1. The subscript s or h indicates whether it is a square lattice or a hexagonal lattice. The statistics gathered are the mean error $\overline{e}$, the maximum error $\max(e)$, and the standard deviation of the error σ.

Visual inspection of the lines reveals several characteristics. Let us denote the lines on the hexagonal and square lattices as l_h and l_s respectively. In Figure 7.5(a), l_h appears less smooth than l_s due to the hexagonal pixel shape. In the cases of other lines, shown in Figures 7.5(b) to 7.5(d), l_h appears less ragged than l_s except for the 45° line. Finally, it is seen from the Figure 7.5(e), that l_h appears to be decidedly worse than l_s. The results are consistent with the earlier expectation, both in terms of appearance and error value. The difficulty for representation of vertical lines is often considered a reason not to perform image processing on a hexagonal lattice. Overall however, the lines which are rendered on a hexagonal lattice exhibit good representations of the lines.

Results of error analysis are shown in Table 7.1. They are fairly consistent with the previous judgements made based on visual inspection. The main exception is that the 45° line has a large average error. This is because it is difficult to represent a 45° line using a set of axes that are 60° apart. The slope in this case turns out to be two which accounts for the large error. The error for the vertical line is quite small despite the ragged appearance of the line itself. Finally, the 30° line appears much smoother but has a slightly higher error. Overall, lines appear to be similar in both the hexagonal and square lattices. This is true despite the error for the lines drawn on the hexagonal lattice having higher relative errors. The reason for this is a perceptual phenomenon

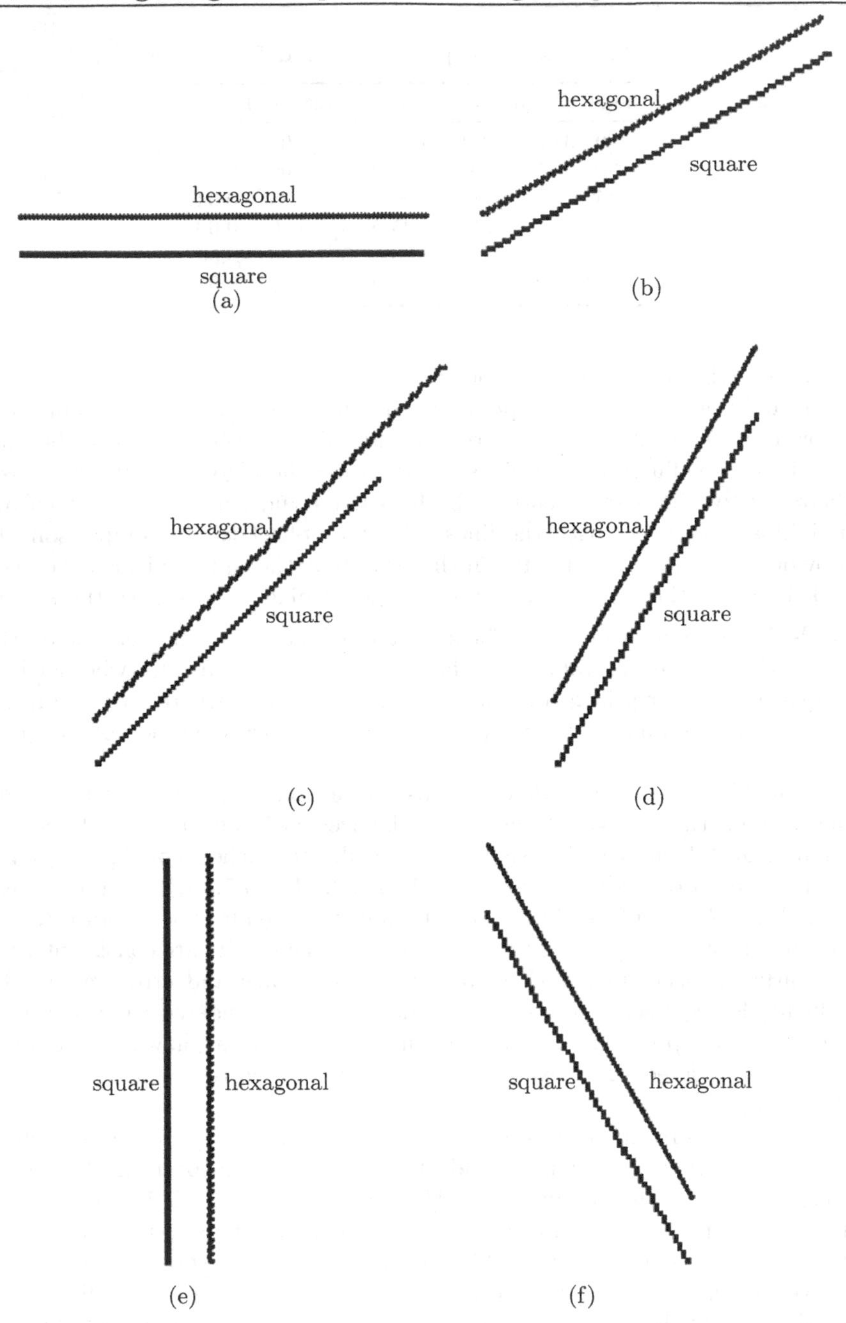

Fig. 7.5. Comparison of lines of different orientation (a) 0° (b) 30° (c) 45° (d) 60° (e) 90° (f) 120°.

known as the oblique effect in human vision [95] wherein obliquely oriented patterns have less visibility compared to horizontal or vertical ones.

Next we turn to a comparison of curve representation on the two lattices. A series of circles of different radii will be considered for this purpose. The basic strategy in depicting a circle on a sampled space would be to approximate the circle with an n-gon whose vertices lie on the circle. Obviously, the higher the value of n the better the approximation to the circle. However, this value is constrained by the desired radius (in pixels). For instance, the best approximation possible for a one-pixel radius circle is with a 6-gon (hexagon) on the hexagonal lattice or a 4-gon (square) on the square lattice and for a larger circle the approximation can be better but still constrained by the lattice geometry.

Let us consider the drawing algorithm which has to find the vertices of this n-gon which are sample points on the circle. The equation for a circle is $x^2 + y^2 = R^2$, where R is the circle's radius. Solving this for y gives $y = \pm\sqrt{R^2 - x^2}$. In order to draw a semi-circle, x can be incremented from 0 to R solving for the allowable value of $+y$ at each step. This approach works reasonably well. However, sample points on the circle are not uniformly spaced and gaps result. This is especially the case with increasing slope or as the value of x approaches R. A similar problem occurs if a trigonometric approach is used to find the points on the circle. There are several algorithms to draw circles on square integer grids, the most famous of which was pioneered by Bresenham [147] .

The approach mentioned in the previous paragraph only discussed computing a segment of the circle. This segment covers an arc of 180°. The rest of the circle can be plotted using symmetry. In fact, using symmetry only a small portion of the circle, say 45°, need be computed. This process is illustrated in Figure 7.6 for both square and hexagonal lattices. In the figure, the axes are labelled as (x, y) in the case of square lattices and (h_1, h_2) in the case of hexagonal lattices, in order to distinguish between the two sets of axes. The radius of the circle drawn in this fashion is given by the value of h_1 while h_2 is zero for the hexagonal lattice and similarly the value of x when $y = 0$ for the square lattice.

The number of symmetric points generated on the circle is different on the two lattices. The circle on a hexagonal lattice has 12 points while that on the square lattice has 8 points. As the points represent the vertices of the n-gon, the value of n will influence the overall shape of the circle. For the hexagonal lattice this will be a slightly dodecagonal shape and for the square case this will be a slightly octagonal shape.

Since, a complete circle can be drawn by exploiting the symmetry, it suffices to detail how to draw one short segment of the circle. Once again,the essence of the approach lies in choosing points which are closer to the actual circle. This is illustrated in Figure 7.7. In the figure, the choice for the next pixel is either $(x + 1, y)$ or $(x + 1, y - 1)$ if continuity of the line is to be maintained. In HIP addressing, these points are $\boldsymbol{g} \oplus \boldsymbol{1}$ and $\boldsymbol{g} \oplus \boldsymbol{6}$. The deci-

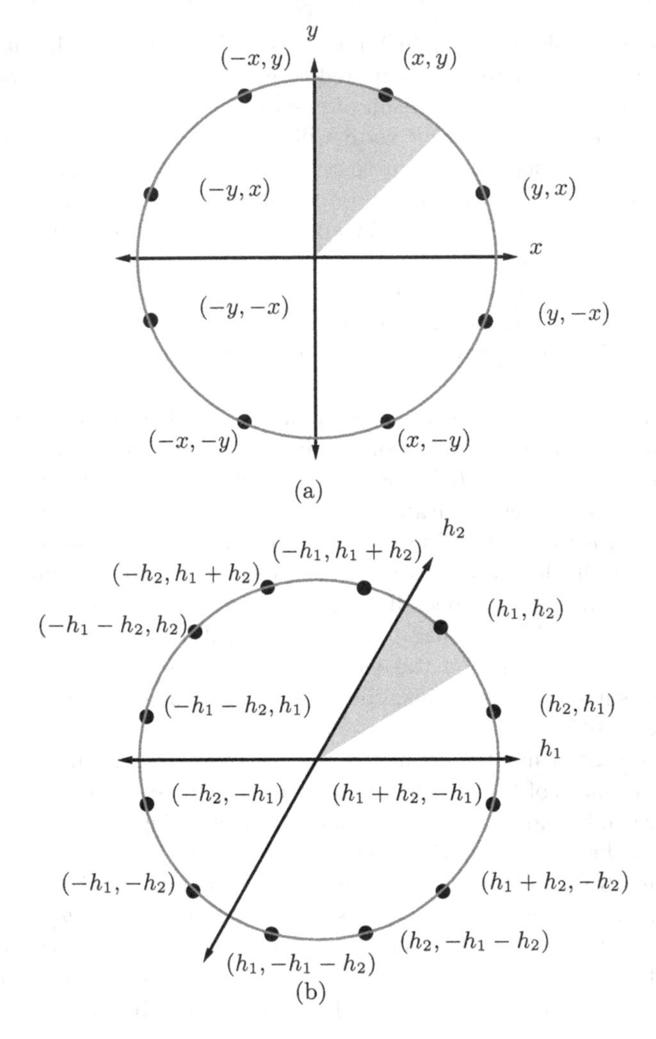

Fig. 7.6. Drawing circles on (a) a square lattice and (b) a hexagonal lattice.

sion on which point to plot can be made using a decision variable which is evaluated at the midpoint between the two competing points. The constraint on the variable is to minimise the error. A procedure to draw the entire circle is as follows. The first point plotted is at $(0, R)$. Subsequent points are plotted based upon the value of the decision variable. After each point is plotted, the decision variable is updated depending on which point is plotted. The arc is considered finished when the x-coordinate and the y-coordinate are equal. The only additional information required to plot the circle is the initial condition, which is the value of the circle at the 30° point (the 45° point for the

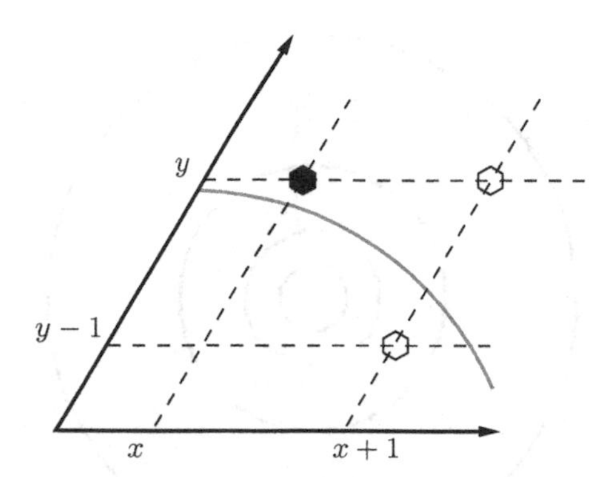

Fig. 7.7. Current point and candidates for the next point in the Bresenham circle drawing algorithm on a hexagonal lattice.

square lattice). More details of the algorithm for the hexagonal case is given in Appendix C.2.

The above algorithm has been used to draw circles of four different diameters for comparison: 100, 50, 25, and 12.5. A wide range of possible diameters is chosen to highlight the best and worst case scenarios. At one extreme, the circle has insufficient pixels and so will lose its fidelity. At the other extreme, the circle will have sufficient points and appear circular. The comparisons between hexagonal and square lattices are carried out first visually and then by computing the deviation from the ideal circle. The errors between ideal and actual circles were computed using equation (7.4), similarly to the case of lines.

Figure 7.8 illustrates the circles on square and hexagonal lattices. The area of the square pixel has been scaled to give the same area as the hexagon (sides of length $\sqrt{\frac{3\sqrt{3}}{2}}$). The computed errors are listed in Table 7.2. The subscript indicates whether the measure refers to a square or a hexagonal lattice. Similar statistics were generated as for the lines.

Several important characteristics of the hexagonal lattice for curve representation can be observed via a visual inspection. Firstly, at a coarse level, the circles drawn on the hexagonal lattice appear smoother. This is especially noticeable in the vertical and horizontal regions of the circle. In the square lattice, these regions appear as thick straight lines. The overall shape of the circles, as the radii reduces, worsens in the square lattice compared to the hexagonal lattice. This is because of the improved angular resolution afforded by having six equidistant neighbouring pixels compared to only four in a square lattice. The oblique effect [95] also plays a part in this perception.

The errors listed in Table 7.2, are consistent with the findings from visual inspection of the images. The average error for the hexagonal lattice is slightly larger, ranging from 0.20 to 0.26 as compared to 0.18 to 0.24 obtained for the square lattice. However, this difference is too small to be significant.

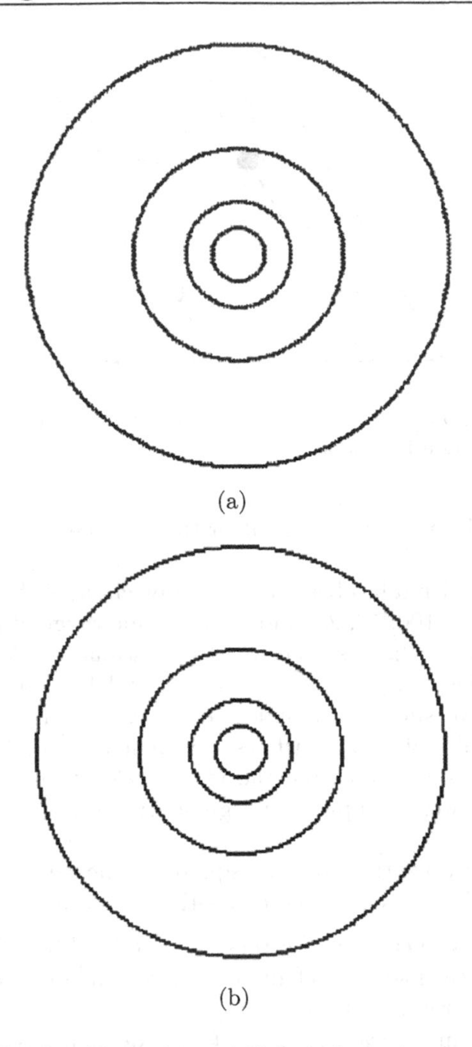

Fig. 7.8. Circles of radii 100, 50, 25, and 12.5 pixels depicted on the (a) hexagonal lattice and (b) square lattice.

Table 7.2. Errors in depicting circles of different radii.

diameter	$\overline{e_s}$	$\max(e_s)$	σ_s	$\overline{e_h}$	$\max(e_h)$	σ_h
12.5	0.18	0.47	0.15	0.20	0.48	0.16
25	0.22	0.50	0.16	0.23	0.48	0.15
50	0.24	0.49	0.15	0.26	0.50	0.15
100	0.24	0.50	0.15	0.26	0.50	0.15

The maximum error and the standard deviation of the error are similar for both the square and hexagonal lattices. Hence, it can be concluded from the quantitative analysis that the hexagonal lattice is on a par with the square lattice for representing curves. However, when this is coupled with the qualitative results, it appears that the human visual system is more tolerant to errors when the underlying representation is a hexagonal lattice. This implies that the hexagonal lattice is a better choice for representing curves than a square lattice.

7.2.2 Down-sampling comparison

A curve and a line are continuous entities in reality, even though we are dealing with discrete entities in image processing. In the previous section, an algorithmic approach was taken to drawing lines and curves based on the minimisation of the error for a particular lattice. In this section, we examine the effect of image resolution on line and curve representation. The assumption is that an image with a suitably high resolution will approximate a continuous image. This effect can be seen when looking at a television set or a newspaper. In both these cases, the images are made up of a large number of individual pixels but the overall perceptual effect is that of a continuous image. By the same

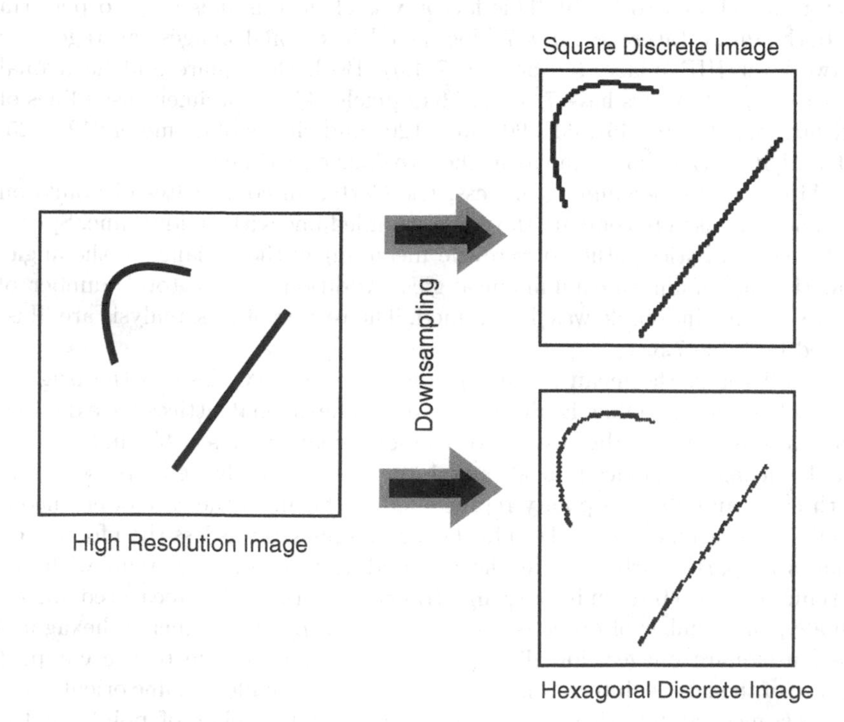

Fig. 7.9. Down-sampling from a large image to a square and hexagonal lattice.

Table 7.3. Error analysis for line depiction after down-sampling.

angle	$\overline{\theta_h}$	$\overline{\theta_s}$	σ_h	σ_s	range$_h$	range$_s$	n_h	n_s
0	0.000	0.000	0.000	0.000	0.000-0.000	0.000-0.000	492	344
30	30.02	29.97	0.003	0.004	29.490-30.600	29.29-30.61	562	514
45	44.990	45.000	0.006	0.000	43.670-46.187	45.000-45.000	468	344
60	60.000	59.980	0.000	0.013	60.000-60.000	53.130-63.430	559	514
90	90.040	90.000	0.005	0.000	89.260-91.050	90.000-90.000	664	344
120	120.000	120.368	0.000	0.013	120.000-120.000	116.365-126.870	556	514

token, if a line or a curve is drawn on a suitably high resolution image then it will also appear continuous. Simple down-sampling of this *continuous* image will allow examination of the tolerance of a lattice to resulting degradation in line and curve representation. This idea is illustrated in Figure 7.9.

In this experiment, we start with high resolution square and hexagonal images of lines and circles to simulate continuous images, the hexagonal image being obtained by resampling the square image. These are then down-sampled. For the purpose of experimentation an initial square image of 2401×2401 was used. This is equivalent to a square image of 5.8 million pixels. Conveniently, this image is equivalent to an eight-layer HIP image. The down-sampling factor was chosen to be 49. This factor was chosen as it is easy to perform in both square (average a 7×7 block) and hexagonal images (average over a two-layer HIP image, see Section 7.4.5). Both the square and hexagonal down-sampled images have 7^6 or 117,649 pixels. The experiment used lines of orientations 0°, 30°, 45°, 60°, 90°, and 120° and circles of diameters 12.5, 25, 50, and 100, which are same as in the previous experiment.

After the down-sampling process, the Cartesian coordinates of points on the line were used to compare the down-sampled line with an ideal line. Specifically, the statistics gathered were the mean angle, the variance of the angle, and the maximum and minimum angles. Additionally, the total number of points on the line/circle was determined. The results of this analysis are illustrated in Table 7.3.

In all cases, the mean angle (θ_h and θ_s) was very close to the original angle of the oriented line. Both the square and hexagonal lattices have distinct orientations at which they excel. For the square lattice this is 45° and 90° and for the hexagonal lattice it is 60° and 120°. 0° was equally well represented in both cases and 30° was poorly represented for both. In the case of erroneous results, the variance (σ_h and σ_s) for hexagonal images was less than for square images as particularly seen in the 60° and 120° lines. The trend with the variance can also be seen in the range (range$_h$ and range$_s$) of measured angles. Finally, the number of observed points (n_h and n_h) was higher in hexagonal than in square images, for all cases. This is expected due to the compact nature of the lattice. Interestingly, this difference depended on line orientation. For instance, at 90° there were nearly twice the number of points in the

Table 7.4. Error analysis for circle depiction after down-sampling.

diameter	$\overline{d_h}$	$\overline{d_s}$	σ_h	σ_s	range$_h$	range$_s$	n_h	n_s
12.5	12.428	12.341	0.365	0.442	10.583-14.000	10.770-14.142	58	49
25.0	24.782	24.653	0.368	0.439	23.324-26.077	23.065-26.457	116	94
50.0	49.310	49.253	0.399	0.432	47.707-50.912	47.286-51.069	232	193
100	98.392	98.388	0.3675	0.4324	96.747-100	96.436-100.240	461	378

hexagonal case as for the square. This is due to the way in which vertical lines are constructed in the hexagonal lattice. This was illustrated in the previous section in Figure 7.5(e).

The results of the same experiment with circles are shown in Table 7.4. The statistics gathered were: mean and variance in diameter value, the range of values obtained for the diameter, and the number of points in the down-sampled circle.

Examining the results shows that there is not much change in the mean value of diameter (d_h and d_s) with change in the lattice. However, the variance (σ_h and σ_s) is less on the hexagonal lattice in all cases. The range (range$_h$ and range$_s$) of measured diameters is marginally less for the hexagonal image. These two factors, taken together, imply that the down-sampled hexagonal image is a better fit for an ideal circle than the square. Finally, circles (of all diameters) on the hexagonal lattice had more points (n_h and n_s) in the down-sampled hexagonal image than in the square image. This is mainly due to the denser nature of the hexagonal lattice which also partly explains the overall improved results on this lattice.

7.2.3 Overview of line and curve experiments

To recap, we described two experiments that were performed to evaluate the effect of lattice change on line and curve representation. The first set of experiments examined this using a well known algorithm for line and curve drawing by Bresenham. The results were analysed qualitatively and quantitatively. The second experiment examined the effect of resolution change by severely down-sampling a high resolution image of lines/curves and quantitatively examined the results. For oriented lines, the results can be summarised as follows. We will use I_H and I_S to denote the hexagonal and square images:

0° line - good in I_H and I_S
30° line - bad in both images
45° line - poor in I_H, good in I_S
60° line - good in I_H, poor in I_S
90° line - poor in I_H, good in I_S
120° line - good in I_H, poor in I_S

Thus, we can conclude that lines can be depicted well, on average, on both the lattices, though some differences can be seen for certain angles due to the lattice geometry. These results also confirm what was expected based on the distance plot in Figure 7.3. Furthermore, we can also infer that the experimental results confirm the expected advantage that a hexagonal lattice has over the square lattice for depicting curves. This is found more in the visual quality than in the error analysis. The reason for this is that the oblique effect enhances the perception of the curves as smooth while it makes the lines look more ragged on a hexagonal lattice.

7.3 General computational requirement analysis

The computational requirements of image processing is an important consideration in general and, specifically, in real-time applications. In this section we address the question of how the computational requirements of an image processing task might change with the choice of the underlying sampling lattice. Computational requirements are very much dependent on the data structure used in the representation for the image. We will use the HIP framework for this study for consistency and thus the discussion in this section is applicable only to HIP and not hexagonal image processing in general. The discussion will start by examining the computational requirements of the HIP framework. This will be followed by a discussion of the fundamental image processing operations discussed in Chapter 3. In each case, the resulting computational requirements will be compared with the corresponding case for square image processing.

A practical implementation of the HIP framework, in the current hardware regime, requires three distinct processing blocks. These are illustrated in Figure 7.10. Underpinning each of these blocks is the exploitation of the HIP addressing scheme. In the figure, the first and last blocks are particularly of interest as they represent the computational overheads in using the HIP framework in conjunction with square pixel-based image acquisition and display hardware. The investigation will begin by evaluating the computational requirements of these two stages.

The cost of the resampling procedure is easy to evaluate. In a typical λ-level HIP image there are 7^λ points. Each of these points is determined

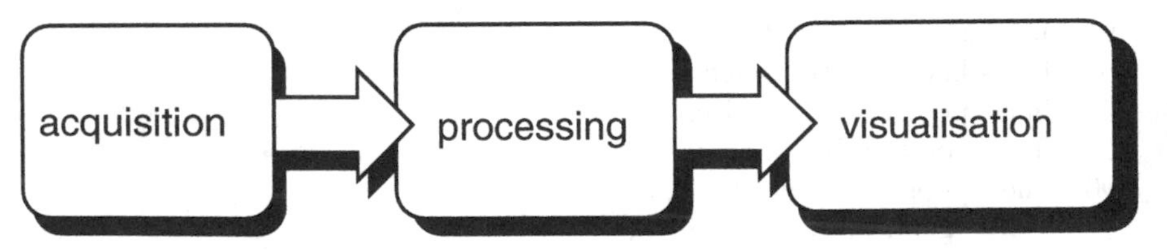

Fig. 7.10. The processing blocks of the HIP framework.

through the use of a resampling kernel which requires a number of multiplications and additions, depending on the specific kernel used. The previous investigation of sampling in Section 6.1, covered three specific methods for interpolation and concluded that the bi-linear kernel was adequate. With this kernel, a hexagonal pixel is computed as a weighted sum of four surrounding points. Also, as the kernel is separable, the computations can be evaluated in the horizontal and vertical directions separately. Thus the number of multiplications at each point is 18 and the number of additions is 8. The figure 18 comes from two multiplications (horizontal and vertical interpolating kernel) for each of the nine points. The additions are required to sum the contributions from the eight-neighbours. The overall number of multiplications for a λ-layer HIP image is:

$$\text{mul} = 18 \times 7^{\lambda}$$

The number of additions is:

$$\text{add} = 8 \times 7^{\lambda}$$

Next, let us examine the computational requirements for displaying a hexagonal image on a display device with a square pixel matrix. In Section 6.2 two methods were described for displaying the resulting hexagonal images. The first method simulated hexagons using a simple aggregate tile and raster graphics. The second method used OpenGL to display hexagons directly. This second method is more versatile and fast because it exploits the video card's polygon generation unit directly. Hence, the second method will be examined here. The HIP coordinate conversion process returns the centre point of the hexagon. To draw a hexagon requires the vertices and the radius of the hexagonal tiles to be known. The vertices relative to the origin and the scale factor for radius can be stored in lookup tables. Furthermore, the coordinate conversion process returns coordinates relative to a particular basis vector. In the frequency domain where the basis vectors rotate this is a particularly useful property as the same algorithm can be employed irrespective of the number of layers. Thus, each point requires an additional two additions and multiplications to be computed. The overall number of multiplications for a λ-level HIP image is thus:

$$\text{mul} = 2 \times 7^{\lambda+1}$$

The number of additions is:

$$\text{add} = 2 \times 7^{\lambda+1}$$

The costs for resampling and display occur only once and can be considerably speeded up using a variety of optimisations. For instance, the resampling process can be optimised for a specific sampling kernel and a simpler algorithm can approximate its behaviour [18].

A common operation that is performed routinely is convolution. It is illustrative to use this to quickly compare hexagonal and square image processing. Let us consider a square image of size $N \times N$ that is to be convolved by a 7×7 mask. This requires 48 additions and 49 multiplications at each point or $48N^2$ additions and $49N^2$ multiplications for the entire image. Comparatively, a HIP image requires $48 \times 7^\lambda$ additions and $49 \times 7^\lambda$ multiplications. This will be in addition to computational overheads for acquisition and visualisation. For simplicity, the cost of performing the looping operations is not mentioned in this example. From these calculations, a simple comparison can be made for a fixed size image. This is illustrated in Figure 7.11 for a square image of resolution 256×256. Three cases of processing are considered: square image (left bar), hexagonal image of a different size but with the same number of computations (middle bar) and finally hexagonal image of equivalent size (right bar). The middle and rightmost bars are made up of three components corresponding to various computational requirements: the lower part for processing, the middle part for visualisation, and the upper part for resampling.

Comparisons between the left and middle bars yield interesting results. The first is that the total overhead associated with HIP due to resampling and display is 20% of the square additions and 40% of the square multiplications respectively. Excluding the overheads, the HIP framework requires roughly 30% less processing. This is equivalent to a reduction of the masking operation containing a maximum of 30 points (5×5 is the closest symmetric mask), or a reduction in image size to 200×200. The right bar shows the total cost for performing HIP processing on an equivalent-sized image. The processing is achieved with an additional overhead of 17% for addition and 29% for multiplication. Depending on the image size, this is certainly achievable on most modern computing platforms. This comparison indicates that the cost of overheads incurred in the acquisition and display stages are not very significant when it comes to image processing problems. In the ideal case, these stages should be replaced by hardware, in which case, the only comparison would be the processing stage.

When performing many operations upon images, it is important to know where the bounds of the image are. This may appear to be a difficult task due to the irregular nature of the HIP image boundary. However, as was shown in equation (3.22) finding the pixels on the image boundary is actually a simple exercise. The equivalent task for a square image of size $M \times N$ where the coordinates start at $(0,0)$ is finding the boundary b_S such that:

$$b_S(x,y) = \begin{cases} 1 & \text{if } (x = 0 \text{ or } x = (M-1)) \text{ and } (y = 0 \text{ or } y = (N-1)) \\ 0 & \text{else} \end{cases} \tag{7.5}$$

Thus, finding boundary points for a square image requires four distinct comparisons whilst for a HIP image it requires six comparisons. This operation

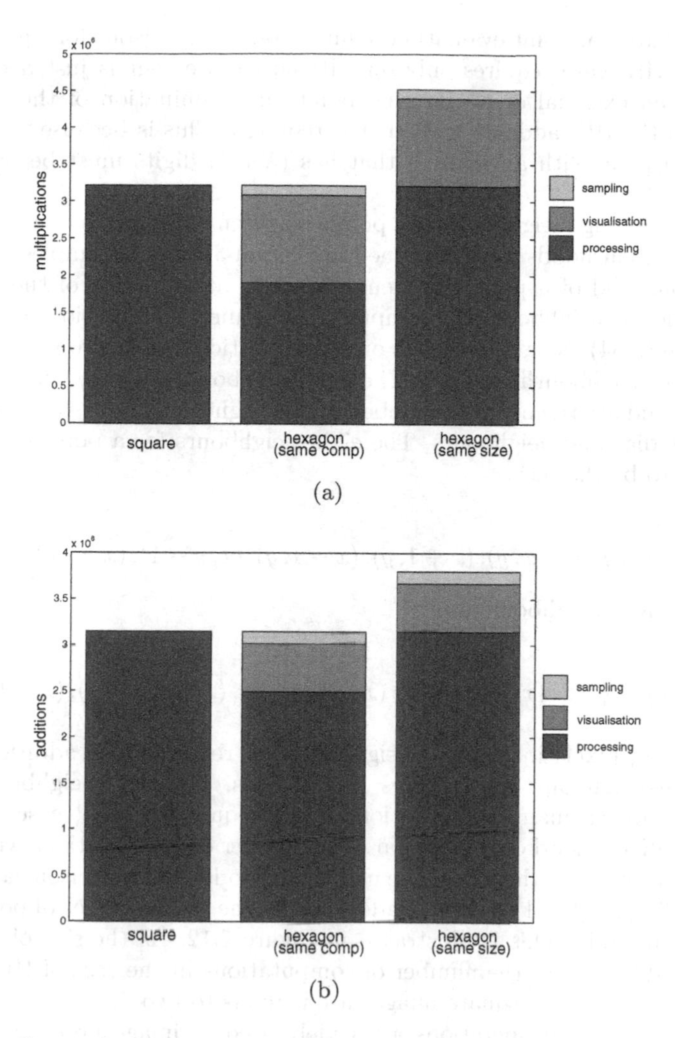

Fig. 7.11. Comparison of computational requirements for HIP and square image processing: (a) multiplications (b) additions.

is slightly more complex for the HIP framework. However, there is a more common need for evaluating if a point is an external point than if it is a boundary point. Using HIP, this is accomplished with equation (3.21). For the square images, this is done as follows:

$$e_S(x,y) = \begin{cases} 1 & \text{if } (x < 0 \text{ or } x > (M-1)) \text{ and } (y < 0 \text{ or } y > (N-1)) \\ 0 & \text{else} \end{cases} \tag{7.6}$$

The external point evaluation requires four comparisons for square images but the HIP case requires only one. If the requirement is just deciding if a pixel is an external or an internal point, an examination of the number of digits in the HIP address is all that is required. This is because for a λ layer image, a pixel with an address that has $(\lambda + 1)$ digits must be outside the image.

Determining neighbourhood points is often required in image processing. Neighbourhoods were examined in Section 3.4.3. The simplest hexagonal neighbourhood of a point was denoted by N_1 and consists of the point and its six closest neighbours. To compute this requires six additions as seen from equation (3.34). Neighbourhoods on square-lattices are of two types: the four- and eight-neighbourhood [118]. The four-neighbours are purely horizontal and vertical and are the nearest neighbours. The eight-neighbours are these points plus the diagonal neighbours. The eight-neighbours for a point (x, y) can be defined to be the set:

$$N_4^S(x,y) = \{(x,y),(x+1,y),(x-1,y),(x,y+1),(x,y-1)\} \tag{7.7}$$

The eight-neighbours are:

$$N_8^S(x,y) = \left\{N_4^S,(x+1,y+1),(x-1,y+1),(x+1,y-1),(x-1,y-1)\right\} \tag{7.8}$$

Finding pixels in the four-neighbourhood requires four additions and in the eight-neighbourhood requires 12 additions. The eight-neighbourhood requires twice as many computations as the equivalent HIP case whilst the four-neighbourhood case is similar. The number of computations will increase with an increase in the size of the neighbourhoods. A comparison can be made by dividing the total number of additions by the total number of points in the neighbourhoods. This is illustrated in Figure 7.12. As the size of the neighbourhood increases, the number of computations in the case of HIP tends to one whereas, for the square image case it tends to two.

Neighbourhood operations are widely used in image processing applications. These operations compute the pixel value at a particular location as the weighted sum of the pixel values in its neighbourhood. Typically, these operations are computed as the convolution of a given image with a mask which represents weights assigned to a point's neighbourhood. For a λ-level HIP image, convolution was defined in equation (3.39). The corresponding equation for a square image using a mask of size $m \times n$ is:

$$M(x,y) * I(x,y) = \sum_{i=-\frac{m}{2}}^{\frac{m}{2}} \sum_{j=-\frac{n}{2}}^{\frac{n}{2}} M(i,j)I(x-i,y-j) \tag{7.9}$$

Computing the convolution at a single point in the square image requires a double summation or two arithmetic loops. This is twice the number of loops

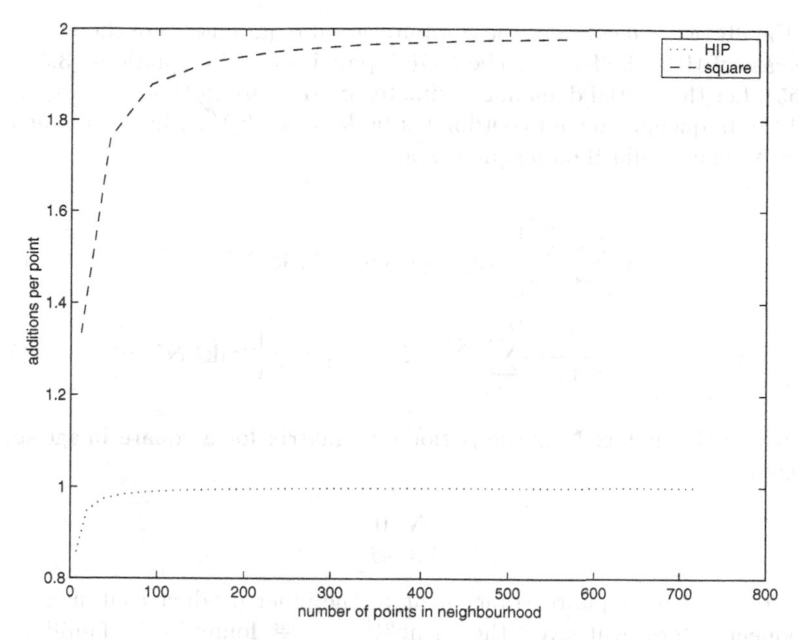

Fig. 7.12. Number of additions/point required to compute neighbourhoods.

required for HIP. It is of course possible to vectorise an image using index mapping. However, this is achieved at the sacrifice of additional computation to compute the new indices. By contrast, a HIP image is naturally vectorised. Thus, an entire image can be convolved in the case of HIP using half the number of loops required for square image processing. The total number of additions and multiplications can now be determined for convolving both types of images. There are three distinct sorts of arithmetic carried out. The first is the multiplications where the image data is multiplied by the mask data. The second is the additions where the multiplied values are summed for the entire mask to compute the current point. The third is a coordinate shift where the current image pixel is translated to the next location. The number of arithmetic operations per point is purely governed by the mask size and, for an entire image, by the mask and image sizes. For a square image of size $X \times Y$ and a mask of size of $M \times N$ the required computations are $XYMN$ multiplications, $XY(MN - 1)$ additions, and $2XYMN$ coordinate shifts. In comparison, a λ-level HIP image with a γ-level mask requires $7^{\lambda+\gamma}$ multiplications, $7^{\lambda+\gamma} - 1$ additions, and $7^{\lambda+\gamma}$ coordinate shifts. The number of multiplications and additions is similar for square image and HIP image convolution excluding the coordinate shift operation which requires half the computations for HIP convolution. This is a direct reflection of the reduction in the number of loops and the efficient single coordinate addressing scheme used in HIP.

Finally, we will consider the computational requirements of the DFT or in the case of HIP, the HDFT. The HDFT pair is given in equations (3.54) and (3.55). Let the spatial domain coordinates be $\mathbf{n} = (n_1, n_2)^T$ for square images and the frequency domain coordinates be $\mathbf{k} = (k_1, k_2)^T$. The DFT pair for a $M \times N$ image defined on a square grid is:

$$X(k_1, k_2) = \sum_{n_1=0}^{M-1} \sum_{n_2=0}^{N-1} x(n_1, n_2) \exp\left[-2\pi \mathrm{j} \mathbf{k}^T \mathbf{N}^{-1} \mathbf{n}\right] \tag{7.10}$$

$$x(n_1, n_2) = \frac{1}{|\det \mathbf{N}|} \sum_{k_1=0}^{M-1} \sum_{k_2=0}^{N-1} X(k_1, k_2) \exp\left[2\pi \mathrm{j} \mathbf{k}^T \mathbf{N}^{-1} \mathbf{n}\right] \tag{7.11}$$

where the matrix $\mathbf{N}$ is the periodicity matrix for a square image and is given as:

$$\begin{bmatrix} N & 0 \\ 0 & M \end{bmatrix}$$

Note that a simple rearrangement of the inner product contained in the exponential term will reveal the term $\frac{n_1 k_1}{M} + \frac{n_2 k_2}{N}$ found in the familiar form for the DFT. As with the convolution, comparison of the sets of equations for HIP and square images shows that HIP requires half the number of loops to compute the DFT than the square image case. Assuming that the exponential term has been precomputed and can be addressed using a lookup table, then the DFT requires M^2N^2 multiplications and $MN(MN-1)$ additions. The HDFT equivalent is $7^{2\lambda}$ multiplications and $(7^{2\lambda}-1)$ additions. The amount of arithmetic in both cases is similar as it is of $O((imagesize)^2)$. However, a significance difference does exist when we examine the lookup table. For a square image, this is a function of four variables and this leads to a complex data structure. A common approach to simplification is to vectorise the result into a two dimensional array using an index mapping as follows:

$$x = k_1 N + k_2, \qquad y = n_1 N + n_2$$

Thus, every time a lookup is required two multiplications and two additions are also required. For an $M \times N$ image this will lead to an extra $2MN$ multiplications and additions. There are ways in which this operation can be performed with fewer computations. In contrast, a lookup table based on HIP addresses is naturally a 2-D array. Hence, no additional computation is required to perform the lookup operation.

Table 7.5 summarises the number of operations required for various processing steps. In the table, N is the total number of points in the image and m_s is the total number of points in the convolution kernel (mask). The overhead incurred in the acquisition and display stages is no worse than performing a masking operation by a 4×4 mask on a square image. Overall, it is seen

Table 7.5. Computational requirements for HIP and square image processing.

	HIP		square	
operation	multiplication	addition	multiplication	addition
boundary	6 comparisons		4 comparisons	
exterior	1 comparison		4 comparisons	
neighbourhood	–	N	–	$2N$
convolution	$m_s^2N^2$	$(m_s^2-1)N^2$	$m_s^2N^2$	$(m_s^2-1)N^2$
coord shift		m_sN		$2m_sN$
DFT	N^4	N^4-1	N^4	N^4-1
DFT lookup	–	–	$2N$	$2N$
acquisition	$18N$	$8N$	–	–
display	$14N$	$14N$	–	–

from the table that the computational complexity is the same for processing square or hexagonal images. However, the number of operations required is different in some instances. For instance, convolution can be computed in quadratic time in both square and HIP images, however, the former requires twice as many addition operations as the latter, for shifting the mask (coordinate shift). The computational complexity of an $N \times N$ DFT is $O(N^4)$ in both cases with the square case requiring further add and multiply operations for table lookup. In practice, of course, this complexity is reduced by using fast algorithms for DFT computation. A comparison of FFT algorithms for square and HIP images is given in the next section.

7.4 Performance of image processing algorithms

In this section, the performance of the algorithms covered in Chapter 4 will be studied for comparison between HIP and square image processing. The aim of this comparison is to see if changing the sampling lattice affects the performance of some of the standard image processing techniques. The comparisons between the various algorithms will be carried out at both computational and qualitative levels. The comparison will start by examining edge detection and skeletonisation, which are spatial domain techniques, followed by the fast Fourier transform and linear filtering in the frequency domain. The final example will be pyramidal decomposition of images.

7.4.1 Edge detection

In Section 4.1.1 three edge detection algorithms within the HIP framework was discussed, namely, the Prewitt edge operator, the Laplacian of Gaussian, and the Canny edge detector. For comparison purposes, these were implemented on square images and tested on the same test images shown in Figures 4.3(a)

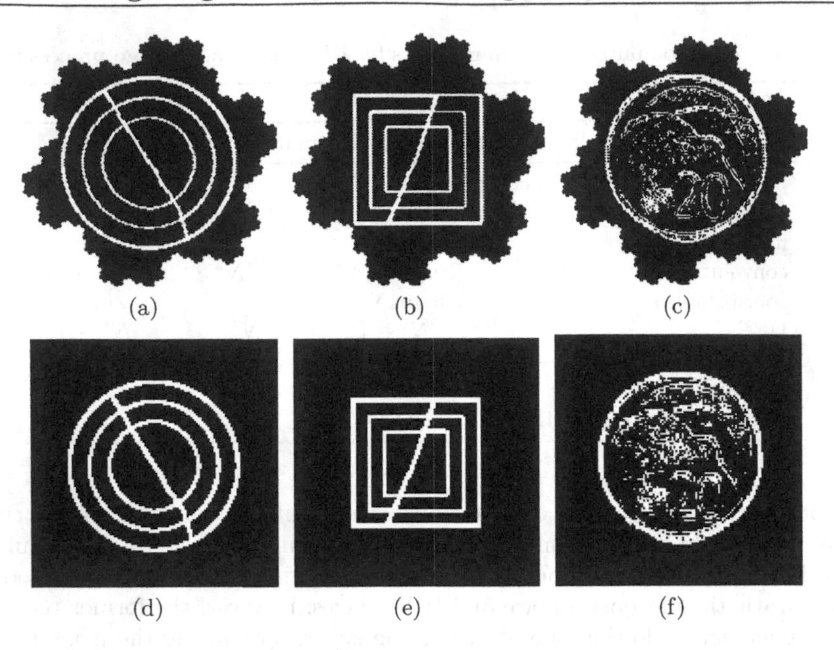

Fig. 7.13. Results of applying the Prewitt edge detector to hexagonal images(a)T1 (b) T2 (c) T3 and square images (d) T1 (e) T2 (f) T3.

to 4.3(c). Of the three images, T1 and T2 were synthetic and one was a real image of a New Zealand coin. The original images were of size 256×256 pixels and contained a mixture of linear and curved segments. The input image was re-sampled onto an equivalent HIP (five-layer) image and a 128×128 pixel square image. These sizes were chosen as they contain roughly the same number of pixels. To compare the results of HIP and square image processing, the threshold was tuned to provide the best qualitative results in both cases.

The Prewitt edge detector [113] is a gradient based edge detector. The detector is considered to be poor due to its bad approximation to the gradient operator. However, it is often used because of its low computational cost. Implementation of the edge operator was covered in Section 4.1.1. Figure 4.1 illustrates the specific masks that were used.

The results of applying the Prewitt edge detector to both square and hexagonal images are shown in Figure 7.13. The results of processing the synthetic images illustrate the superiority of representation of circles in a hexagonal image by way of the smoothness of the circles in the edge detected image. Furthermore, the diagonal dividing lines are also smoother in the hexagonal image. Examination of the edge detected image T3 shows that the shape of the coin is much more circular in the hexagonal image, and the kiwi bird, the fern, and the number '20' all appear more clearly in the hexagonal image. Overall, the appearance of the hexagonal image is less noisy. This is due to

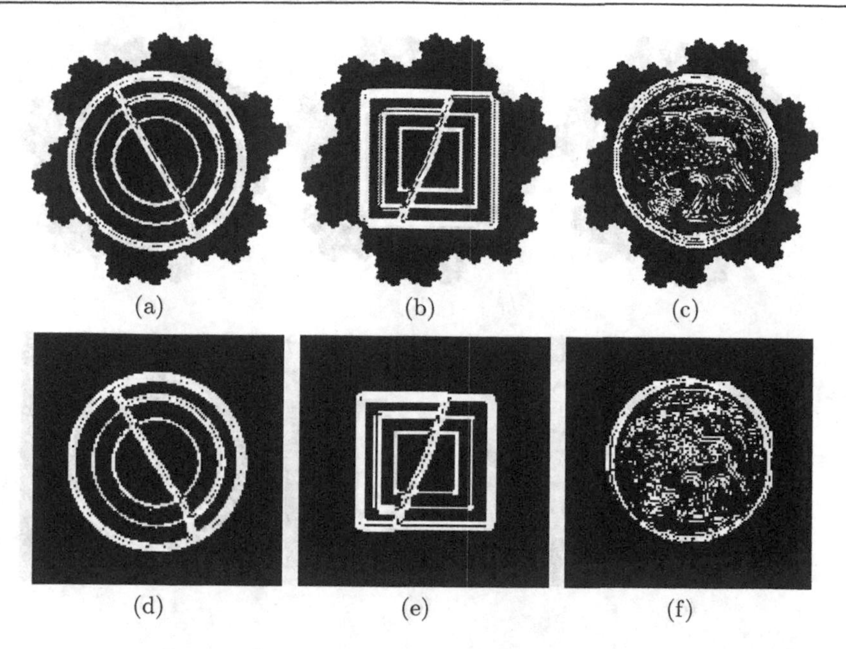

Fig. 7.14. Results of applying the LoG edge detector to hexagonal images (a) T1 (b) T2 (c) T3 and square images (d) T1 (e) T2 (f) T3.

the edges in the square image being thicker. The ratio of edge pixels to image size for T1 is 11.5% for the hexagonal image and 11.3% for the square image. For T2, the ratios are 11.2% for the hexagonal image and 9.5% for the square image. In T3 the ratio of edge pixels in the two cases are 13.3% for the hexagonal image and 11.9% for the square image.

The Laplacian of Gaussian (LoG) edge detector was first proposed by Marr [114]. The detection regime is, as mentioned in Section 4.1.1, isotropic and consequently should perform well on hexagonal images. The masks used in implementation to approximate the LoG function are made up of 49 pixels in square and hexagonal images. The results of edge detection with the LoG operator, with the same test images, are shown in Figure 7.14. For the hexagonal images, the curves are shown with good clarity. In T3, the leaves of the fern are also revealed. The poor performance of edge detection in the square image is due to the fact that the mask is a poor fit for the LoG function. To improve the performance of the square image would require a bigger mask which would also result in an increased computational cost. Referring to T1, the ratio of edge pixels for the hexagonal image is 19.4% and 14.3% for the square image. For T2, the ratios are 18.3% for the hexagonal image and 13.1% for the square image. T3 gives ratios of 18.0% and 20.1% for the hexagonal image and square image respectively.

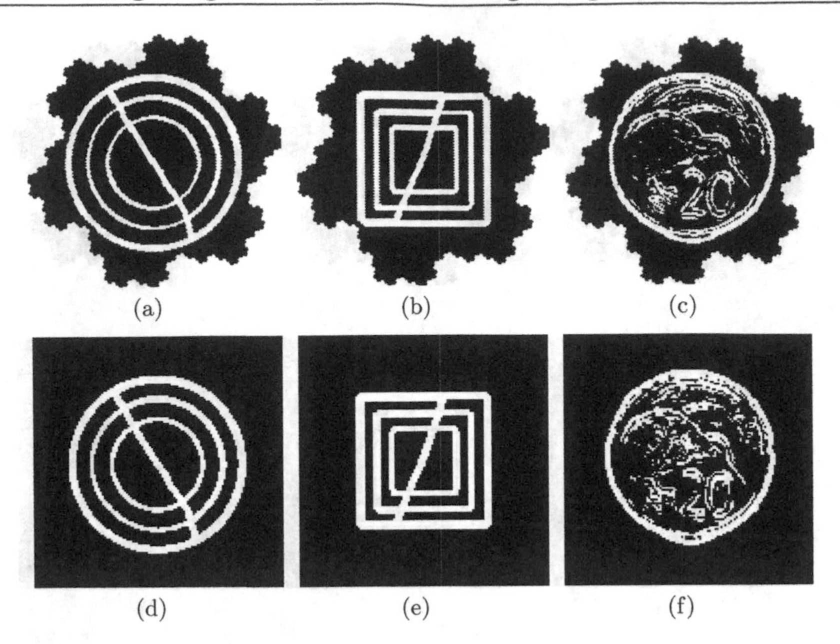

Fig. 7.15. Results of applying the Canny edge detector to hexagonal images (a) T1 (b) T2 (c) T3 and square images(d) T1 (e) T2 (f) T3.

The Canny edge detector [113] combines two Gaussians to produce the gradient estimates in the horizontal and vertical directions. For the experiment, the masks used are of the same size as the Laplacian of Gaussian. The results are illustrated in Figure 7.15. As expected, this edge detector shows improved performance (such as edges are less noisy) over the previous two detectors. This is especially dramatic in the case of the square image. The improvement is mainly due to the maximal suppression step. The interesting point to note is that this step appears to be more effective in the hexagonal case, for images containing curved features as seen in T1 and T3. Additionally, fewer computations are required to achieve good results with HIP as was the case with the LoG detector. For T1, the ratios of edge pixels are 18.7% and 15.1% for the hexagonal and the square images, respectively. For T2, these ratios are 19.9% and 15.3% for the hexagonal and the square images, respectively. Finally, the ratios for T3 are 18.2% and 14.2% for the hexagonal and the square images, respectively.

Edge detection under noisy conditions is important for many applications. Hence, the performance of the Prewitt Edge detector for this condition was examined. The LoG and Canny detectors were not examined as they have in-built smoothing functions which would bias the results. Gaussian noise of standard deviation 20% was added to the original images before processing as illustrated in Figures 7.16(a) and 7.16(d). The results indicate a degradation

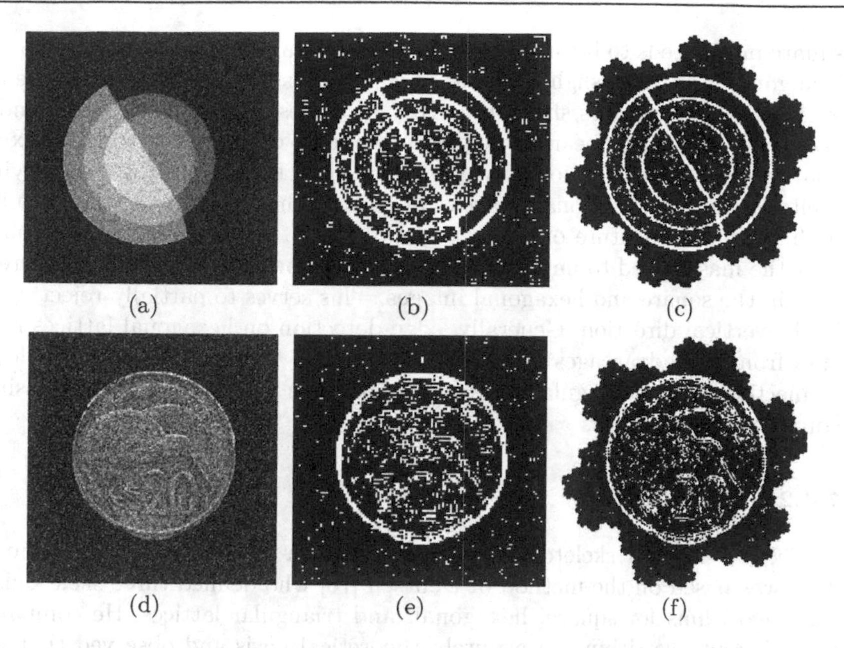

Fig. 7.16. Results of Prewitt edge detection on noisy images. (a) Test image T1 and results for (b)square image T1 (c) hexagonal image T1. (d) Test image T3 and results for (e) square image T3 (f) hexagonal image T3.

in performance over the noise free case. However, in relative comparison, the result with the hexagonal image is superior to the square image as indicated by fewer parts of the image being incorrectly classified as edges. The level of noise retained in the interior of the circle in T1 in particular is significantly different. For T3, the outlines of the kiwi and the number '20' have better clarity in the hexagonal image as well. In the case of T1, the ratio of edge pixels are 18.9% for the hexagonal image and 15.4% for the square image. In T3, the ratio of edge pixels for the hexagonal image are 13.9% and 12.9% for the square image.

There are a number of conclusions that can be drawn from the above comparisons. In all cases, the number of edge pixels was roughly the same but the edge-detected hexagonal image appears to be qualitatively better. This stems from the consistent connectivity of the pixels in hexagonal images which aids edge detection. Pixels without consistent connectivity show up as discontinuities in the contour (in the input image) and breaks in the edge image. This is especially a problem when it comes to finding the edges of curved objects in square images. This is related to the issue of adjacency mentioned in Section 7.2. Another issue, which has a direct bearing on the computational efficiency, is the mask size. For the HIP Prewitt operator similar performance is achieved with 22% fewer points in the mask. For the LoG operator, the square mask needs to be much bigger to produce a performance similar to the hexagonal mask. The slight increase in edge pixels in the hexagonal image for all three edge detectors shows that HIP produces a richer edge map. Under noisy conditions, it appears to be preferable to do edge detection using hexagonal images. Noise rejection was also reported by Deutsch [76] when studying skeltonisation on hexagonal images. More study has to be done to determine if this a general feature of the hexagonal lattice. A plausible explanation is that the masks used to implement the Prewitt operators align along different axes in the square and hexagonal images. This serves to partially reject noise in the vertical direction. Generally, edge detection on hexagonal lattices benefits from the advantages mentioned in sections 7.1 to 7.3 namely, uniform connectivity, better angular resolution, and with the use of HIP addressing, computational savings.

7.4.2 Skeletonisation

In Section 4.1.2, a skeletonisation algorithm for HIP images was outlined. This was based on the method of Deutsch [76] who defined three skeletonisation algorithms for square, hexagonal, and triangular lattices. He compared the different algorithms on a purely theoretical basis and observed that the hexagonal lattice had good noise rejection and was computationally efficient. Later, Staunton [67,68] derived an alternative thinning algorithm using mathematical morphology and the hit-or-miss transform. The approach was based on that of Jang and Chin [148, 149] with which the hexagonal skeletonisation algorithm was compared. The algorithm was found to be efficient and to produce rich skeletal descriptions. In this section, a comparison of square and hexagonal skeletonisations is performed. The methodologies used are the ones described in Section 4.1.2 for the HIP image and the method described by Deutsch for square images. The comparison is first at a qualitative level and then at the quantitative level.

For the purpose of comparison, four different images were examined. The first (Figures 7.17(a) and 7.17(e)) was a ring of known radius. The second (Figures 7.17(b) and 7.17(f)) was a box with a horizontal major axis. The third (figures 7.17(c) and 7.17(g)) was a box with the major axis aligned in a vertical direction. The final image (Figures 7.17(d) and 7.17(h)) was a thresholded image of a chest x-ray. The results are illustrated in Figure 7.17 with the skeleton in white and the original image in grey. The theoretical

Table 7.6. Errors in derived skeletons.

image	min(e_s)	max(e_s)	$\overline{e_s}$	σ_s	min(e_h)	max(e_h)	$\overline{e_h}$	σ_h
ring	-1.11	0.50	-0.25	0.40	-1.10	0.67	-0.25	0.40
horizontal box	0.00	0.00	0.00	0.00	0.00	0.00	0.00	0.00
vertical box	-0.25	0.25	0.00	0.25	0.00	0.00	0.00	0.00

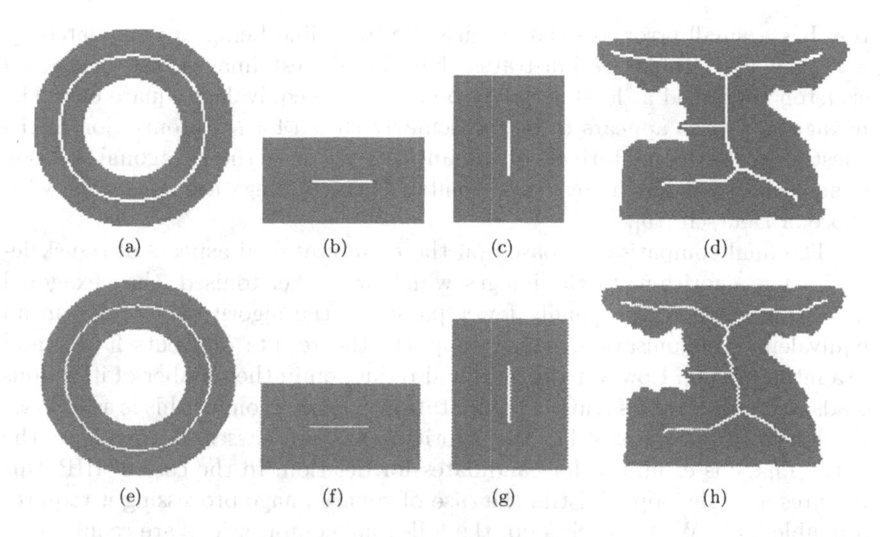

Fig. 7.17. Results for skeletonisation using square images (top row) and hexagonal images (bottom row).

skeleton was known for the first three pairs of images, hence the computed results could be compared with the ideal. The comparison performed was similar to that in Section 7.2 for lines and curves. The results are given in Table 7.6. Two other metrics are also of interest in the comparison. The first is the number of iterations required to perform the skeletonisation algorithm and the second is the number of points in the resulting skeleton. These are given in Table 7.7.

The visual appearance of the resulting skeletons is fairly similar. The main difference appears to be that the lines look thinner in the hexagonal compared to the square case. The raggedness of the vertical line for the hexagonal case is noteworthy in Figure 7.17(g). The ring appears more circular in the hexagonal than in the square skeleton and has more points as well (see Table 7.6). These results are consistent with the finding for line and curve representation in Section 7.2. The hexagonal skeleton of the ring has a smaller range of errors than the square case which is reflected in its better visual quality. The vertical

Table 7.7. Iterations and size of the derived skeletons.

	hexagonal			square		
image	iterations	skeleton	original	iterations	skeleton	original
ring	14	246	6798	14	227	5740
horizontal	25	30	4000	26	30	4000
vertical	25	30	4000	26	30	4000
real	23	324	6155	25	237	6325

box has a small error associated with the ideal line being approximated by a ragged line. Table 7.7 illustrates that for all test images, the hexagonal skeleton contained at least as many points as the equivalent square one. The hexagonal lattice appears to be particularly suited for skeletonisation of the chest image both qualitatively and quantitatively since the hexagonal skeleton is smoother and has many more points. These findings are consistent with those of Deutsch [76].

The final comparison is based on the computational aspects of the skeletonisation algorithm. In the images which were skeletonised, the hexagonal algorithm required marginally fewer passes of the algorithm to perform an equivalent skeletonisation, which supports the results of Deutsch [76] and Staunton [67,68]. However the marginal reduction in the number of iterations leads to a substantial saving in computations. The reason for this is as follows. As described in Section 4.1.2, the algorithm has two passes. In each pass, the entire image is examined for candidates for deletion. In the case of HIP, this requires a single loop whilst in the case of square image processing it requires a double loop. Within each loop, the following computations are required:

operation	HIP	square
crossing number	12	18
non zero neighbours	6	8
neighbourhood	6	8
total	24	34

Thus, there is a 30% computational saving for one pass of the algorithm using hexagonal (HIP) rather than square lattices.

In summary, the comparison of skeletonisation algorithms on square or hexagonal images indicates the following: the skeletons appear similar, regardless of the lattice used, however, the hexagonal lattice enables a richer structural description and, when HIP framework is used, increased computational efficiency. Finally, hexagonal lattices are preferable for accurate skeletonisation of curved structures and these are widely found in most real images.

7.4.3 Fast Fourier transform

The history of fast Fourier transform (FFT) algorithms is extremely long, with some evidence that it dates back to Gauss [150]. Generally, the algorithms are attributed to Brigham and Morrow [151] in the 1960s. Their method involved reducing the number of multiplications by observing that an L-length sequence where L is an even number, could be split into two sequences - one with odd terms and one with even terms - to reduce the numbers of multiplications. The process could be repeated to drastically reduce the number of computations required. There are a variety of other FFT algorithms each based upon different reduction techniques. The HFFT was presented in a previous chapter and used a similar factorisation method to decimation in space. The method is

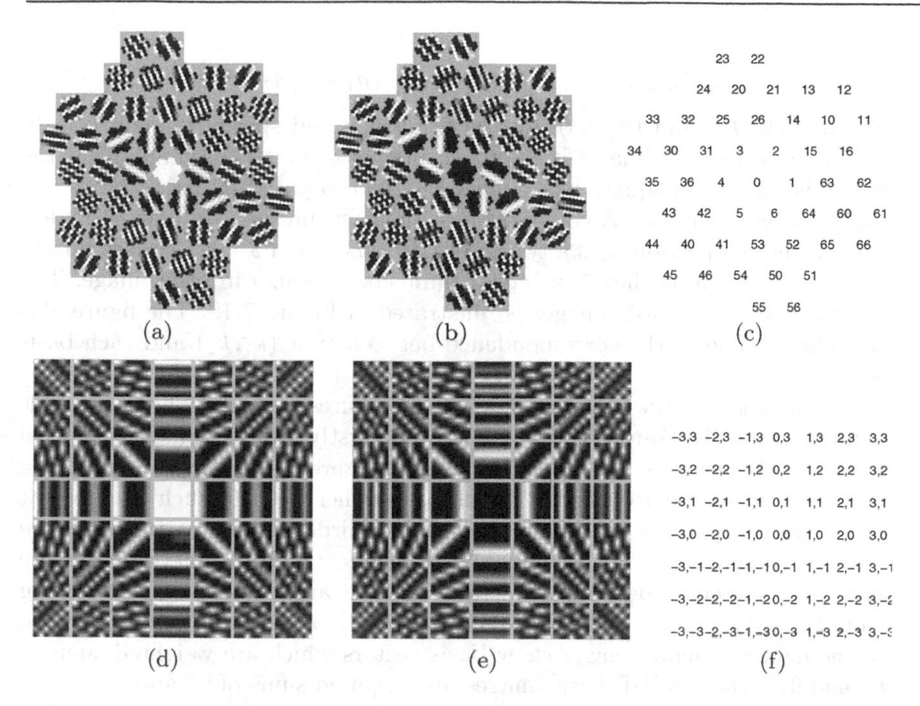

Fig. 7.18. Basis images for the discrete Fourier Transform (a) HIP-real (b) HIP-imaginary (c) HIP addresses (d) square-real (e) square-imaginary (f) square addresses.

based on work by Mersereau who derived a general FFT for multidimensional periodic signals [59]. In this section, a comparison of the FFT algorithm is undertaken between that proposed for the HIP framework and the one used for square image processing.

The Fourier transform, along with all general image transforms [152], decomposes an image in terms of a set of basis images. In the case of the Fourier transform, these basis images are sinusoidal components. Given the HDFT (from equation (3.54), Section 3.4.5):

$$X(\boldsymbol{k}) = \sum_{\boldsymbol{n} \in \mathbb{G}^\lambda} x(\boldsymbol{n}) \exp\left[-2\pi \mathrm{j} H(\boldsymbol{k})^T \mathbf{N}_\lambda^{-1} h(\boldsymbol{n})\right]$$

the set of all basis images can be defined to be :

$$B_{\boldsymbol{k}}(\boldsymbol{n}) = \exp\left[-2\pi \mathrm{j} H(\boldsymbol{k})^T \mathbf{N}_\lambda^{-1} h(\boldsymbol{n})\right] \tag{7.12}$$

Here, $\boldsymbol{k}$ and $\boldsymbol{n}$ are the frequency and spatial domain HIP addresses, respectively, while $\mathbf{N}_\lambda$ is the periodicity matrix for a λ-level HIP image. A particular basis image can be computed by keeping $\boldsymbol{k}$ fixed and varying through all possible values of $\boldsymbol{n}$. A similar formula can be found for the square case:

$$B_{k_1,k_2}(n_1,n_2) = \exp\left[-2\pi \mathrm{j}(k_1 n_1 + k_2 n_2)/N\right] \tag{7.13}$$

Here (k_1, k_2) and (n_1, n_2) are the frequency and spatial domain coordinates, respectively, while N is the total number of pixels (assuming that the image is square). A square basis image can be computed by fixing k_1 and k_2 and varying n_1 and n_2. A comparison of the basis images can now be undertaken. The comparison uses a two-level HIP image and a 7×7 square image. These are chosen as they have equal numbers of points in each image. The resulting set of all basis images is illustrated in Figure 7.18. The figure also includes a guide to the correspondence between $\boldsymbol{k}$ or (k_1, k_2) and each basis image.

Examination of the square and hexagonal figures show the following trends conforming to the Fourier transform theory. Firstly, the DC image (0 spatial frequency) which lies in the centre of each figure is identical in all cases. Secondly, at spatial frequencies which are further from 0, such as those at the edges, the pattern has more repetitions. Thirdly, as the basis images are based upon sinusoids, the individual basis images exhibit symmetry. There are however, differences between the square and hexagonal basis images arising from the nature of the lattices. This is most noticeable in the central portion of the figures. Square images show basis vectors which are weighted sums of $0°$ and $90°$ whereas HIP basis images are weighted sums of $0°$ and $120°$.

The Fourier transform of some images with predictable responses are illustrated in Figure 7.19. The Fourier images in this figure illustrate the logarithm of the magnitude. The rectangular image has two dominant features, namely the major and minor axes of the rectangle. These are evident in the magnitude spectra of both the DFT and the HDFT. Additionally, there are bands of energy which are distributed along the horizontal and vertical axes in the square image. This banding is also present in the hexagonal case but the energy is aligned along axes at multiples of $60°$. This is because the HDFT is aligned along the axes of symmetry of the hexagon. This trend was also observed in the basis images in Figure 7.18. The second test image being a circle, should have a DFT and HDFT where the spectral energy spreads across many frequency bands. This is illustrated in Figure 7.19 for both the HDFT and the DFT. The low frequency information should yield the major and minor axes of the image. In the DFT, this is seen as a cross at the centre of the magnitude spectrum. The HDFT has a star showing the major axes to be along each of the axes of symmetry of the hexagon. Furthermore, there are a number of rings in both the square and the HIP image. This is due to the circle being solid, resulting in its energy being spread out in the frequency domain. The rings in the HIP image look much more circular than for the square image especially at lower frequencies.

Thus, the observed responses of the HDFT are similar to the case for the DFT. The same is seen by looking at the transform of the images of rectangle and circle. In the case of HIP, images which result in a spread of energy across

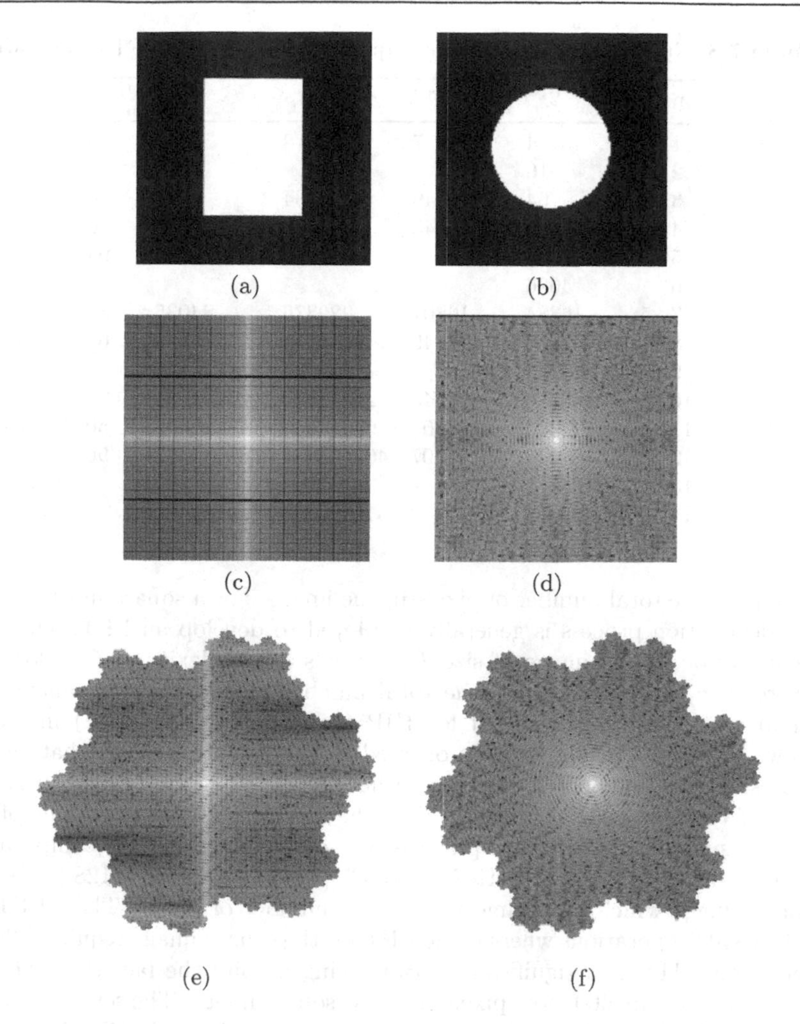

Fig. 7.19. Examples of the Fourier magnitude spectra. (a) and (b) are the test images and (c), (e) and (d), (f) are the corresponding magnitude spectra of the square and hexagonal versions of the test images respectively.

many frequency bands are better represented. This is due to the basis images being in terms of the three axes of symmetry of the hexagon. Furthermore, the advantage of the lattice for spatial representation of curved features is also carried over into the frequency domain.

Analysis of the HFFT algorithm showed it to be a decimation-by-seven algorithm. The computational requirements were described in a previous chapter. The overall order of complexity for a λ-layer HIP image is $\lambda 7^{\lambda+1} + (\lambda - 1)7^{\lambda}$. Due to redundancies, the overall complexity is of $O(N \log_7 N)$, where

Table 7.8. Number of multiplications required for HFFT and FFT computation.

n	k	2^{2n}	7^k	M_S	M_H	$\frac{M_H}{M_S}$%
1	1	4	7	8	7	88
2		16				
3	2	64	49	384	98	26
4	3	256	343	2048	1029	50
5	4	1024	2401	10240	9604	94
6		4096				
7	5	16384	16807	229376	84035	37
8	6	65536	117649	1048576	705894	67
9		262144				
10	7	1048576	823543	20971520	5764801	27
11	8	4194304	5764801	92274688	46118408	50
12	9	16777216	40353607	402653184	363182463	90
13		67108864				
14	10	268435456	282475249	7516192768	2824752490	38

$N = 7^\lambda$ is the total number of pixels in the image. For a square image, a similar decimation process is generally employed to develop an FFT. The FFT computation for an image of size $L \times L$, has a complexity of $O(N \log_2 N)$, where $N = L^2$ is once again the total number of pixels in the image. The number of operations required for HIP (M_H) and square (M_H) images is shown in table 7.8. The table is ordered to show nearest sized (that can be expressed as a power of 2) square and hexagonal image near each other. The sizes of these images are illustrated in the second and third columns of the table, respectively. As an example, it is possible to compare DFT computation on a five-layer HIP image (with $7^5 = 16807$ pixels) and a 128×128 ($= 16384$) square image which have almost the same number of pixels. The HFFT requires 84035 operations whereas the FFT for the square image requires 229376 operations. This is a significant (63%) saving despite the fact that the HIP image has roughly 400 more pixels than the square image. The seventh column in the table shows the savings to be had in computing a DFT using HFFT instead of FFT. The extent of savings is dependent on the relative sizes of the square and hexagonal images. For a fair comparison we consider the case when the two types of images are almost of the same size (see the seventh and fourteenth rows). In these cases the savings are roughly 60%.

7.4.4 Linear Filtering

The linear filtering example in Section 4.2.2 was chosen as it is a simple demonstration of how the HIP framework can be used to implement image processing in the frequency domain. Linear filtering in the frequency domain exploits the result of the convolution theorem. For a square image $f(n_1, n_2)$, with Fourier transform, $F(k_1, k_2)$, the convolution theorem can be applied by

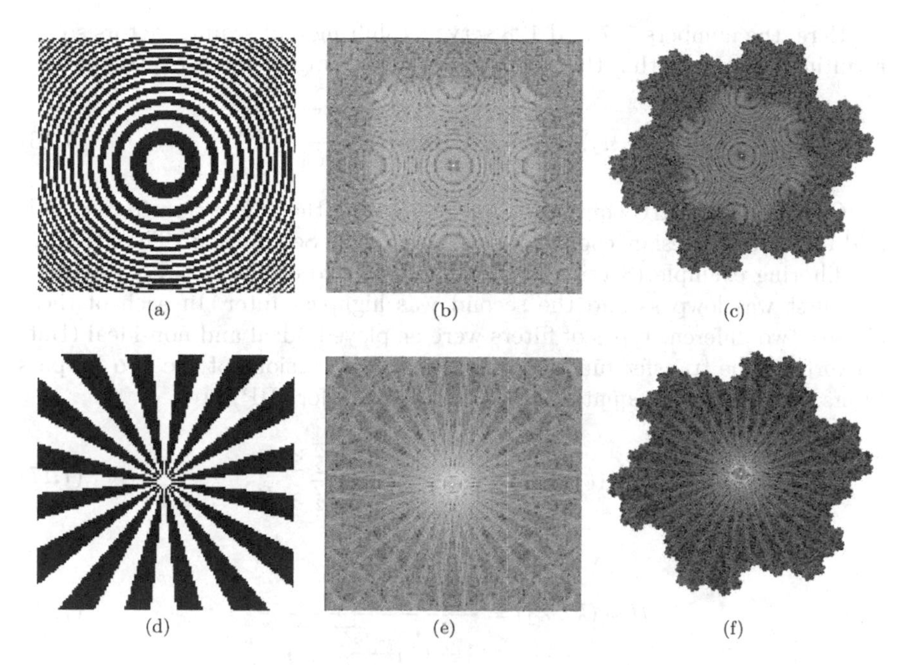

Fig. 7.20. Fourier magnitude spectra of the test images (a) ring R1 and (d) star S1. (b), (c) and (e), (f) are the magnitude spectra of the square and hexagonal versions of R1 and S1 respectively.

selecting a filter, $H(k_1, k_2)$, which when multiplied with F will produce the desired result, $g(n_1, n_2)$. This can be written concisely as:

$$g(n_1, n_2) = \mathcal{F}^{-1}\left[H(k_1, k_2)F(k_1, k_2)\right] \tag{7.14}$$

This is equivalent to equation (4.26) but using Cartesian coordinates rather than HIP addresses.

It is now possible to compare linear filtering applications in HIP and square image processing. This will be performed using a pair of images with known characteristics. The images are thresholded versions of two of the images that make up F1 in Figure 6.5 namely the star and rings. These are illustrated in Figure 7.20 along with the associated DFT and HDFT magnitude spectra. The ring was also used in the HIP case study for linear filtering in Section 4.2.2. Binary versions of the images were chosen as the frequency domain behaviour under filtering is more predictable. Another feature of these images is that they can be expressed analytically. The star image defined on a square lattice is given as:

$$s(n_1, n_2) = 127 + 128 \cos\left(16 \tan^{-1} \frac{n_2}{n_1}\right) \tag{7.15}$$

Here, the numbers 127 and 128 serve as shifting and scaling factors so the resulting image is within the range $[0, 255]$. The ring image is expressed as:

$$r(n_1, n_2) = 127 + 128 \cos\left(\frac{\sqrt{n_1^2 + n_2^2}}{64}\right) \tag{7.16}$$

Of course, the HIP equivalent images require the conversion from a HIP address to a Cartesian coordinate as covered in Section 3.3. The HIP linear filtering example (Section 4.2.2) examined two distinct classes of filters. The first was lowpass and the second was highpass filter. In each of these classes, two different types of filters were employed: ideal and non-ideal (Butterworth). The transfer functions for the square versions of the two lowpass filters (compare with equations (4.27) and (4.28) for HIP) are:

$$H_{IL}(k_1, k_2) = \begin{cases} 1, & \sqrt{k_1^2 + k_2^2} \le R \\ 0, & \sqrt{k_1^2 + k_2^2} > R \end{cases} \tag{7.17}$$

and

$$H_{BL}(k_1, k_2) = \frac{1}{1 + C\left(\frac{\sqrt{k_1^2+k_2^2}}{R}\right)^{2n}} \tag{7.18}$$

The two highpass filters for square images (compare with equations (4.29) and (4.30)) are:

$$H_{IH}(k_1, k_2) = \begin{cases} 1, & \sqrt{k_1^2 + k_2^2} > R \\ 0, & \sqrt{k_1^2 + k_2^2} \le R \end{cases} \tag{7.19}$$

and

$$H_{HB}(k_1, k_2) = \frac{1}{1 + C\left(\frac{R}{\sqrt{k_1^2+k_2^2}}\right)^{2n}} \tag{7.20}$$

The filtered images can be compared by tuning the cutoff frequencies of the filter to attain the same reduction in total power. This was the same approach used in Section 4.2.2. Using the same labels as in equation (7.14) the power transfer ratio is:

$$\beta = \frac{\sum_{k_1,k_2} G(k_1, k_2)}{\sum_{k_1,k_2} F(k_1, k_2)} \tag{7.21}$$

Here the summations are performed over all possible spatial frequencies. This is similar to equation (4.31) for the HIP framework. Results of each of the classes of filters will be examined next, starting with the lowpass filter (LPF).

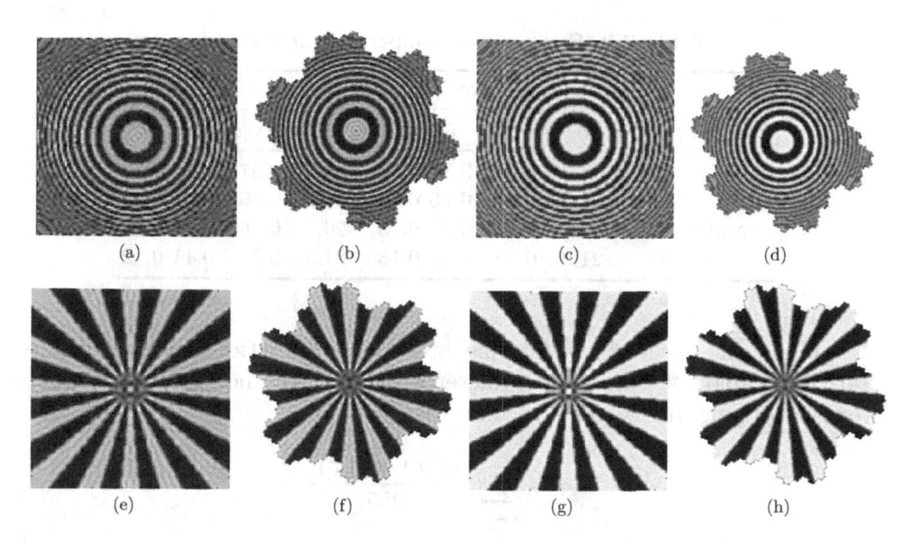

Fig. 7.21. Results of lowpass filtering of R1 and S1 test images sampled on square and hexagonal lattices: (a), (b) and (e), (f) are outputs of ideal LPF; (c), (d) and (g), (h) are outputs of the Butterworth LPF.

Filtering with a LPF rejects all spatial frequencies above a certain cutoff frequency. Since the frequency content in R1 increases radially, there should be a degradation closer to the edges of the image after lowpass filtering. Likewise, the frequency content in S1 decreases radially and hence there should be a loss of definition towards the centre after lowpass filtering. An initial comparison was performed by setting $\beta = 0.5$. The results for both images are given in Figure 7.21. In all cases, the results are as expected. There is more aliasing noticeable in the filtered rings around the edges of the images, more so in the case of the square image. Note that the aliasing in the case of the square images is aligned with the horizontal and vertical axes whilst for HIP it is aligned with the hexagon's axes of symmetry. This spreading of the aliasing across the three axes of symmetry is the reason that the HIP image exhibits less aliasing distortion. Additionally, the number of clearly visible rings is fewer in the square image than the HIP image. In the filtered star image, the centre has degraded and more so in the square image. When an ideal LPF is used, distinct ripples can be observed in the white portions of the star due to Gibbs phenomenon. Inspection of the central portion of the star indicates that the square image seems to preserve the information in the horizontal and vertical directions, whilst the HIP image serves to degrade the information evenly in all directions. As a result, the central region of the image looks smoother for HIP than for the square image.

The filtering experiment can be evaluated quantitatively in terms of difference from the original image after the filtering process. This metric is employed as, ideally, with no filtering the difference should be 0. The larger the number

Table 7.9. Performance comparison for the LPF.

		square				HIP			
filter	image	R	max(e)	$\overline{e}$	σ_e	R	max(e)	$\overline{e}$	σ_e
ideal	R1	38.7	0.93	0.31	0.20	33.5	0.97	0.28	0.19
	S1	28.7	0.78	0.20	0.15	28.5	0.86	0.18	0.14
non-ideal	R1	30.0	0.65	0.24	0.13	26.5	0.66	0.21	0.13
	S1	21.9	0.69	0.11	0.13	21.5	0.74	0.11	0.12

the greater the effect of the filtering procedure. Statistics for this are illustrated in Table 7.9 with the parameter e meaning the normalised difference from the original image. This is computed for HIP as:

$$e_H = \sum_{\boldsymbol{x} \in \mathbb{G}^\lambda} \frac{|g(\boldsymbol{x}) - f(\boldsymbol{x})|}{255}$$

Similarly, for the square image the error is computed as:

$$e_S = \sum_{n_1, n_2 \in \mathbb{R}^2} \frac{|g(n_1, n_2) - f(n_1, n_2)|}{255}$$

There are several observations that can be made by examination of the figures in Table 7.9. The first is that, for a given cutoff frequency, the value of R is consistently smaller for HIP. This is due to the compact nature of the hexagonal lattice. Due to the higher sampling density of the hexagonal lattice, a filled circular region on a hexagonal image will contain more points than a square lattice. As a result, there should be relatively more energy concentrated in the central portion of the HIP image than a square image. This was illustrated in the Fourier magnitude spectra in Figure 7.20. The next feature is that the maximum error is consistently higher for the hexagonal case, though this is partially offset by the fact that the mean error is lower. In conjunction with the fact that the standard deviation is similar for both the square and HIP images, it can be concluded that the HIP image has fewer erroneous points, however, the error magnitude is large. This is again due to the compact nature of the hexagonal lattice.

Next, processing with a highpass filter (HPF) was examined. This rejects all spatial frequencies below a certain minimum frequency. Hence, as a result of this operation the inner rings in R1 should disappear. For image S1, the inner portion of the image should be enhanced. Furthermore the result on the arms of the star should appear similar to an edge detected result due to the presence of sharp transitions in intensity. As with the LPF example, the value of β was set to 0.5 leading to rejecting 50% of the frequency spectra. This should result in a severe reduction of the image data and thus a degradation in the resulting image quality. The results are illustrated in Figure 7.22.

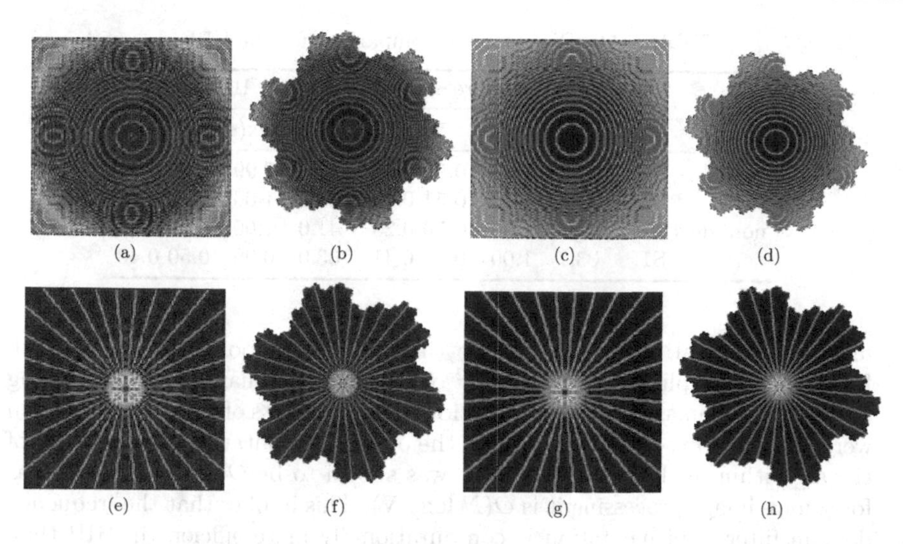

Fig. 7.22. Results of highpass filtering of R1 and S1 test images sampled on square and hexagonal lattices: (a), (b) and (e), (f) are outputs of ideal HPF; (c), (d) and (g), (h) are outputs of the Butterworth HPF.

The qualitative findings reflect the theoretical expectations. The ring image loses most of the large outer rings. The filtered star image retains edges and emphasises the central region. Further examination of the filtered ring image reveals several trends. The first is the reduction in the number of rings. After filtering however, more rings remain in the HIP images. The second is that the artifacts around the edges in the original image have been emphasised though less so in the HIP case. This is due to the artifacts being spread over three symmetric axes for HIP rather than two for square images. The filtering operation also performs better for the HIP image. The central region which contains the highest frequency features suffers from less distortion in HIP images than in the square images. Furthermore, the edges are clearer and more distinct for the HIP. Finally, with ideal filtering, the HIP image has fewer ringing artifacts than the square image.

Similar statistics as with the LPF example, were collected from these images. These are presented in Table 7.10. Overall, the statistics illustrate that most of the image data was lost in this filtering process however they do illustrate some interesting trends. Once again the average error is less for the HIP filtering examples when compared to the square image examples. Also, the radius for a given required cutoff frequency is less in the HIP case.

The second comparison to perform is in terms of the computational requirements of the filtering algorithm. The algorithm requires the use of three FFT algorithms and one complete point by point multiplication. The three FFT algorithms are required to be applied, one each, for the image and the

Table 7.10. Performance comparison for the HPF.

		square				HIP			
filter	image	R	max(e)	$\bar{e}$	σ_e	R	max(e)	$\bar{e}$	σ_e
ideal	R1	38.7	1.00	0.50	0.33	33.5	0.99	0.49	0.29
	S1	28.7	1.00	0.51	0.38	28.5	1.00	0.49	0.37
non-ideal	R1	46.2	1.00	0.50	0.25	41.0	1.00	0.49	0.25
	S1	33.7	1.00	0.51	0.41	33.0	0.99	0.50	0.40

filter, and a final inverse FFT after they have been point-wise multiplied. Point-wise multiplication requires N^2 operations for square image processing and N for HIP images. The computational requirements of the FFT algorithm were discussed in section 7.4.3 where the overall computational complexity of the algorithm in the HIP framework was shown to be $O(N \log_7 N)$ whereas for square image processing it is $O(N \log_2 N)$. This implies that the frequency domain filter implementation is computationally more efficient in HIP than for the equivalent square image case.

This section provided a comparison of the image processing techniques presented in Section 4.2.2 for HIP images against square images. The overall results are promising. Generally, processing using HIP gave a better performance than processing square images. The HIP images exhibited less aliasing and yielded better qualitative results. An interesting point is that the equivalent cutoff frequency radius is less for HIP than for square images. This is due to the denser packing of the hexagonal lattice and greater packing of spectral energy in the centre of the image. Related to this observation are the findings of Mersereau [41] who compared the design of finite impulse response (FIR) filters on hexagonal and square lattices. It was noted that imposing symmetry conditions on the impulse response of the hexagonal filter leads to the filter having zero phase and a frequency response with twelve-fold symmetry (in contrast to the eight-fold symmetry possible for square filters). This leads to computational efficiency as, due to symmetry, it is possible to compute the entire frequency response of the filter from the knowledge of the response over a width of $\frac{\pi}{6}$. This has important implications in physical realisation of systems, especially when it comes to memory usage and implementation of the filtering algorithms within the systems. Mersereau reported that for large cutoff frequencies the savings is as much as 50%. The results in this section also imply that to achieve similar performance to the hexagonal case, a square image must have a higher resolution.

7.4.5 Image pyramids

Section 4.3 covered several algorithms for deriving image pyramids in the HIP framework. Two distinct methodologies were covered, the first of which exploited the hierarchical nature of the HIP indexing scheme, and the second employed averaging followed by down-sampling. The first method is efficient as it requires only selecting from within the existing image data and no further processing. Furthermore, by combining different pyramids generated in this fashion, an approach similar to the averaging method can be achieved, though it is memory-intensive. This first method however, has no equivalent in square images due to the use of Cartesian coordinates. Hence, we will discuss only the second method. The method of averaging for the HIP framework was given in equation (4.33). An equivalent formulation for square images is:

$$f(x,y) = \sum_{j=0}^{J-1} \sum_{k=0}^{K-1} a(j,k) o(x-j, y-k) \tag{7.22}$$

In this equation, $a(j,k)$ is an averaging function and $o(x,y)$ is the original image. The averaging function is defined over a range of $(0,0)$ to $(J-1, K-1)$. The image pyramid is a multiresolution representation for a given image, preferably, without significant loss of information. Hence, the size of the averaging kernel must be large enough so that information is integrated over a wide neighbourhood. Additionally, the down-sampling rate or reduction in resolution must be small enough to enable gradual loss of information. Typically, in square image processing, images of size $N \times N$ are used where $N = 2^n$. It is thus logical that a reduction of order by four is carried out at each step and the resulting averaging function, $a(j,k)$ is defined over some range, the simplest of which is $(-1,-1)$ to $(1,1)$, or a 3×3 neighbourhood. Larger sizes increase processing time but integrate more information.

To perform a visual comparison of the square image and HIP pyramids a simple example was employed. An original eight-bit gray scale square image of size 128×128 was decomposed into a simple image pyramid. The original image was also resampled into a five-layer HIP image as described previously in Section 6.1 for the experiment. Reduction orders of four and seven were employed for the square and HIP images, respectively, for convenience. The averaging functions that were used had nine points (3×3 mask) for the square image and seven points (defined over $\mathbb{G}^1$) for the HIP image. The pyramids generated have five layers for the HIP image and seven for the square image. The first three layers in the pyramid produced in each case are illustrated in Figure 7.23.

The image pyramids appear similar for the top two layers. The third layer of the HIP pyramid shows significantly less detail than the square image pyramid due to the larger reduction in order at each stage. A logical consequence of this is that the HIP pyramid will always have fewer layers than the square image pyramid whose implications depend upon the application that is being examined. For instance, the HIP pyramid exhibits more compression between layers than the square image pyramid which is advantageous for image coding applications. For image analysis applications, the fact that the square image pyramid generates more representations may be an advantage. Finally, it should be noted that the reduction of resolution by seven holds only when processing is done using HIP. By using skewed and other coordinate systems proposed for hexagonal image representation, pyramids with resolution reductions in powers of two are also possible.

The algorithms for pyramid generation using square and HIP images are similar. Yet, due to the HIP data structure there is an inherent advantage in using HIP in terms of implementation. For the square image pyramid, the computation of the first order of reduction (from 128×128 to 64×64) requires 4,096 masking operations each with a mask size of 3×3. The computations associated with this was covered in Section 7.3. Thus there are a total of 36,864 multiplications required. In the HIP pyramid, the equivalent first reduction (from 7^5 points to 7^4 points) requires 2401 masking operations with an overall number of multiplications of 16,807. The number of required multiplications in each stage is shown in table 7.11. The HIP pyramid is computed with less than

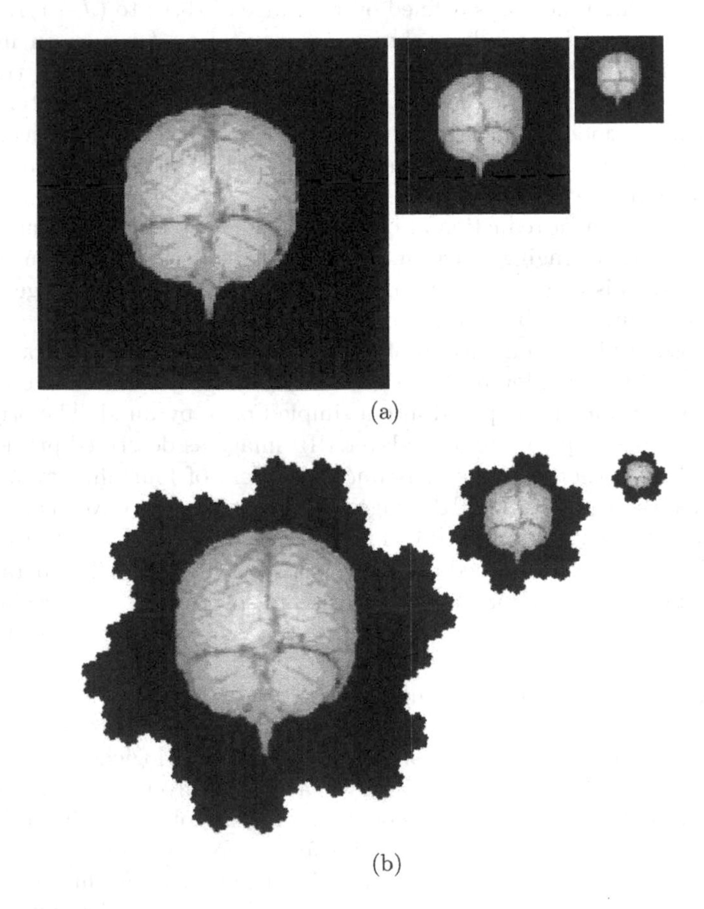

Fig. 7.23. First three layers of pyramid decomposition of a brain image: (a) square image b) HIP image.

Table 7.11. Number of multiplications required for image pyramid computation.

order	HIP	square
1	16807	36864
2	2401	9216
3	343	2304
4	49	576
5	0	144
6	0	36
total	19600	49140

half the number of computations required for the square image pyramid. Even if only the first four reductions are considered, the computational requirements for the HIP case is still far less than that of the square image pyramid.

7.5 Concluding Remarks

This chapter was concerned with a comparison of square image and HIP processing. Three major themes have run throughout the entire chapter. These are sampling density, curved structure representation, and computational efficiency. The packing efficiency of the hexagonal and square lattices were studied. A reduction (by 13.4%) in the number of sample points required to represent a continuous signal is possible when the sampling lattice used is hexagonal instead of a square. This might not seem very significant since the camera and display technologies (which use square grids) are tending to support increasingly higher resolution images. However, in keeping with this trend, there is a corresponding increase in processing load and communication bandwidth requirement, neither of which is desirable. The increased sampling density and uniform connectivity of the hexagonal lattice lead to interesting advantages in image processing as found in the denser skeletons and edge maps obtained with hexagonal images, both of which are beneficial for image analysis.

The hexagonal lattice has non-orthogonal basis vectors unlike the square lattice. This is a boon in some cases and a bane in others. Enhanced curve representation and noise rejection in edge detection are examples of the former. In the case of DFT, the basis images are aligned along the three axes of symmetry of the hexagon as compared to two axes of symmetry for the square, which has important implications when it comes to the spread of energy within the frequency domain. Two shortcomings of the non-orthogonality of the basis vectors are that it inhibits the use of natural coordinates such as Cartesian which would lead to a non-integer grid, and that the DFT is no longer separable. The square lattice on the other hand permits the use of

Cartesian coordinates and a separable DFT. It also offers an advantage in representing vertical and horizontal lines.

Changing the sampling lattice has an interesting impact on computational requirements of image processing. An obviously negative one is the overhead cost incurred to acquire (via resampling) and display hexagonal images. This is due to the unavailability of suitable hardware at present. Nevertheless, the computational complexity is comparable for processing hexagonal and square images, and in many of the cases we studied, the number of required operations was found to be higher for square than for hexagonal images. Smaller mask sizes for filtering and more efficient implementation of filtering and morphological operations for hexagonal images were such examples.

Some of the computational advantages offered by the hexagonal lattice for image processing arise specifically from the HIP framework which uses a single index to address pixels. Firstly, the HIP framework offers an advantage for implementing the convolution (or masking) operation due to the data structure used in the framework. Secondly, morphological algorithms such as skeletonisation also require fewer computations despite using an algorithm not tailored to the hexagonal lattice. Thirdly, the FFT developed for the HIP framework is also more efficient than the equivalent algorithms for square images. For instance, the FFT computation for a five-layer HIP image (equivalent to a 128×128 square image) requires 63% fewer operations. Overall, the HFFT requires about 60% fewer operations than the FFT when the square and hexagonal images are almost of equal size. In the case of image pyramids, the natural order of reduction between adjacent levels in the square case is four, while the corresponding figure for a HIP pyramid is seven. This can be an advantage in image compression. HIP image pyramids are easier to construct and have lower storage requirements than square image pyramids.

The HIP framework does have some drawbacks. The first is the restriction on the image sizes to be powers of seven which means that while resampling arbitrary sized square images, the resulting HIP image (stored as a vector) can be much larger albeit sparse. The second is the ragged boundary of the image as compared to the clean edges of a square image which will cause the screen of a display devise to be under-utilised for full image display. And finally, the HIP framework requires the use of modulo-7 arithmetic for address manipulations. However, these we believe are offset by the benefits (listed above) of using the HIP framework.

Overall, the findings of our comparative study suggest that the impact of changing the sampling lattice from square to hexagonal on image processing is, by and large, quite positive.

8

Conclusion

This monograph focused on the use of hexagonal lattices for defining digital images and developed a practical framework, called HIP, within which these images can be processed. As pointed out in Chapter 2, interest in using hexagonal sampling grids for images is as old as the field of digital image processing itself. However, the development in this area of research is no match for the vigour or advances in the research on processing images defined on a square lattice. Does this imply that hexagonal image processing is an opportunity lost forever? We began this monograph by observing that nature and mathematical sciences favour the hexagonal lattice and our studies revealed that image processing can also benefit from using this lattice for defining images. Hence, the answer to the question appears to be a simple *No*. As increased computational power becomes available, there seems to be much energy and time directed at devising more complex algorithms whereas it may be equally worthwhile to examine alternate sampling regimes which can lead to more efficiencies and thus permit even more complicated algorithms.

The work presented in this monograph included practical solutions for processing hexagonal images and an appreciation of the variety of applications and algorithms that could be implemented. Thus, it sets the stage for further development of hexagonal image processing. In this chapter we will outline some avenues for such development.

There are several distinct areas in which future work could be concentrated. These include study of key areas in general and improvements to the HIP framework. One of the hurdles in using hexagonal image processing is the lack of supporting hardware for image acquisition and display. This scenario needs to change. Prototypes for hexagonal sensors have been developed and reported in the literature [30,32], which are positive signals. Software solutions in the form of resampling were presented in Section 6.1. The sampling part could also be significantly improved via a hardware implementation. There are two ways to do this. The first is to modify existing hardware, such as a flatbed scanner, to produce the alternate offset lines of a hexagonal lattice.

Another method is to use an existing CCD sensor and build some custom hardware to perform the resampling process. This could be performed using simple reprogrammable hardware such as Field Programmable Gate Arrays (FPGA).

The work reported in this monograph covers many of the basic techniques in image processing implemented using the HIP framework. However, there is a wide scope for the study of many other important problems. Two of these that require immediate and in-depth study are spatio-temporal techniques and wavelets. The former study is needed to examine the likelihood of enhancing the performance of algorithms and the development of more efficient techniques for video processing. The higher sampling density of the hexagonal lattice should prove to be a positive asset in its use in spatio-temporal processing applications where the high processing loads deter real-time applications at present. An in-depth study of wavelets is desirable given the important role wavelets play in a wide range of applications from image compression to analysis and fusion. The compactness of the hexagonal lattice and the ability to decompose images into oriented subbands are attractive features that can be exploited.

As far as the HIP framework is concerned, there a few improvements that need to be made. The first point that needs attention is the ability to handle arbitrary size images. As mentioned at the end of Chapter 7, an image whose size is not a power of seven ends up being stored as a much larger image (where many of the pixel values are zero) in the HIP framework. An image is stored as a long vector within this framework, with the centre of the image being mapped to the first element of the vector and the pixels towards the periphery of the image being mapped to the elements at the end of the vector as shown in Figure 3.11. Hence, the hexagonal image will be mapped into a sparse vector where most of the elements at one end are zero-valued. Special techniques have to be devised to handle such a *sparse vector* efficiently for both storage and manipulation. A second area which can be examined in the future is speeding up the addressing part of the scheme. At the moment, the algorithm is implemented in software using C++ and Python. The C++ code is reasonably optimal but could be significantly improved via a hardware implementation. Additionally, a hardware system could be built which exploits the HIP framework and then converts the result to square images to help implement mixed system designs, as mentioned in Chapter 6. For a full hexagonal image processing system, a hardware solution needs to be developed for resampling (if an appropriate hardware for direct acquisition is unavailable), implementation of modulo-7 operations, and for display to further improve the efficiency of the HIP framework.

An entirely new direction of investigation that is also of interest is the extension of alternative sampling regimes to 3-D space. The body-centred cubic grid (with truncated octahedral voxel) is the optimal covering for 3-D space while the densest packing is provided by the face-centred cubic grid (with rhombic dodecahedral voxel) [144]. These can be shown to be generalisations of the hexagonal sampling in 3-D. Non-cubic voxels such as those listed above are viable alternatives to the currently used cubic voxels and has drawn some interest recently [153]. It is noteworthy that changing the sampling regime in volume imagery is not difficult as raw tomographic data from X-ray or magnetic imaging is used to compute the volume image. Furthermore, there is an incentive to do so with many of the application areas for 3-D reconstruction being in medical imaging. Here, the natural structures, such as various organs, being imaged contain an abundance of curved features which are best represented on a hexagonal lattice.

A

Mathematical derivations

This appendix provides proofs and derivations of some of the equations in the main body of this monograph.

A.1 Cardinality of A_λ

In Section 3.2.1 the HIP addressing scheme was introduced. The set of all possible points in a λ-level aggregate was given in equation (3.8) and was denoted by A_λ. A simple proof of the cardinality of this set follows.

By definition, the 0-th level aggregate has only one hexagonal cell. Hence, the cardinality of the 0-th level aggregate is:

$$\mathrm{card}(A_0) = 1$$

From the rule of aggregation, the first level aggregate consists of the point in A_0 plus six more hexagonal cells which are translated versions of A_0. Therefore, the cardinality of the first-level aggregate is:

$$\begin{aligned}\mathrm{card}(A_1) &= \mathrm{card}(A_0) + 6\,\mathrm{card}(A_0) \\ &= 7\end{aligned}$$

Now consider a general case, with the aggregate level being λ. The cardinality of A_λ for $\lambda = 0$ is $7^0 = 1$. Assume that this is true for some value k. This gives:

$$\mathrm{card}(A_k) = 7^k \tag{A.1}$$

The A_{k+1} layer aggregate has seven copies of A_k as it is built by taking the A_k aggregate and surrounding it with six more such aggregates. Hence cardinality of A_{k+1} can be found from the cardinality of A_k as follows:

$$\begin{aligned} \text{card}(A_{k+1}) &= \text{card}(A_k) + 6\,\text{card}(A_k) \\ &= 7^k + 6 \times 7^k \\ &= 7^{k+1} \end{aligned} \tag{A.2}$$

The trivial case ($k = 0$) and equations (A.1) and (A.2) are necessary and sufficient to show that it is true $\forall k \in \mathbb{Z}$.

A.2 Locus of aggregate centres

In the construction of the HIP structure through a process of aggregation, the centre of an aggregate can be shown to rotate at an angle which increases with the level of aggregation. Consider a vector joining the 0-th level aggregate and the centre of a λ-level aggregate. Let the length of this vector be r and the angle subtended by this vector with the horizontal be θ_λ. In Figure A.1 showing the top quarter of the HIP structure, this angle is shown for the cases $\lambda = 2, 3$, along with the HIP addresses of these centres. Note that for $\lambda = 0$, this length and angle are zero as the aggregate consists of only one cell. For all other values of λ, the expression for the length and angle of the vector are given as:

$$\begin{aligned} r_k &= (\sqrt{7})^{k-1} \\ \theta_k &= (k-1)\tan^{-1}\frac{\sqrt{3}}{2} \end{aligned} \tag{A.3}$$

We will now prove these relationships. The centre of the aggregate A_λ, where $\lambda > 0$, is related to the centre of A_1, as expressed in equation (3.8) by a translation matrix. This matrix is given as:

$$\mathbf{N}_{\lambda-1} = \begin{bmatrix} 3 & -2 \\ 2 & 1 \end{bmatrix}^{\lambda-1} \tag{A.4}$$

The translation matrix can be used to compute the centres $\mathbf{x}_\lambda$ of each aggregate. This computation is with respect to a pair of skewed axes where one axis is aligned to the horizontal and the other at 120° to the first. For $\lambda = 2, 3$ these centres can be found as follows:

$$\begin{aligned} \mathbf{x}_2 &= \mathbf{N}_1 \begin{bmatrix} 1 \\ 0 \end{bmatrix} \\ &= \begin{bmatrix} 3 & -2 \\ 2 & 1 \end{bmatrix} \begin{bmatrix} 1 \\ 0 \end{bmatrix} \\ &= \begin{bmatrix} 3 \\ 2 \end{bmatrix} \end{aligned}$$

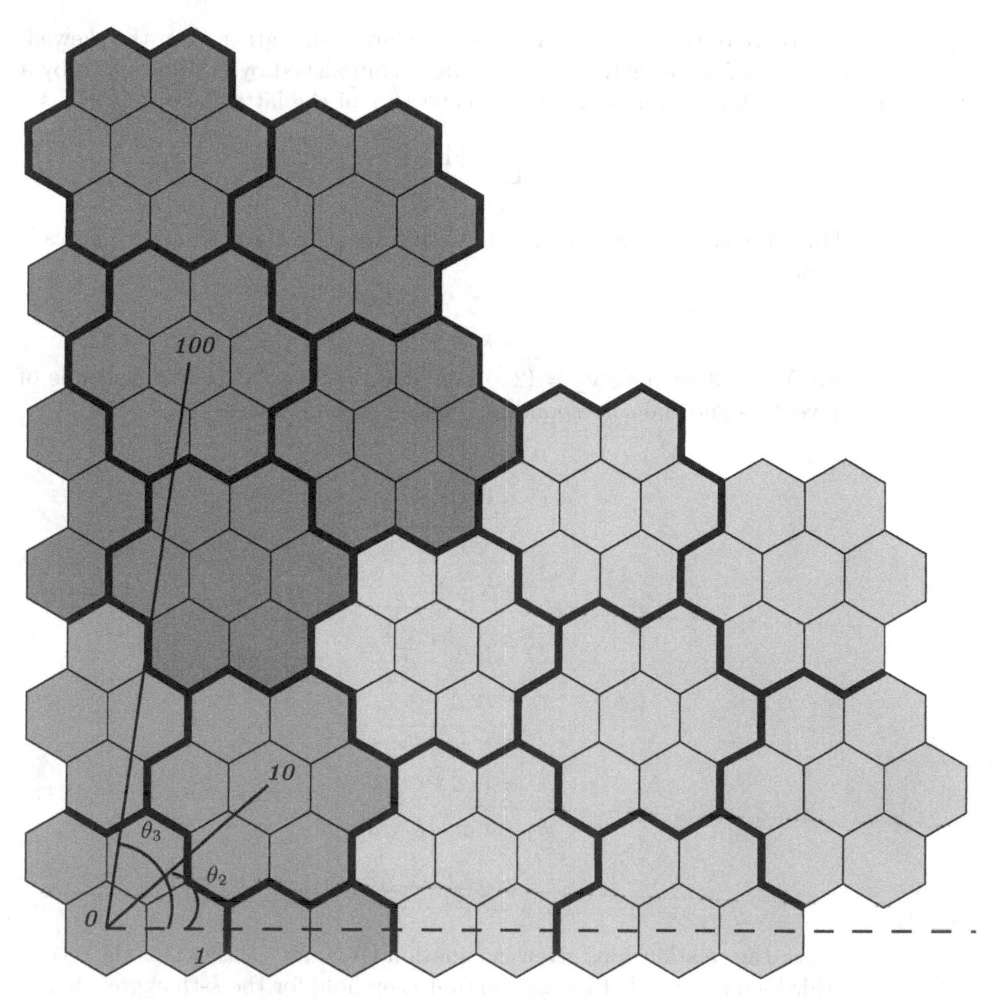

Fig. A.1. The increase in angle from the horizontal for A_2 and A_3.

$$\begin{aligned}
\mathbf{x}_3 &= \mathbf{N}_2 \begin{bmatrix} 1 \\ 0 \end{bmatrix} \\
&= \begin{bmatrix} 5 & -8 \\ 8 & -3 \end{bmatrix} \begin{bmatrix} 1 \\ 0 \end{bmatrix} \\
&= \begin{bmatrix} 5 \\ 8 \end{bmatrix}
\end{aligned}$$

To compute the angle θ, we need the Cartesian, rather than the skewed, coordinates of these centres. This can be accomplished by multiplication by a matrix which has as its rows the basis vectors of the lattice. This matrix is:

$$\mathbf{B} = \begin{bmatrix} 1 & -\frac{1}{2} \\ 0 & \frac{\sqrt{3}}{2} \end{bmatrix}$$

Thus, the centre of any aggregate with respect to Cartesian coordinates is found as:

$$\mathbf{y}_\lambda = \mathbf{B}\mathbf{x}_\lambda \tag{A.5}$$

For $\lambda = 2, 3$ we have $\mathbf{y}_2 = (2, \sqrt{3})$ and $\mathbf{y}_3 = (1, 4\sqrt{3})$. The magnitude of these vectors and the corresponding angles can be found as:

$$\begin{aligned} r_2 &= |\mathbf{y}_2| \\ &= \sqrt{(4+3)} \\ &= \sqrt{7} \\ \theta_2 &= \tan^{-1}\frac{\sqrt{3}}{2} \\ r_3 &= |\mathbf{y}_3| \\ &= \sqrt{1+48} = 7 \\ &= (\sqrt{7})^{3-1} \\ \theta_3 &= \tan^{-1}(4\sqrt{3}) \\ &= 2\tan^{-1}\frac{\sqrt{3}}{2} \end{aligned}$$

Thus, the relationships given in equation (A.3) are satisfied for the second and third aggregates. Let us assume that they hold for the k-th aggregate A_k. The Cartesian coordinate of this point is given by:

$$\mathbf{y}_k = \mathbf{B}\mathbf{N}_{k-1}\mathbf{e}_0$$

where $\mathbf{e}_0 = (1, 0)$. Next, let us consider the aggregate A_{k+1}. The centre of this aggregate can be found as:

$$\begin{aligned} \mathbf{y}_{k+1} &= \mathbf{B}\mathbf{N}_k\mathbf{e}_0 \\ &= \mathbf{B}\mathbf{N}_1\mathbf{N}_{k-1}\mathbf{e}_0 \end{aligned}$$

Examination of this equation shows that the $(k+1)$-th coordinate can be computed directly from the k-th coordinate with an extra scaling and rotation attributable to the matrix $\mathbf{N}_1$. These have been previously computed to be r_2 and θ_2. Thus we have the following:

$$\begin{aligned}
r_{k+1} &= r_2(\sqrt{7})^{k-1} \\
&= \sqrt{7}(\sqrt{7})^{k-1} \\
&= (\sqrt{7})^{k} \\
\theta_{k+1} &= (k-1)\tan^{-1}\frac{\sqrt{3}}{2} + \theta_2 \\
&= (k-1)\tan^{-1}\frac{\sqrt{3}}{2} + \tan^{-1}\frac{\sqrt{3}}{2} \\
&= k\tan^{-1}\frac{\sqrt{3}}{2}
\end{aligned}$$

Thus, so long as these statements are true for A_k then it is also true for A_{k+1}. This, along with the trivial case is sufficient to prove that the relationships in equation (A.3) are true for all values of λ so long as $\lambda > 0$.

Finally, the locus of the centres of the successive aggregates in the HIP structure can be shown to be a spiral. The spiral is illustrated for the first three layers in Figure A.2.

This is an exponential spiral with general equation $r = a\exp(b\theta)$. The parameters a, b can be simply estimated by substitution. We can find a using point ***1*** for which $\theta = 0$, as follows:

$$\begin{aligned}
ae^{b0} &= 1 \\
\Leftrightarrow a &= 1
\end{aligned}$$

Similarly, the parameter b can be estimated using point ***10***:

$$\begin{aligned}
\exp\left(b\tan^{-1}\frac{\sqrt{3}}{2}\right) &= \sqrt{7} \\
\Rightarrow b &= \frac{\log\sqrt{7}}{\tan^{-1}\frac{\sqrt{3}}{2}} \\
&= \frac{\log 7}{2\tan^{-1}\frac{\sqrt{3}}{2}}
\end{aligned}$$

Together these give the equation:

$$r = \exp(\theta\frac{\log\sqrt{7}}{\tan^{-1}\frac{\sqrt{3}}{2}})$$

This is nothing but equation (3.10) in Section 3.2.1.

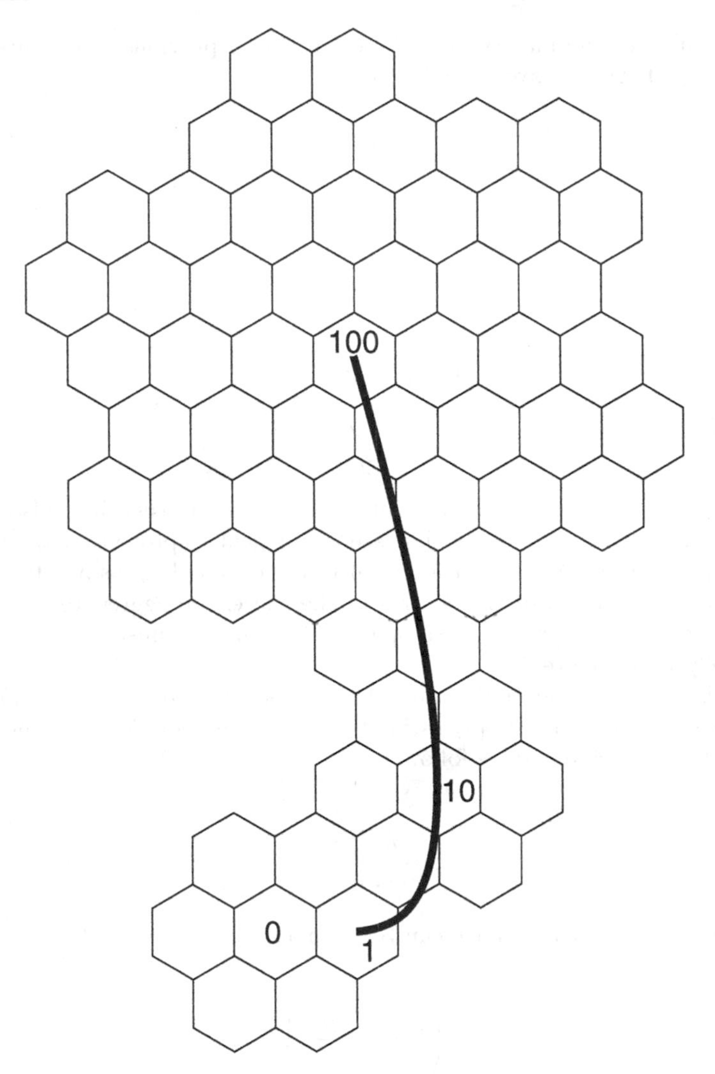

Fig. A.2. The spiral through ***1***, ***10***, and ***100***.

A.3 Conversion from HIP address to Her's 3-tuple

As mentioned in Section 2.2.2 there are several coordinate schemes used for addressing pixels in a hexagonal lattice. Conversion from a HIP address to Her's [20] 3-tuple coordinate scheme is used widely in this monograph for various purposes. We will derive a mapping function in this section to achieve this conversion. The desired mapping function is $c : \mathbb{G}^1 \rightarrow \mathbb{R}^3$.

First we note that all HIP addresses from ***1*** to ***6*** can be found by a suitable rotation of address ***1***. A rotation matrix that effects a rotation of a lattice point about the origin by an angle θ (measured anticlockwise) was defined by Her as follows:

$$R_\theta = \frac{1}{3}\begin{bmatrix} 1+2\cos\theta & 1-\cos\theta+\sqrt{3}\sin\theta & 1-\cos\theta-\sqrt{3}\sin\theta \\ 1-\cos\theta-\sqrt{3}\sin\theta & 1+2\cos\theta & 1-\cos\theta+\sqrt{3}\sin\theta \\ 1-\cos\theta+\sqrt{3}\sin\theta & 1-\cos\theta-\sqrt{3}\sin\theta & 1+2\cos\theta \end{bmatrix}$$

Now, the HIP addresses $\boldsymbol{n}$, $\boldsymbol{n} = \boldsymbol{1}, \boldsymbol{2}, \boldsymbol{3}, \boldsymbol{4}, \boldsymbol{5}, \boldsymbol{6}$ denote pixels at unit distance from the origin and at $\theta = 0, \frac{5\pi}{3}, \frac{4\pi}{3}, \pi, \frac{2\pi}{3}, \frac{\pi}{3}$. Hence, the corresponding rotation matrices are:

$$R_0 = \begin{bmatrix} 1 & 0 & 0 \\ 0 & 1 & 0 \\ 0 & 0 & 1 \end{bmatrix} \qquad R_\pi = \frac{1}{3}\begin{bmatrix} -1 & 2 & 2 \\ 2 & -1 & 2 \\ 2 & 2 & -1 \end{bmatrix}$$

$$R_{\frac{5\pi}{3}} = \frac{1}{3}\begin{bmatrix} 2 & -1 & 2 \\ 2 & 2 & -1 \\ -1 & 2 & 2 \end{bmatrix} \qquad R_{\frac{2\pi}{3}} = \begin{bmatrix} 0 & 1 & 0 \\ 0 & 0 & 1 \\ 1 & 0 & 0 \end{bmatrix}$$

$$R_{\frac{4\pi}{3}} = \begin{bmatrix} 0 & 0 & 1 \\ 1 & 0 & 0 \\ 0 & 1 & 0 \end{bmatrix} \qquad R_{\frac{\pi}{3}} = \frac{1}{3}\begin{bmatrix} 2 & 2 & -1 \\ -1 & 2 & 2 \\ 2 & -1 & 2 \end{bmatrix}$$

Since, Her's coordinate system has the property that the coordinates sum to zero, we can easily add a constant to each column of the matrix to obtain matrix entries of 0 or ± 1 to which are more convenient to use. Thus we obtain:

$$R_0 = \begin{bmatrix} 1 & 0 & 0 \\ 0 & 1 & 0 \\ 0 & 0 & 1 \end{bmatrix} \qquad R_\pi = \begin{bmatrix} -1 & 0 & 0 \\ 0 & -1 & 0 \\ 0 & 0 & -1 \end{bmatrix}$$

$$R_{\frac{5\pi}{3}} = \begin{bmatrix} 0 & -1 & 0 \\ 0 & 0 & -1 \\ -1 & 0 & 0 \end{bmatrix} \qquad R_{\frac{2\pi}{3}} = \begin{bmatrix} 0 & 1 & 0 \\ 0 & 0 & 1 \\ 1 & 0 & 0 \end{bmatrix}$$

$$R_{\frac{4\pi}{3}} = \begin{bmatrix} 0 & 0 & 1 \\ 1 & 0 & 0 \\ 0 & 1 & 0 \end{bmatrix} \qquad R_{\frac{\pi}{3}} = \begin{bmatrix} 0 & 0 & -1 \\ -1 & 0 & 0 \\ 0 & -1 & 0 \end{bmatrix}$$

It is apparent that any matrix R_θ can be obtained by repeated multiplication by $R_{\frac{\pi}{3}}$. Now we have the desired mapping function. The mapping function for the trivial case of address $\boldsymbol{0}$ is a zero vector. For all other $\boldsymbol{n}$ we have a rotation matrix of appropriate power multiplying the Her coordinate of $\boldsymbol{1}$ which is $(1, 0, -1)^T$ as shown below:

$$c(\boldsymbol{n}) = \begin{cases} \begin{bmatrix} 0 \\ 0 \\ 0 \end{bmatrix} & \text{if } \boldsymbol{n} = \boldsymbol{0} \\ \begin{bmatrix} 0 & 0 & -1 \\ -1 & 0 & 0 \\ 0 & -1 & 0 \end{bmatrix}^{(7-n)} \begin{bmatrix} 1 \\ 0 \\ -1 \end{bmatrix} & \text{if } \boldsymbol{1} \leq \boldsymbol{n} < \boldsymbol{6} \end{cases}$$

A.4 Properties of the HDFT

In Section 3.4.5 the HIP discrete Fourier transform was introduced. In equations (3.54) and (3.55) the HDFT pair were defined to be:

$$X(\boldsymbol{k}) = \sum_{\boldsymbol{n} \in \mathbb{G}^{\lambda}} x(\boldsymbol{n}) \exp\left[-2\pi \mathrm{j} H(\boldsymbol{k})^T \mathbf{N}_{\lambda}^{-1} h(\boldsymbol{n})\right] \tag{A.6}$$

$$x(\boldsymbol{n}) = \frac{1}{|\det \mathbf{N}_{\lambda}|} \sum_{\boldsymbol{k} \in \mathbb{G}^{\lambda}} X(\boldsymbol{k}) \exp\left[2\pi \mathrm{j} H(\boldsymbol{k})^T \mathbf{N}_{\lambda}^{-1} h(\boldsymbol{n})\right] \tag{A.7}$$

where the spatial and frequency variables are specified as HIP addresses. The HDFT possesses several useful properties. These are verified in this section.

A.4.1 Linearity

Given the HDFT pair

$$x(\boldsymbol{n}) \overset{\text{HDFT}}{\longleftrightarrow} X(\boldsymbol{k}) \quad \text{and} \quad y(\boldsymbol{n}) \overset{\text{HDFT}}{\longleftrightarrow} Y(\boldsymbol{k})$$

then

$$ax(\boldsymbol{n}) + by(\boldsymbol{n}) \overset{\text{HDFT}}{\longleftrightarrow} aX(\boldsymbol{k}) + bY(\boldsymbol{k})$$

This holds for any scalars $a, b \in \mathbb{C}$. This is verified easily as follows. Taking HDFT of the left hand side of this equation yields:

$$\begin{aligned} \text{HDFT}\{ax(\boldsymbol{n}) + by(\boldsymbol{n}\} &\triangleq \sum_{\boldsymbol{n} \in \mathbb{G}^{\lambda}} (ax(\boldsymbol{n}) + by(\boldsymbol{n})) \exp\left[-2\pi \mathrm{j} H(\boldsymbol{k})^T \mathbf{N}_{\lambda}^{-1} h(\boldsymbol{n})\right] \\ &= a \sum_{\boldsymbol{n} \in \mathbb{G}^{\lambda}} x(\boldsymbol{n}) \exp\left[-2\pi \mathrm{j} H(\boldsymbol{k})^T \mathbf{N}_{\lambda}^{-1} h(\boldsymbol{n})\right] \\ &\quad + b \sum_{\boldsymbol{n} \in \mathbb{G}^{\lambda}} y(\boldsymbol{n}) \exp\left[-2\pi \mathrm{j} H(\boldsymbol{k})^T \mathbf{N}_{\lambda}^{-1} h(\boldsymbol{n})\right] \\ &= aX(\boldsymbol{k}) + bY(\boldsymbol{k}) \end{aligned}$$

.4.2 Shift/translation

ıere are two possibilities for a shift. The first is a spatial shift:

$$x(\boldsymbol{n} \ominus \boldsymbol{a}) \stackrel{\text{HDFT}}{\longleftrightarrow} \exp\left[-2\pi \mathrm{j} H(\boldsymbol{k})^T \mathbf{N}_\lambda^{-1} h(\boldsymbol{a})\right] X(\boldsymbol{k})$$

This is the linear phase condition whereby a spatial shift results in a lin-
r change in phase in the frequency domain. This is also verified using the
finition of HDFT as follows:

$$\begin{aligned}
\text{)FT}\{x(\boldsymbol{n} \ominus \boldsymbol{a})\} &= \sum_{\boldsymbol{n} \in \mathbb{G}^\lambda} x(\boldsymbol{n} \ominus \boldsymbol{a}) \exp\left[-2\pi \mathrm{j} H(\boldsymbol{k})^T \mathbf{N}_\lambda^{-1} h(\boldsymbol{n})\right] \qquad \text{let } \boldsymbol{m} = \boldsymbol{n} \ominus \boldsymbol{a} \\
&= \sum_{\boldsymbol{m} \in \mathbb{G}^\lambda \ominus \boldsymbol{a}} x(\boldsymbol{m}) \exp\left[-2\pi \mathrm{j} H(\boldsymbol{k})^T \mathbf{N}_\lambda^{-1} h(\boldsymbol{m} \oplus \boldsymbol{a})\right] \\
&= \sum_{\boldsymbol{m} \in \mathbb{G}^\lambda \ominus \boldsymbol{a}} x(\boldsymbol{m}) \exp\left[-2\pi \mathrm{j} H(\boldsymbol{k})^T \mathbf{N}_\lambda^{-1} h(\boldsymbol{m})\right] \exp\left[-2\pi \mathrm{j} H(\boldsymbol{k})^T \mathbf{N}_\lambda^{-1} h(\boldsymbol{a})\right] \\
&= \exp\left[-2\pi \mathrm{j} H(\boldsymbol{k})^T \mathbf{N}_\lambda^{-1} h(\boldsymbol{a})\right] \sum_{\boldsymbol{m} \in \mathbb{G}^\lambda} x(\boldsymbol{m}) \exp\left[-2\pi \mathrm{j} H(\boldsymbol{k})^T \mathbf{N}_\lambda^{-1} h(\boldsymbol{m})\right] \\
&= \exp\left[-2\pi \mathrm{j} H(\boldsymbol{k})^T \mathbf{N}_\lambda^{-1} h(\boldsymbol{a})\right] X(\boldsymbol{k})
\end{aligned}$$

Note the use of $h(\boldsymbol{n} \oplus \boldsymbol{a}) = h(\boldsymbol{n}) + h(\boldsymbol{a})$. This property follows from the
t that HIP addition is a complex addition.

A second shift possible is in the frequency domain. This property states
ıt:

$$x(\boldsymbol{n}) \exp\left[2\pi \mathrm{j} H(\boldsymbol{a})^T \mathbf{N}_\lambda^{-1} h(\boldsymbol{n})\right] \stackrel{\text{HDFT}}{\longleftrightarrow} X(\boldsymbol{k} \ominus \boldsymbol{a})$$

This follows from the duality of HDFT. However, we can also verify this
ng the definitions as follows:

$$\begin{aligned}
\text{DFT}\{X(\boldsymbol{k} \ominus \boldsymbol{a})\} &= \frac{1}{|\det \mathbf{N}_\lambda|} \sum_{\boldsymbol{k} \in \mathbb{G}^\lambda} X(\boldsymbol{k} \ominus \boldsymbol{a}) \exp\left[2\pi \mathrm{j} H(\boldsymbol{k})^T \mathbf{N}_\lambda^{-1} h(\boldsymbol{n})\right] \qquad \text{let } \boldsymbol{l} = \boldsymbol{k} \ominus \boldsymbol{a} \\
&= \frac{1}{|\det \mathbf{N}_\lambda|} \sum_{\boldsymbol{l} \in \mathbb{G}^\lambda \ominus \boldsymbol{a}} X(\boldsymbol{l}) \exp\left[2\pi \mathrm{j} H(\boldsymbol{l} \oplus \boldsymbol{a})^T \mathbf{N}_\lambda^{-1} h(\boldsymbol{n})\right] \\
&= \frac{1}{|\det \mathbf{N}_\lambda|} \sum_{\boldsymbol{l} \in \mathbb{G}^\lambda \ominus \boldsymbol{a}} X(\boldsymbol{l}) \exp\left[2\pi \mathrm{j} H(\boldsymbol{l})^T \mathbf{N}_\lambda^{-1} h(\boldsymbol{n})\right] \exp\left[2\pi \mathrm{j} H(\boldsymbol{a})^T \mathbf{N}_\lambda^{-1} h(\boldsymbol{n})\right] \\
&= \exp\left[2\pi \mathrm{j} H(\boldsymbol{a})^T \mathbf{N}_\lambda^{-1} h(\boldsymbol{n})\right] \frac{1}{|\det \mathbf{N}_\lambda|} \sum_{\boldsymbol{l} \in \mathbb{G}^\lambda} X(\boldsymbol{l}) \exp\left[2\pi \mathrm{j} H(\boldsymbol{l})^T \mathbf{N}_\lambda^{-1} h(\boldsymbol{n})\right] \\
&= \exp\left[2\pi \mathrm{j} H(\boldsymbol{a})^T \mathbf{N}_\lambda^{-1} h(\boldsymbol{n})\right] x(\boldsymbol{n})
\end{aligned}$$

A.4.3 Convolution theorem

Given two spatial domain images $x(\boldsymbol{n})$ and $y(\boldsymbol{n})$ with Fourier transforms $X(\boldsymbol{k})$ and $Y(\boldsymbol{k})$ respectively, the following relationship holds:

$$x(\boldsymbol{n}) \circledast y(\boldsymbol{n}) \overset{\text{HDFT}}{\longleftrightarrow} X(\boldsymbol{k})Y(\boldsymbol{k})$$

Using equation (A.6) and the definition of the HIP convolution (equation (3.39)) this can be proven as follows:

$$\begin{aligned}
\text{HDFT}\{x(\boldsymbol{n}) \circledast y(\boldsymbol{n})\} &\triangleq \sum_{\boldsymbol{n}\in\mathbb{G}^\lambda} (x(\boldsymbol{n}) \circledast y(\boldsymbol{n})) \exp\left[-2\pi \mathrm{j} H(\boldsymbol{k})^T \mathbf{N}_\lambda^{-1} h(\boldsymbol{n})\right] \\
&\triangleq \sum_{\boldsymbol{n}\in\mathbb{G}^\lambda} \left(\sum_{\boldsymbol{m}\in\mathbb{G}^\lambda} x(\boldsymbol{m}) y(\boldsymbol{n} \ominus \boldsymbol{m}) \right) \exp\left[-2\pi \mathrm{j} H(\boldsymbol{k})^T \mathbf{N}_\lambda^{-1} h(\boldsymbol{n})\right] \\
&= \sum_{\boldsymbol{m}\in\mathbb{G}^\lambda} x(\boldsymbol{m}) \sum_{\boldsymbol{n}\in\mathbb{G}^\lambda} y(\boldsymbol{n} \ominus \boldsymbol{m}) \exp\left[-2\pi \mathrm{j} H(\boldsymbol{k})^T \mathbf{N}_\lambda^{-1} h(\boldsymbol{n})\right] \\
&= \sum_{\boldsymbol{m}\in\mathbb{G}^\lambda} x(\boldsymbol{m}) \exp\left[-2\pi \mathrm{j} H(\boldsymbol{k})^T \mathbf{N}_\lambda^{-1} h(\boldsymbol{m})\right] Y(\boldsymbol{k}) \\
&= X(\boldsymbol{k})Y(\boldsymbol{k})
\end{aligned}$$

This proof makes use of the spatial domain shifting property. The frequency domain convolution can be proved in a similar fashion.

$$\begin{aligned}
\text{IHDFT}\{X(\boldsymbol{k}) \circledast Y(\boldsymbol{k})\} &\triangleq \frac{1}{|\det \mathbf{N}_\lambda|} \sum_{\boldsymbol{k}\in\mathbb{G}^\lambda} (X(\boldsymbol{k}) \circledast Y(\boldsymbol{k})) \exp\left[2\pi \mathrm{j} H(\boldsymbol{k})^T \mathbf{N}_\lambda^{-1} h(\boldsymbol{n})\right] \\
&\triangleq \frac{1}{|\det \mathbf{N}_\lambda|} \sum_{\boldsymbol{k}\in\mathbb{G}^\lambda} \left(\sum_{\boldsymbol{m}\in\mathbb{G}^\lambda} X(\boldsymbol{m}) Y(\boldsymbol{k} \ominus \boldsymbol{m}) \right) \exp\left[2\pi \mathrm{j} H(\boldsymbol{k})^T \mathbf{N}_\lambda^{-1} h(\boldsymbol{n})\right] \\
&= \frac{1}{|\det \mathbf{N}_\lambda|} \sum_{\boldsymbol{m}\in\mathbb{G}^\lambda} X(\boldsymbol{m}) \sum_{\boldsymbol{k}\in\mathbb{G}^\lambda} y(\boldsymbol{k} \ominus \boldsymbol{m}) \exp\left[2\pi \mathrm{j} H(\boldsymbol{k})^T \mathbf{N}_\lambda^{-1} h(\boldsymbol{n})\right] \\
&= \frac{1}{|\det \mathbf{N}_\lambda|} \sum_{\boldsymbol{m}\in\mathbb{G}^\lambda} X(\boldsymbol{m}) \exp\left[2\pi \mathrm{j} H(\boldsymbol{m})^T \mathbf{N}_\lambda^{-1} h(\boldsymbol{n})\right] y(\boldsymbol{n}) \\
&= x(\boldsymbol{n})y(\boldsymbol{n})
\end{aligned}$$

B

Derivation of HIP arithmetic tables

For the HIP framework to be useful it is essential that the arithmetic operations be defined. This work was covered in Section 3.2.2 but due to space constraints the derivations were left incomplete. This appendix will provide some details about the derivations. The vectorial nature of the HIP addresses is useful for this purpose.

B.1 HIP addition

The derivation of the addition property starts with an already populated HIP structure as illustrated in Figure B.1. Taking the individual hexagonal cells in the figure to be unit distance apart, the centre of every hexagonal cell can be expressed in terms of a unique vector. For the addresses ***0*** to ***6*** these are:

$$
\begin{aligned}
\mathit{0} &\equiv \begin{bmatrix}0 & 0\end{bmatrix}^T & & \\
\mathit{1} &\equiv \begin{bmatrix}1 & 0\end{bmatrix}^T & \mathit{4} &\equiv \begin{bmatrix}-1 & 0\end{bmatrix}^T \\
\mathit{2} &\equiv \frac{1}{2}\begin{bmatrix}1 & \sqrt{3}\end{bmatrix}^T & \mathit{5} &\equiv -\frac{1}{2}\begin{bmatrix}1 & \sqrt{3}\end{bmatrix}^T \\
\mathit{3} &\equiv \frac{1}{2}\begin{bmatrix}-1 & \sqrt{3}\end{bmatrix}^T & \mathit{6} &\equiv \frac{1}{2}\begin{bmatrix}1 & -\sqrt{3}\end{bmatrix}^T
\end{aligned}
$$

In order to generate the addition table (Table 3.1, reproduced here as Table B.2) we add pairwise combinations of these seven vectors. It should be noted that when the angle between the vectors is 0 or $\frac{\pi}{3}$, they will add to produce a vector pointing to a cell outside the original seven. The set of such cells form the outermost layer in the N_2 neighbourhood with addresses: (***15***, ***14***, ***26***, ***25***, ***31***, ***36***, ***42***, ***41***, ***53***, ***52***,***64*** and ***63***). The vector equivalent of these addresses are:

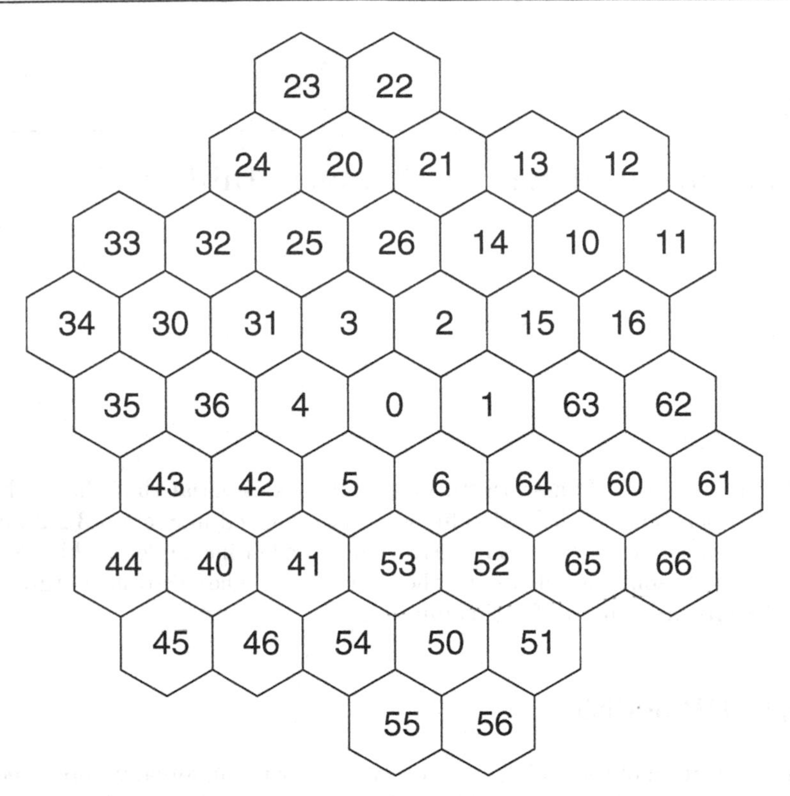

Fig. B.1. A fully populated level 2 HIP aggregate.

$$\mathit{14} \equiv \left[1\ \sqrt{3}\right]^T \qquad \mathit{41} \equiv -\left[1\ \sqrt{3}\right]^T$$
$$\mathit{15} \equiv \frac{1}{2}\left[3\ \sqrt{3}\right]^T \qquad \mathit{42} \equiv -\frac{1}{2}\left[3\ \sqrt{3}\right]^T$$
$$\mathit{25} \equiv \left[-1\ \sqrt{3}\right]^T \qquad \mathit{52} \equiv \left[1\ -\sqrt{3}\right]^T$$
$$\mathit{26} \equiv \left[0\ \sqrt{3}\right]^T \qquad \mathit{53} \equiv \left[0\ -\sqrt{3}\right]^T$$
$$\mathit{31} \equiv \frac{1}{2}\left[-3\ \sqrt{3}\right]^T \qquad \mathit{64} \equiv \frac{1}{2}\left[3\ -\sqrt{3}\right]^T$$
$$\mathit{36} \equiv \left[-2\ 0\right]^T \qquad \mathit{63} \equiv \left[2\ 0\right]^T$$

The entire table for HIP addition can now be generated by adding the vectors pairwise. The result of this pairwise addition is given in Table B.1.

By substituting the equivalent HIP addresses for the individual vectors, the desired HIP addition table can be derived. This is given in Table B.2.

$+$	$\begin{bmatrix}0\\0\end{bmatrix}$	$\begin{bmatrix}1\\0\end{bmatrix}$	$\frac{1}{2}\begin{bmatrix}1\\\sqrt{3}\end{bmatrix}$	$\frac{1}{2}\begin{bmatrix}-1\\\sqrt{3}\end{bmatrix}$	$\begin{bmatrix}-1\\0\end{bmatrix}$	$-\frac{1}{2}\begin{bmatrix}1\\\sqrt{3}\end{bmatrix}$	$\frac{1}{2}\begin{bmatrix}1\\-\sqrt{3}\end{bmatrix}$
$\begin{bmatrix}0\\0\end{bmatrix}$	$\begin{bmatrix}0\\0\end{bmatrix}$	$\begin{bmatrix}1\\0\end{bmatrix}$	$\frac{1}{2}\begin{bmatrix}1\\\sqrt{3}\end{bmatrix}$	$\frac{1}{2}\begin{bmatrix}-1\\\sqrt{3}\end{bmatrix}$	$\begin{bmatrix}-1\\0\end{bmatrix}$	$-\frac{1}{2}\begin{bmatrix}1\\\sqrt{3}\end{bmatrix}$	$\frac{1}{2}\begin{bmatrix}1\\-\sqrt{3}\end{bmatrix}$
$\begin{bmatrix}1\\0\end{bmatrix}$	$\begin{bmatrix}1\\0\end{bmatrix}$	$\begin{bmatrix}2\\0\end{bmatrix}$	$\frac{1}{2}\begin{bmatrix}3\\\sqrt{3}\end{bmatrix}$	$\frac{1}{2}\begin{bmatrix}1\\\sqrt{3}\end{bmatrix}$	$\begin{bmatrix}0\\0\end{bmatrix}$	$\frac{1}{2}\begin{bmatrix}1\\-\sqrt{3}\end{bmatrix}$	$\frac{1}{2}\begin{bmatrix}3\\-\sqrt{3}\end{bmatrix}$
$\frac{1}{2}\begin{bmatrix}1\\\sqrt{3}\end{bmatrix}$	$\frac{1}{2}\begin{bmatrix}1\\\sqrt{3}\end{bmatrix}$	$\frac{1}{2}\begin{bmatrix}3\\\sqrt{3}\end{bmatrix}$	$\begin{bmatrix}1\\\sqrt{3}\end{bmatrix}$	$\begin{bmatrix}0\\\sqrt{3}\end{bmatrix}$	$\frac{1}{2}\begin{bmatrix}-1\\\sqrt{3}\end{bmatrix}$	$\begin{bmatrix}0\\0\end{bmatrix}$	$\begin{bmatrix}1\\0\end{bmatrix}$
$\frac{1}{2}\begin{bmatrix}-1\\\sqrt{3}\end{bmatrix}$	$\frac{1}{2}\begin{bmatrix}-1\\\sqrt{3}\end{bmatrix}$	$\frac{1}{2}\begin{bmatrix}1\\\sqrt{3}\end{bmatrix}$	$\begin{bmatrix}0\\\sqrt{3}\end{bmatrix}$	$\begin{bmatrix}-1\\\sqrt{3}\end{bmatrix}$	$\frac{1}{2}\begin{bmatrix}-3\\\sqrt{3}\end{bmatrix}$	$\begin{bmatrix}-1\\0\end{bmatrix}$	$\begin{bmatrix}0\\0\end{bmatrix}$
$\begin{bmatrix}-1\\0\end{bmatrix}$	$\begin{bmatrix}-1\\0\end{bmatrix}$	$\begin{bmatrix}0\\0\end{bmatrix}$	$\frac{1}{2}\begin{bmatrix}-1\\\sqrt{3}\end{bmatrix}$	$\frac{1}{2}\begin{bmatrix}-3\\\sqrt{3}\end{bmatrix}$	$\begin{bmatrix}-2\\0\end{bmatrix}$	$-\frac{1}{2}\begin{bmatrix}3\\\sqrt{3}\end{bmatrix}$	$-\frac{1}{2}\begin{bmatrix}1\\\sqrt{3}\end{bmatrix}$
$-\frac{1}{2}\begin{bmatrix}1\\\sqrt{3}\end{bmatrix}$	$-\frac{1}{2}\begin{bmatrix}1\\\sqrt{3}\end{bmatrix}$	$\frac{1}{2}\begin{bmatrix}1\\-\sqrt{3}\end{bmatrix}$	$\begin{bmatrix}0\\0\end{bmatrix}$	$\begin{bmatrix}-1\\0\end{bmatrix}$	$-\frac{1}{2}\begin{bmatrix}3\\\sqrt{3}\end{bmatrix}$	$-\begin{bmatrix}1\\\sqrt{3}\end{bmatrix}$	$\begin{bmatrix}0\\-\sqrt{3}\end{bmatrix}$
$\frac{1}{2}\begin{bmatrix}1\\-\sqrt{3}\end{bmatrix}$	$\frac{1}{2}\begin{bmatrix}1\\-\sqrt{3}\end{bmatrix}$	$\frac{1}{2}\begin{bmatrix}3\\-\sqrt{3}\end{bmatrix}$	$\begin{bmatrix}1\\0\end{bmatrix}$	$\begin{bmatrix}0\\0\end{bmatrix}$	$-\frac{1}{2}\begin{bmatrix}1\\\sqrt{3}\end{bmatrix}$	$\begin{bmatrix}0\\-\sqrt{3}\end{bmatrix}$	$\begin{bmatrix}1\\-\sqrt{3}\end{bmatrix}$

Table B.1. Vectorial addition

$\oplus$	*0*	*1*	*2*	*3*	*4*	*5*	*6*
0	*0*	*1*	*2*	*3*	*4*	*5*	*6*
1	*1*	*63*	*15*	*2*	*0*	*6*	*64*
2	*2*	*15*	*14*	*26*	*3*	*0*	*1*
3	*3*	*2*	*26*	*25*	*31*	*4*	*0*
4	*4*	*0*	*3*	*31*	*36*	*42*	*5*
5	*5*	*6*	*0*	*4*	*42*	*41*	*53*
6	*6*	*64*	*1*	*0*	*5*	*53*	*52*

Table B.2. Addition table for HIP addresses.

B.2 HIP multiplication

We once again start with the fully populated HIP structure in Figure B.1 for the derivation of the HIP multiplication table. We can use the vectorial nature of the HIP addresses as in the previous section. However, since the operation of interest is multiplication, a polar representation for the vectors is more appropriate now. Given $\boldsymbol{a}$, $\boldsymbol{b}$ and their product $\boldsymbol{a} \otimes \boldsymbol{b}$, the corresponding vectors are expressed as:

$\times$	0	1	$e^{\jmath\frac{\pi}{3}}$	$e^{\jmath\frac{2\pi}{3}}$	-1	$e^{\jmath\frac{4\pi}{3}}$	$e^{\jmath\frac{5\pi}{3}}$
0	0	0	0	0	0	0	0
1	0	1	$e^{\jmath\frac{\pi}{3}}$	$e^{\jmath\frac{2\pi}{3}}$	-1	$e^{\jmath\frac{4\pi}{3}}$	$e^{\jmath\frac{5\pi}{3}}$
$e^{\jmath\frac{\pi}{3}}$	0	$e^{\jmath\frac{\pi}{3}}$	$e^{\jmath\frac{2\pi}{3}}$	-1	$e^{\jmath\frac{4\pi}{3}}$	$e^{\jmath\frac{5\pi}{3}}$	1
$e^{\jmath\frac{2\pi}{3}}$	0	$e^{\jmath\frac{2\pi}{3}}$	-1	$e^{\jmath\frac{4\pi}{3}}$	$e^{\jmath\frac{5\pi}{3}}$	1	$e^{\jmath\frac{\pi}{3}}$
-1	0	-1	$e^{\jmath\frac{4\pi}{3}}$	$e^{\jmath\frac{5\pi}{3}}$	1	$e^{\jmath\frac{\pi}{3}}$	$e^{\jmath\frac{2\pi}{3}}$
$e^{\jmath\frac{4\pi}{3}}$	0	$e^{\jmath\frac{4\pi}{3}}$	$e^{\jmath\frac{5\pi}{3}}$	1	$e^{\jmath\frac{\pi}{3}}$	$e^{\jmath\frac{2\pi}{3}}$	-1
$e^{\jmath\frac{5\pi}{3}}$	0	$e^{\jmath\frac{5\pi}{3}}$	1	$e^{\jmath\frac{\pi}{3}}$	$e^{\jmath\frac{2\pi}{3}}$	-1	$e^{\jmath\frac{4\pi}{3}}$

(a)

⊗	*0*	*1*	*2*	*3*	*4*	*5*	*6*
0	*0*	*0*	*0*	*0*	*0*	*0*	*0*
1	*0*	*1*	*2*	*3*	*4*	*5*	*6*
2	*0*	*2*	*3*	*4*	*5*	*6*	*1*
3	*0*	*3*	*4*	*5*	*6*	*1*	*2*
4	*0*	*4*	*5*	*6*	*1*	*2*	*3*
5	*0*	*5*	*6*	*1*	*2*	*3*	*4*
6	*0*	*6*	*1*	*2*	*3*	*4*	*5*

(b)

Table B.3. Multiplication tables in terms of (a) polar coordinates and (b) HIP addresses.

$$
\begin{aligned}
\boldsymbol{a} &\in \mathbb{G}, & \boldsymbol{a} &\to (r_a, \theta_a) \\
\boldsymbol{b} &\in \mathbb{G}, & \boldsymbol{b} &\to (r_b, \theta_b) \\
\boldsymbol{a} \otimes \boldsymbol{b} &= r_a r_b e^{\jmath(\theta_a + \theta_b)}
\end{aligned}
$$

The polar equivalent of addresses ***0*** to ***6*** are:

$$
\begin{aligned}
\boldsymbol{0} &\equiv 0e^{\jmath 0} = 0 & & \\
\boldsymbol{1} &\equiv e^{\jmath 0} = 1 & \boldsymbol{4} &\equiv e^{\jmath\pi} = -1 \\
\boldsymbol{2} &\equiv e^{\jmath\frac{\pi}{3}} & \boldsymbol{5} &\equiv e^{\jmath\frac{4\pi}{3}} \\
\boldsymbol{3} &\equiv e^{\jmath\frac{2\pi}{3}} & \boldsymbol{6} &\equiv e^{\jmath\frac{5\pi}{3}}
\end{aligned}
$$

The multiplication table is generated by a pairwise multiplication of the vectors corresponding to the above set of addresses. Unlike HIP addition, this set is sufficient to generate the entire HIP multiplication table. This is because the vectors under consideration are of unit magnitude and consequently, their pairwise multiplication also yields a vector of unit magnitude. In other words, multiplication of two vectors of unit magnitude each, is only a rotation of one of the vectors by an appropriate angle. The resulting multiplication tables are given in Tables B.3(a) (in terms of polar coordinates) and B.3(b) (in terms of HIP addresses).

C

Bresenham algorithms on hexagonal lattices

Chapter 7 contained a discussion of representation of lines and curves on different lattices. The aim was to compare the effect of changing the lattice on representing curves. Central to the discussion was the use of a hexagonal form of the Bresenham line and circle drawing algorithms. The main body only provided a brief discussion of the actual algorithms. This appendix provides the equivalent algorithms for the HIP framework.

C.1 HIP Bresenham line algorithm

For a hexagonal lattice, it is sufficient to provide an algorithm for angles measured with respect to the horizontal which are less than or equal to 30°. This is due to the reflectional symmetry of a hexagon about a line at 30°. The Bresenham line drawing algorithm can be described in terms of drawing a line from address $\boldsymbol{g_1}$ to address $\boldsymbol{g_2}$. The details are given in Algorithm 1.

The algorithm requires two conversion functions ($getX()$ and $getY()$) which return the Cartesian coordinates of a HIP address. The details of this conversion were covered in Section 3.3.

Algorithm 1 will draw lines of orientation less than 30°. Using symmetry, we can draw lines which are outside this range. For example, a line between 30° and 60° can be drawn by swapping the x- and y-coordinates. This means that all the conditions involving Δx become ones involving Δy and vice versa. Lines of arbitrary angles can be drawn by a similar process. Usually, the code will take care of this automatically and assign the various variables appropriately.

C.2 Hexagonal Bresenham circle algorithm

Once again due to symmetry, it suffices to develop an algorithm for drawing a 30° arc of a circle. An example of the relationship between coordinates on a skewed frame of reference is shown in Figure 7.6(b) in Section 7.2. Unlike

Algorithm 1 Hexagonal version of the Bresenham line algorithm

```
Δg ← g₂ ⊖ g₁
Δx ← getX(Δg)
Δy ← getY(Δg)
g ← g₁
ε ← 0
while g ≠ g₂ do
  plot point g
  ε ← ε + Δy
  if 2(ε + Δy) > Δx then
    g ← g ⊕ 1
    ε ← ε − Δx
  else
    g ← g ⊕ 2
  end if
end while
```

the case of line drawing, it is not possible to describe a circle directly in terms of HIP addresses. An example of this difficulty is the trouble with specifying a circle's radius using HIP addresses. Hence, it is assumed that the radius would be specified as a real number. The algorithm developed here is centred about the origin. Due to the nature of HIP addressing, a circle centred about another point just requires the addition of the new centre to all the points in the circle.

Algorithm 2 Hexagonal version of the Bresenham circle algorithm

```
g ← R ⊗ 2
ε ← 1/4
c_x ← 0, c_y ← R
while c_y ≥ c_x do
  plot point g
  if ε < 0 then
    ε ← ε + 2c_x + c_y + 5/2
    g ← g ⊕ 1
  else
    ε ← ε + c_x − c_y + 11/4
    g ← g ⊕ 6
    c_y ← c_y − 1
  end if
  c_x ← c_x + 1
end while
```

The first line of Algorithm 2 uses the previously defined HIP multiplication (see Section 3.2) to define the starting point for the arc. The remaining points on the circle can be generated by replacing $\boldsymbol{g}$ by the 12 symmetric points shown in Figure 7.6(b).

D

Source code

Algorithms that were presented throughout the book were usually implemented within the HIP framework. The purpose of this appendix is to provide some simple source code that implements HIP addressing and the HIP data structure to allow readers to experiment on hexagonal algorithms for themselves. Additionally, resampling and visualisation routines are also provided. However, providing the source code for *all* algorithms covered in this book is outside the scope of this section. The specific implementations have been left as an exercise for the reader.

The algorithms in this appendix are implemented using the Python programming language (`http://www.python.org`). This language was chosen as it is readily available on all platforms and contains a variety of high-level data structures that make programming easy.

The rest of this appendix is split into several sections which provide annotated code along with some examples of the code usage. The code for HIP addressing is presented first followed by that for the HIP data structure, resampling and visualisation.

D.1 HIP addressing

This code provides a class that implements HIP arithmetic via a new data type known as a `Hexint`.

```
#!/usr/bin/env python

import math

class Hexint:

    # class constructor, provides conversion from a
    # normal integer if required
```

```
    def __init__( self, value=0, base7=True  ):
        # error is less than zero
        if value<0:
            raise 'BadHexintError','must be greater 
                than or equal to 0'
        if base7==True:
            # any non 0-6 digit will raise error
            temp = '%ld'%value
            for i in range(len(temp)):
                if int(temp[i])>6:
                    raise 'BadHexintError','cant have
                         digit > 6'
            self.numval = value
        else:
            temp = 0
            mul = 1
            while value>0:
                temp += (value%7)*mul
                value /= 7
                mul *= 10
            self.numval = temp
        self.numstr = '%ld'%self.numval
        self.ndigits = len( self.numstr )

    # convert to a string for displaying result
    def __str__( self ):
        return '('+self.numstr+')'

    __repr__ = __str__

    # comparison ==
    def __eq__( self, object ):
        if isinstance( object,Hexint):
            return self.numval==object.numval
        else:
            return False

    # Hexint addition
    def __add__( self, object ):
        # lookup table
        A = [ [ 0, 1, 2, 3, 4, 5, 6 ] ,
              [ 1,63,15, 2, 0, 6,64 ] ,
              [ 2,15,14,26, 3, 0, 1 ] ,
              [ 3, 2,26,25,31, 4, 0 ] ,
              [ 4, 0, 3,31,36,42, 5 ] ,
```

```
            [ 5, 6, 0, 4,42,41,53 ] ,
            [ 6,64, 1, 0, 5,53,52 ] ]

      # pad out with zeros to make strs same length
      slen = self.ndigits-object.ndigits
      if slen>0:
          numa = '0'+self.numstr
          numb = '0'+('0'*slen)+object.numstr
      else:
          numa = '0'+('0'*(-slen))+self.numstr
          numb = '0'+object.numstr

      maxlen = len(numa)

      total = 0
      mul = 1
      for i in range(maxlen):
          ii = maxlen-i-1
          t = A[ int(numa[ii]) ][ int(numb[ii]) ]
          total += (t%10)*mul
          carry = t/10
          for j in range(i+1,maxlen):
              jj = maxlen-j-1
              if carry>0:
                  t = A[ int(numa[jj]) ][ carry ]
                  numa = numa[:jj]+str(t%10)+numa[
                      jj+1:]
                  carry = t/10
          mul *=10

      return Hexint( total )

  # Hexint multiplication works for another Hexint
  # or a scalar
  def __mul__( self, object ):
      if isinstance(object,Hexint):
          # lookup table
          M = [ [ 0,0,0,0,0,0,0 ] ,
                [ 0,1,2,3,4,5,6 ] ,
                [ 0,2,3,4,5,6,1 ] ,
                [ 0,3,4,5,6,1,2 ] ,
                [ 0,4,5,6,1,2,3 ] ,
                [ 0,5,6,1,2,3,4 ] ,
                [ 0,6,1,2,3,4,5 ] ]
```

```
            # pad out with zeros to make strs
            # same length
            slen = self.ndigits-object.ndigits
            if slen>0:
                numa = '0'+self.numstr
                numb = '0'+('0'*slen)+object.numstr
            else:
                numa = '0'+('0'*(-slen))+self.numstr
                numb = '0'+object.numstr

            maxlen = len(numa)
            powers = [10**i for i in range(maxlen)]

            sum = Hexint( 0 )

            for i in range(maxlen):
                ii = maxlen-i-1
                partial = long(0)
                mul = powers[i]
                for j in range(maxlen):
                    jj = maxlen-j-1
                    if numa[ii]!=0:
                        partial += M[ int(numa[ii])
                            ][ int(numb[jj]) ]*mul
                    mul *= 10
                sum += Hexint(partial)
            return sum
        # scalar multiplication
        elif isinstance(object,int):
            if object>0:
                num = Hexint(self.numval)
            else:
                num = -Hexint(self.numval)

            total = Hexint(0)
            for i in range(abs(object)):
                total += num
            return total

    __rmul__ = __mul__

    # negate a Hexint
    def __neg__( self ):
        total = 0
        mul = 1
```

```
        for i in range(self.ndigits-1,-1,-1):
            if self.numstr[i]=='1':
                total += (4*mul)
            elif self.numstr[i]=='2':
                total += (5*mul)
            elif self.numstr[i]=='3':
                total += (6*mul)
            elif self.numstr[i]=='4':
                total += (1*mul)
            elif self.numstr[i]=='5':
                total += (2*mul)
            elif self.numstr[i]=='6':
                total += (3*mul)
            mul *= 10

        return Hexint( total )

    # Hexint subtraction
    def __sub__( self, object ):
        return self + (-object)

    # get the digit of the Hexint at position pos
    def __getitem__( self, pos ):
        if pos>=self.ndigits:
            raise 'HexIndexError','not␣that␣many␣
                layers'
        else:
            return int( self.numstr[self.ndigits-pos
                -1] )

    # get a Hexint that is some part of the original
    def __getslice__( self, low,high ):
        return Hexint( int(self.numstr[low:high]) )

    # number of digits in the Hexint
    def __len__( self ):
        return self.ndigits

    # return spatial integer coord pair
    def getSpatial( self ):
        xi,yi,zi = self.getHer( )
        # return result
        return ( (xi + yi - 2*zi)/3, (-xi + 2*yi - zi
            )/3 )
```

```
    # return fequency integer coord pair
    def getFrequency( self ):
        xi,yi,zi = self.getHer( )
        # return result
        return ( (-xi + 2*yi - zi)/3, (2*xi - yi - zi
            )/3 )

    # return integer coord of skewed 60 degree axis
    def getSkew( self ):
        xi,yi,zi = self.getHer( )
        # return result
        return ( (2*xi-yi-zi)/3, (-xi+2*yi-zi)/3 )

    # return polar coords for a hexint
    def getPolar( self ):
        if self.numval==0:
            return (0,0)
        (x,y) = self.getReal( )
        r = math.sqrt(x*x+y*y)
        t = math.atan2(y,x)

        return (r,t)

    # return cartesian coords of hexint
    def getReal( self ):
        xc,yc = 1.0,0.0
        x,y = 0.0,0.0
        sqrt3 = math.sqrt(3)

        for i in range(self.ndigits-1,-1,-1):
            if i<self.ndigits-1: # compute key points
                xc,yc = 2*xc - sqrt3*yc, sqrt3*xc +
                    2*yc
            # compute rotation
            if self.numstr[i]=='1':
                x += xc
                y += yc
            elif self.numstr[i]=='2':
                x += (xc/2) - (sqrt3*yc/2)
                y += (sqrt3*xc/2) + (yc/2)
            elif self.numstr[i]=='3':
                x += -(xc/2) - (sqrt3*yc/2)
                y += (sqrt3*xc/2) - (yc/2)
            elif self.numstr[i]=='4':
                x -= xc
```

```
                y -= yc
            elif self.numstr[i]=='5':
                x += -(xc/2) + (sqrt3*yc/2)
                y += -(sqrt3*xc/2) - (yc/2)
            elif self.numstr[i]=='6':
                x += (xc/2) + (sqrt3*yc/2)
                y += -(sqrt3*xc/2) + (yc/2)
        return (x,y)

    # returns a 3-tuple using Her's coord system
    def getHer( self ):
        xc,yc,zc = 1,0,-1
        x,y,z = 0,0,0

        for i in range(self.ndigits-1,-1,-1):
            if i<self.ndigits-1: # compute key points
                xc,yc,zc = (4*xc - 5*yc + zc)/3, (xc
                    + 4*yc - 5*zc)/3, \
                          (-5*xc + yc + 4*zc)/3
            # compute the rotation
            if self.numstr[i]=='1':
                x += xc
                y += yc
                z += zc
            elif self.numstr[i]=='2':
                x -= yc
                y -= zc
                z -= xc
            elif self.numstr[i]=='3':
                x += zc
                y += xc
                z += yc
            elif self.numstr[i]=='4':
                x -= xc
                y -= yc
                z -= zc
            elif self.numstr[i]=='5':
                x += yc
                y += zc
                z += xc
            elif self.numstr[i]=='6':
                x -= zc
                y -= xc
                z -= yc
        # return result
```

```
        return (x,y,z)

    # returns a base 10 integer corresponding
    # to a Hexint
    def getInt( self ):
        total = 0
        mul = 1
        for i in range(self.ndigits-1,-1,-1):
            total += int(self.numstr[i])*mul
            mul *= 7
        return total

    # find the nearest Hexint to Cartesian coords
    def getNearest( self, x,y ):
        sqrt3 = math.sqrt(3)
        o1 = Hexint(1)
        o2 = Hexint(2)

        h = Hexint(0)

        r1 = x -y/sqrt3
        r2 = 2*y/sqrt3

        if r1<0:
            o1 = Hexint(4)
            if r1+math.floor(r1)>0.5:
                h += o1
        elif r1-math.floor(r1)>0.5:
            h += o1

        if r2<0:
            o2 = Hexint(5)
            if r2+math.floor(r2)>0.5:
                h += o2
        elif r2-math.floor(r2)>0.5:
            h += o2
        h += abs(int(r1))*o1
        h += abs(int(r2))*o2
        return h
```

Here are some examples of the object being used from within a Python interpreter.

```
>>> from Hexint import Hexint
>>> h1 = Hexint(1)
>>> h2 = Hexint(2)
```

```
>>> h1+h2
(15)
>>> -Hexint(15)
(42)
>>> h2*Hexint(15)
(26)
>>> h1-h2
(6)
>>> h = Hexint(654321)
>>> h[3],h[1]
(4,2)
>>> Hexint(42).getSpatial()
(-2, -1)
>>> Hexint(42).getFrequency()
(-1, -1)
>>> Hexint(42).getPolar()
(1.7320508075688772, -2.6179938779914944)
>>> Hexint(42).getReal()
(-1.5, -0.8660254037844386)
>>> Hexint(42).getHer()
(-1, -1, 2)
```

D.2 HIP data structure

This code implements the HIP data structure, which is stored internally using a Python list. It provides accessors that use either a `Hexint` or an int.

```
#!/usr/bin/env python

from Hexint import Hexint
import math

class Hexarray:

    # class constructor, creates a list which has
    # 7^{size} elements
    def __init__( self, size ):
        self.nhexs = 7**size
        self.hexdata = [ 0 ]*self.nhexs
        self.layers = size

    # provided to print out a Hexarray in the form
    #       <0,0,...0>
    def __str__( self ):
```

```
        output = '<'
        for i in self.hexdata:
            output += str(i)+', '
        output += '>'
        return output

    __repr__ = __str__

    # return the number of elements in the hexarray
    def __len__( self ):
        return self.nhexs

    # return the value of the Hexarray at pos.
    # pos can be either a Hexint or an integer
    def __getitem__( self, pos ):
        if isinstance( pos, Hexint ):
            p = pos.getInt()
        else:
            p = pos
        return self.hexdata[p]

    # set the value of the Hexarray at pos.
    # pos can be either a Hexint or an integer
    def __setitem__( self, pos, value ):
        if isinstance( pos, Hexint ):
            p = pos.getInt()
        else:
            p = pos
        self.hexdata[p] = value

    # get a list of data from within a Hexarray
    # low,high can be either a Hexint or a int
    def __getslice__( self, low,high ):
        if isinstance( low, Hexint ) and isinstance(
           high,Hexint):
            p1,p2 = low.getInt(), high.getInt()
        else:
            p1,p2 = low,high
        return self.hexdata[low:high]

    # assign a list of data to the Hexarray
    def __setslice__( self, low,high, values ):
        if isinstance( low, Hexint ) and isinstance(
           high,Hexint):
            p1,p2 = low.getInt(), high.getInt()
```

```
        else:
            p1,p2 = low,high
        self.hexdata[low:high] = values

    # return the number of layers in the HIP
       structure
    # represented by this Hexarray
    def getLayers( self ):
        return self.layers

    # save the current Hexarray in an XML file
    # which is called filename
    def save( self, filename ):
        data = open( filename, 'w' )
        output = self.getXML( )
        data.write( output )
        data.close( )

    # generate XML that represents the Hexarray
    def getXML( self ):
        data = ''
        cmplxflag = False
        tupleflag = False
        if isinstance(self.hexdata[0],int):
            dname = 'int'
        elif isinstance(self.hexdata[0],float):
            dname = 'double'
        elif isinstance(self.hexdata[0],complex):
            dname = 'complex'
            cmplxflag = True
        elif isinstance(self.hexdata[0],tuple):
            dname = 'rgb'
            tupleflag = True
        data += '<struct>\n'
        data += '<member>\n'
        data += '<name>layers</name>\n'
        data += '<value><int>%d</int></value>\n'%self
            .layers
        data += '</member>\n'
        data += '<member>\n'
        data += '<name>datatype</name>\n'
        data += '<value><string>%s</string></value>\n
            '%dname
        data += '</member>\n'
        data += '</struct>\n'
```

```
        data += '<array>\n'
        data += '<data>\n'
        for i in self.hexdata:
            if cmplxflag==True:
                data += '<value><%s>%d,%d</%s></value
                    >\n' \
                        % (dname,i.real,i.imag,dname)
            else:
                if tupleflag==True:
                    data += '<value><%s>%d,%d,%d</%s
                        ></value>\n' % (dname,i[0],i
                        [1],i[2],dname)
                else:
                    data += '<value><%s>%d</%s></
                        value>\n'%(dname,i,dname)
        data += '</data>\n'
        data += '</array>\n'

        return data
```

Here are some examples of the object being used from within a python interpreter.

```
>>> h = Hexarray(1)
>>> h
<0, 0, 0, 0, 0, 0, 0, >
>>> for i in range(len(h)):
...   h[i] = i
...
>>> h
<0, 1, 2, 3, 4, 5, 6, >
>>> h[Hexint(3)]=42
>>> h
<0, 1, 2, 42, 4, 5, 6, >
>>> h.getLayers()
1
```

D.3 HIP resampling

This code implements resampling to and from the HIP data structure. It uses the Python Imaging Library (PIL) to handle images. Details of this library can be found at `http://www.pythonware.com/library/index.htm`. The code provides two different versions, for colour and grayscale images respectively.

```
#!/usr/bin/env python

import math
import sys
import types
import Image # python imaging library (PIL)
try:
    from Hexint import Hexint
    from Hexarray import Hexarray
    import Hexdisp
except:
    print 'ERROR:␣Hex␣libs␣are␣not␣installed'
    sys.exit( )

# different sampling techniques
BLINEAR = 1
BCUBIC = 2

# the sampling kernel
def kernel( x,y, tech ):
    xabs,yabs = abs(x), abs(y)
    if tech==BLINEAR:
        if xabs>=0 and xabs<1:
            xf = 1-xabs
        else:
            xf = 0.0
        if yabs>=0 and yabs<1:
            yf = 1-yabs
        else:
            yf = 0.0
    else: # BCUBIC
        a = 0
        if xabs>=0 and xabs<1:
            xf = (a+2)*(xabs**3) - (a+3)*(xabs**2) +
                1
        elif xabs>=1 and xabs<2:
            xf = a*(xabs**3) - 5*a*(xabs**2) + 8*a*
                xabs - 4*a
        else:
            xf = 0.0
        if yabs>=0 and yabs<1:
            yf = (a+2)*(yabs**3) - (a+3)*(yabs**2) +
                1
        elif yabs>=1 and yabs<2:
```

```
            yf = a*(yabs**3) - 5*a*(yabs**2) + 8*a*
                yabs - 4*a
        else:
            yf = 0.0
    return xf*yf

# resample from a square image into a hip image for
# gray scale images
# order : the size of the hip image
# scale : the spacing between points in the hex
#         lattice
def hipsampleGray( image, order, sc, technique ):
    (w,h) = image.size
    ox,oy = w/2.0,h/2.0
    scale = sc
    himage = Hexarray( order )

    for i in range(len(himage)):
        x,y = Hexint(i,False).getReal( )
        y = -y # y direction for PIL is inverted
        xa,ya = ox+scale*x, oy+scale*y
        out = 0.0
        for m in range( int(round(xa-3)),int(round(xa
            +4)) ):
            for n in range( int(round(ya-3)),int(
                round(ya+4)) ):
                if m>=0 and m<w and n>=0 and n<h:
                    pixel = image.getpixel( (m,n) )
                    out += pixel*kernel(xa-m,ya-n,
                        technique)
            himage[i] = round(out)
    return himage

# resample from a square image into a hip image for
# colour images
# order : the size of the hip image
# scale : the spacing between points in the hex
#         lattice
def hipsampleColour( image, order, sc, technique ):
    (w,h) = image.size
    ox,oy = w/2.0,h/2.0
    himage = Hexarray( order )
    for i in range(len(himage)):
        x,y = Hexint(i,False).getReal( )
        y = -y # y direction for PIL is inverted
```

```
        xa,ya = ox+sc*x, oy+sc*y
        out = [0.0,0.0,0.0]
        for m in range( int(round(xa-3)),int(round(xa
            +4)) ):
            for n in range( int(round(ya-3)),int(
                round(ya+4)) ):
                if m>=0 and m<w and n>=0 and n<h:
                    pixel = image.getpixel( (m,n) )
                    out[0] += pixel[0]*kernel(xa-m,ya
                        -n,technique)
                    out[1] += pixel[1]*kernel(xa-m,ya
                        -n,technique)
                    out[2] += pixel[2]*kernel(xa-m,ya
                        -n,technique)
        himage[i] = ( out[0],out[1],out[2] )
    return himage

# resample from a hip image to a square image
# this version is for grayscale images
# rge : radius to resample from
# sc : the spacing between points in the hex lattice
def sqsampleGray( himage, rge,sc, technique ):
    # find size of image
    mh = Hexint(len(himage)-1,False)
    xvs = [Hexint(i,False).getReal()[0] for i in
        range(len(himage))]
    yvs = [Hexint(i,False).getReal()[1] for i in
        range(len(himage))]
    mxx,mnx = round(max(xvs)),round(min(xvs))
    mxy,mny= round(max(yvs)),round(min(yvs))
    rx = int( round((mxx-mnx)/sc) )+1
    ry = int( round((mxy-mny)/sc) )+1
    # create a square image of the right size
    image = Image.new( 'L',(rx,ry) )
    # offset table
    sizes = [ Hexint(7**o-1,False) for o in range(8)
        ]
    order = [ o for o in range(len(sizes)) if sizes[o
        ].getPolar()[0]>rge ]
    offsets = [ Hexint(h,False) for h in range(7**
        order[0]) if Hexint(h,False).getPolar()[0]<=
        rge ]
    for i in range(ry):
        for j in range(rx): # for the points in the
            image
```

```
            xa,ya = mnx+j*sc, mny+i*sc
            # find hex lattice points near the point
              (xa,ya)
            list = []
            hn = Hexint().getNearest(xa,ya)
            for h in offsets:
                hi = hn + h
                (x,y) = hi.getReal()
                if abs(x-xa)<=1 and abs(y-ya)<=1:
                  list.append( hi )
            # compute the colour of the square pixel
            out = 0.0
            for h in list:
                if h.getInt()<=mh.getInt():
                    (x,y) = h.getReal()
                    pixel = himage[h]
                    out += pixel*kernel(xa-x,ya-y,
                       technique)
                    image.putpixel((j,ry-i-1),int(
                       round(out)) )
    return image

# resample from a hip image to a square image.
# this version is for colour images
# rge : radius to resample from
# sc : the spacing between points in the hex lattice
def sqsampleColour( himage, rge,sc, technique ):
    # find size of image
    mh = Hexint(len(himage)-1,False)
    xvs = [Hexint(i,False).getReal()[0] for i in
       range(len(himage))]
    yvs = [Hexint(i,False).getReal()[1] for i in
       range(len(himage))]
    mxx,mnx = round(max(xvs)),round(min(xvs))
    mxy,mny = round(max(yvs)),round(min(yvs))
    rx = int( round((mxx-mnx)/sc) )+1
    ry = int( round((mxy-mny)/sc) )+1
    # create a square image
    image = Image.new( 'RGB',(rx,ry) )
    # offset table
    sizes = [ Hexint(7**o-1,False) for o in range(8)
       ]
    order = [ o for o in range(len(sizes)) if sizes[o
       ].getPolar()[0]>rge ]
```

```
    offsets = [ Hexint(h,False) for h in range(7**
        order[0]) if Hexint(h,False).getPolar()[0]<=
        rge ]
    for i in range(ry):
        for j in range(rx): # for the points in the
            image
            xa,ya = mnx+j*sc, mny+i*sc
            # find hex lattice points near the point
                (xa,ya)
            list = []
            hn = Hexint().getNearest(xa,ya)
            for h in offsets:
                hi = hn + h
                (x,y) = hi.getReal()
                if abs(x-xa)<=1 and abs(y-ya)<=1:
                    list.append( hi )
            # compute the colour of the square pixel
            out = [0.0,0.0,0.0]
            for h in list:
                if h.getInt()<=mh.getInt():
                    (x,y) = h.getReal()
                    pixel = himage[h]
                    out[0] += pixel[0]*kernel(xa-x,ya
                        -y,technique)
                    out[1] += pixel[1]*kernel(xa-x,ya
                        -y,technique)
                    out[2] += pixel[2]*kernel(xa-x,ya
                        -y,technique)
                    image.putpixel((j,ry-i-1),
                                   (int(round(out[0])
                                       ),
                                   int(round(out[1]))
                                       ,
                                   int(round(out[2]))
                                       ) )
    return image

# perform sampling from a square sampled image to a
# HIP image
# image : a valid PIL image
# order : how many layers in the HIP image
# sc : spacing between points in the hex lattice
# technique : which kernel to use
def hipsample( image, order=5, sc=1.0, technique=
    BLINEAR ):
```

```
    if image.mode=='L':
        return hipsampleGray( image, order, sc,
            technique )
    elif image.mode=='RGB':
        return hipsampleColour( image, order, sc,
            technique )
    else:
        raise Exception('hex␣sample␣:␣do␣not␣support␣
            this␣colour␣model')

# perform sampling from a HIP image to a
# square image
# image : a valid PIL image
# rad : range on which to perform interpolation
# sc : spacing between points in the hex lattice
# technique : which kernel to use
def sqsample( himage, rad=1.0, sc=1.0, technique=
    BLINEAR):
    if type(himage[0])==types.FloatType or type(
        himage[0])==types.IntType:
        return sqsampleGray( himage, rad,sc,
            technique )
    elif type(himage[0])==types.TupleType or type(
        himage[0])==types.ListType:
        return sqsampleColour( himage, rad,sc,
            technique )
    else:
        raise Exception('square␣sample␣:␣do␣not␣
            suport␣this␣colour␣model')
```

It is hard to demonstrate this function in operation, but for educational purposes here is a simple example of how to use it:

```
>>> image = Image.open( 'line.png' ) # load an image
>>> h = hipsample( image,4,8.0,BCUBIC ) # hip -> sq image
>>> image2 = sqsample( h, 4.0,0.25,BCUBIC ) # sq -> hip
```

D.4 HIP visualisation

This code implements a display algorithm to visualise HIP images. The methodology is as described in Section 6.2.3. This requires that the Python OpenGL extension be installed as part of your Python installation. The provided code is a simple viewer that allows rotation and scaling of the resulting images. This will handle both spatial and frequency domain HIP images.

```
#!/usr/bin/env python

import math
import sys
try:
    from OpenGL.GLUT import *
    from OpenGL.GL import *
    from OpenGL.GLU import *
except:
    print 'ERROR:␣PyOpenGL␣not␣installed␣properly.'
    sys.exit()

try:
    from Hexint import Hexint
    from Hexarray import Hexarray
except:
    print 'ERROR:␣Hex␣libs␣are␣not␣installed'
    sys.exit( )

# default rotation and scale
xrot,yrot,zrot = 0.0,0.0,0.0
scale = -7.5
max = -9.99e99
scf = 1.0

# initialise the open GL context
def initGL( hdata, domain ):
    glClearColor(0.0, 1.0, 1.0, 0.0)
    glShadeModel( GL_SMOOTH )
    glPolygonMode( GL_FRONT, GL_FILL )
    glPolygonMode( GL_BACK,  GL_LINE )
    glEnable( GL_DEPTH_TEST )
    #  be efficient--make display list
    global hexList
    hexList = glGenLists(1)
    glNewList( hexList, GL_COMPILE )
    compute2DDisplayList( hdata,domain )
    glEndList ()

# what to do when window is resized
def reshapeGL( w,h ):
    if w>h:
        w=h
    elif h>w:
        h=w
```

```
    glViewport( 0,0, w,h )
    glMatrixMode( GL_PROJECTION )
    glLoadIdentity( )
    glFrustum(-1.0, 1.0, -1.0, 1.0, 5.0, 10.0)
    glMatrixMode( GL_MODELVIEW )

# the display function
# uses a display list to store current
# HIP image
def displayGL():
    global scf
    glClear(GL_COLOR_BUFFER_BIT | GL_DEPTH_BUFFER_BIT
        )

    glLoadIdentity( )
    glTranslatef( 0.0,0.0, scale )
    glScalef( scf,scf,scf )
    glRotatef( xrot, 1.0, 0.0, 0.0 )
    glRotatef( yrot, 0.0, 1.0, 0.0 )
    glRotatef( zrot, 0.0, 0.0, 1.0 )
    glCallList(hexList)
    glFlush()

# keyboard controls
# escape : exit
# x/X,y/Y,z/Z : rotate about x,y,z axis
# s/S : scale the image
def keyboardGL(key, x, y):
    global xrot,yrot,zrot
    global scale
    # quit
    if key == chr(27):
        sys.exit(0)
    # change rotation
    if key==chr(88): # X
        xrot += 0.5
    if key==chr(120): # x
        xrot -= 0.5
    if key==chr(89): # Y
        yrot += 0.5
    if key==chr(121): # y
        yrot -= 0.5
    if key==chr(90): # Z
        zrot += 0.5
    if key==chr(122): # z
```

```
        zrot -= 0.5
    # change scale
    if key==chr(83): # S
        scale -= 0.1
    if key==chr(115): # s
        scale += 0.1
    # reset all vals
    if key==chr(82) or key==chr(114):
        scale = -10
        xrot,yrot,zrot = 0.0,0.0,0.0
    xrot,yrot,zrot = xrot%360,yrot%360,zrot%360
    if scale>-5.0: scale = -5.0
    if scale<-10.0: scale = -10.0
    displayGL( )

# find the basis for plotting
def findBasis( domain,order):
    sqrt3 = math.sqrt(3.0)

    if domain: # spatial
        N = [ [1.0,-0.5],[0.0,sqrt3/2.0] ]
    else: # frequency
        if order==1:
            N = [ [1.0/7.0, -2.0/7.0],
                  [5.0/(7*sqrt3), 4.0/(7*sqrt3) ] ]
        elif order==2:
            N = [ [-3.0/49.0, -8.0/49.0],
                  [13.0/(49.0*sqrt3), 2.0/(49.0*sqrt3
                      )] ]
        elif order==3:
            N = [ [ -19.0/343.0, -18.0/343.0],
                  [17.0/(343.0*sqrt3), -20.0/(343.0*
                      sqrt3)] ]
        elif order==4:
            N = [ [ -55.0/2401.0, -16.0/2401.0],
                  [-23.0/(2401.0*sqrt3),
                      -94.0/(2401.0*sqrt3)] ]
        elif order==5:
            N = [ [-87.0/16807.0, 62.0/16807.0],
                  [-211.0/(16807.0*sqrt3),
                      -236.0/(16807.0*sqrt3)] ]
        elif order==6:
            N = [ [37.0/117649.0, 360.0/117649.0],
                  [-683.0/(117649.0*sqrt3),
                      -286.0/(117649.0*sqrt3)] ]
```

```
        elif order==7:
            N = [ [757.0/823543.0, 1006/823543.0],
                  [-1225.0/(823543*sqrt3),
                     508/(823543*sqrt3)] ]
        else:
            N = [ [ 1.0, 0.0], [0.0, 1.0] ]
    return N

# compute the coords for a single hexagon
def doHex( ox,oy, r,o ):
    glBegin( GL_POLYGON )
    for i in range(7):
        x = r*math.cos( i*math.pi/3.0 + math.pi/2.0 +
            o )
        y = r*math.sin( i*math.pi/3.0 + math.pi/2.0 +
            o )
        glVertex3f(ox+x,oy+y,0.0)
    glEnd( )

# compute the display list
def compute2DDisplayList( hdata,domain ):
    global max,scf
    N = findBasis( domain,hdata.layers )
    if domain: # spatial
        radius = 1/math.sqrt(3.0)
        offset = 0
    else: # frequency
        radius = math.sqrt( N[0][1]*N[0][1] + N
            [1][1]*N[1][1] )/math.sqrt(3.0)
        offset = math.atan2( (N[1][0]-N[1][1]),(N
            [0][0]-N[0][1]) )
    max = -9.99e99
    for i in range( len(hdata) ):
        if domain:
            xa,ya = Hexint(i,False).getSpatial()
        else:
            xa,ya = Hexint(i,False).getFrequency()
        hx = xa*N[0][0] + ya*N[0][1]
        hy = xa*N[1][0] + ya*N[1][1]
        if hx>max: max = hx
        if hy>max: max = hy
        if not isinstance( hdata[i],tuple ):
            glColor3f( hdata[i]/255.0, hdata[i
                ]/255.0, hdata[i]/255.0 )
        else:
```

```
            glColor3f( hdata[i][0]/255.0,hdata[i
                ][1]/255.0,hdata[i][2]/255.0 )
        doHex( hx,hy, radius, offset )
    scf = 1.0/(max*(1+0.8/hdata.layers))

# display the image
def display( hdata, domain=True, ang=0,sc=-7.5 ):
    global zrot,scale
    zrot = ang
    scale = sc
    glutInit(sys.argv)
    glutInitDisplayMode(GLUT_SINGLE | GLUT_RGB |
        GLUT_DEPTH)
    glutInitWindowSize(500, 500)
    glutInitWindowPosition(50,50)
    glutCreateWindow("heximage␣layers=%d" % hdata.
        layers )
    initGL( hdata, domain )
    glutReshapeFunc(reshapeGL)
    glutDisplayFunc(displayGL)
    glutKeyboardFunc(keyboardGL)
    glutMainLoop()
```

To use the code presented here, issue the command `Hexdisp.display(hipimage)` from the Python interpreter or your code.

References

1. T. C. Hales, "Cannonballs and honeycombs," *Notices of the American Mathematical Society*, vol. 47, no. 4, pp. 440–449, 2000.
2. T. C. Hales, "The Honeycomb Conjecture," *Discrete Computational Geometry*, vol. 25, pp. 1–22, 2001.
3. E. R. Kandel, J. H. Schwartz, and T. M. Jessell, *Principles of neural science*. McGraw-Hill, Health Professions Division (New York), 2000.
4. A. Meyer, *Historical Aspects of Cerebral Anatomy*. Oxford University Press, 1971.
5. M. H. Pirenne, *Optics, Painting and Photography*. Cambridge University Press, 1970.
6. R. Descartes, *Discourse on Method, Optics, Geometry, and Meteorology*. Hackett Pub., 2001.
7. H. von Helmholtz, *Treatise on Physiological Optics*. The Optical Society of America, 1924.
8. G. Osterberg, "Topography of the layer of rods and cones in the human retina," *Acta Ophthalmology*, vol. 6, pp. 1–103, 1935.
9. M. H. Pirenne, *Vision and the Eye*. Chapman and Hall (London), 2nd ed., 1967.
10. C. A. Curcio, K. R. Sloan, O. Packer, A. E. Hendrickson, and R. E. Kalina, "Distribution of cones in human and monkey retina: Individual variability and radial asymmetry," *Science*, vol. 236, no. 4801, pp. 579–582, 1987.
11. D. Hubel and T. Weisel, "Receptive fields and functional architecture of monkey striate cortex," *Journal of Physiology*, vol. 195, pp. 215–243, 1968.
12. R. C. Gonzalez and R. E. Woods, *Digital Image Processing*, p. 5. Prentice Hall (New Jersey), 2nd ed., 2001.
13. B. H. McCormick, "The Illinois pattern recognition computer: ILLIAC III," *IEEE Transactions on Electronic Computers*, vol. EC-12, pp. 791–813, 1963.
14. D. P. Petersen and D. Middleton, "Sampling and Reconstruction of Wave-Number-Limited Functions in N-Dimensional Euclidean Spaces," *Information and Control*, vol. 5, pp. 279–323, 1962.
15. N. P. Hartman and S. L. Tanimoto, "A Hexagonal Pyramid data structure for Image Processing," *IEEE Transactions on Systems, Man, and Cybernetics*, vol. SMC-14, pp. 247–256, Mar/Apr 1984.

16. A. B. Watson and A. J. Ahumada, Jr., "A hexagonal orthogonal-oriented pyramid as a model of image representation in the visual cortex," *IEEE Transactions on Biomedical Engineering*, vol. BME-36, pp. 97–106, Jan 1989.
17. A. P. Fitz and R. J. Green, "Fingerprint classification using a Hexagonal Fast Fourier Transform," *Pattern Recognition*, vol. 29, no. 10, pp. 1587–1597, 1996.
18. I. Overington, *Computer Vision : a unified, biologically-inspired approach.* Elsevier Science Publishing Company, 1992.
19. A. F. Laine, S. Schuler, W. Huda, J. C. Honeyman, and B. Steinbach, "Hexagonal wavelet processing of digital mammography," in *Proceedings of SPIE*, vol. 1898, pp. 559–573, 1993.
20. I. Her and C.-T. Yuan, "Resampling on a Pseudohexagonal Grid," *CVGIP: Graphical Models and Image Processing*, vol. 56, pp. 336–347, July 1994.
21. D. Van De Ville, R. Van de Walle, W. Philips, and I. Lemahieu, "Image resampling between orthogonal and hexagonal lattices," in *Proceedings of the 2002 IEEE International Conference on Image Processing (ICIP '02)*, pp. 389–392, 2002.
22. D. Van De Ville, W. Philips, I. Lemahieu, and R. Van de Walle, "Suppression of sampling moire in color printing by spline-based least-squares prefiltering," *Pattern Recognition Letters*, vol. 24, pp. 1787–1794, 2003.
23. D. Van De Ville, T. Blu, M. Unser, W. Philips, I. Lemahieu, and R. Van de Walle, "Hex-splines: A novel spline family for hexagonal lattices," *IEEE Transactions on Image Processing*, vol. 13, no. 6, pp. 758–772, 2004.
24. R. C. Staunton and N. Storey, "A comparison between square and hexagonal sampling methods for pipeline image processing," *Proc. SPIE*, vol. 1194, pp. 142–151, 1989.
25. R. C. Staunton, "The processing of hexagonally sampled images," *Advances in imaging and electron physics*, vol. 119, pp. 191–265, 2001.
26. W. Snyder, H. Qi, and W. Sander, "A coordinate system for hexagonal pixels," in *Proceedings of the International Society for Optical Engineering*, vol. 3661, pp. 716–727, 1999.
27. C. Mead, *Analog VLSI and neural systems.* Addison-Wesley, 1989.
28. L. Gibson and D. Lucas, "Vectorization of raster images using hierarchical methods," *Computer Graphics and Image Processing*, vol. 20, pp. 82–89, 1982.
29. M. Tremblay, S. Dallaire, and D. Poussart, "Low level Segmentation using CMOS Smart Hexagonal Image Sensor," in *Proc. of IEEE / Computer Architecture for Machine Perception Conference (CAMP '95)*, pp. 21–28, 1995.
30. R. Hauschild, B. J. Hosticka, and S. Müller, "A CMOS Optical Sensor System Performing Image Sampling on a Hexagonal Grid," in *Proc of ESSCIRC (European Solid-State Circuits Conference)*, 1996.
31. M. Schwarz, R. Hauschild, B. Hosticka, J. Huppertz, T. Kneip, S. Kolnsberg, L. Ewe, and H. K. Trieu, "Single-chip CMOS image sensors for a retina implant system," *Circuits and Systems II: Analog and Digital Signal Processing*, vol. 46, no. 7, pp. 870–877, 1999.
32. M. D. Purcell, D. Renshaw, J. E. D. Hurwitz, K. M. Findlater, S. G. Smith, A. A. Murray, T. E. R. Bailey, A. J. Holmes, P. Mellot, and B. Paisley, "CMOS sensors using hexagonal pixels," in *ACIVS2002: Advanced concepts in Intelligent Systems*, pp. 214–223, 2002.
33. S. Jung, R. Thewes, T. Scheiter, K. F. Goser, and W. Weber, "Low-power and high-performance CMOS fingerprint sensing and encoding architecture," *IEEE Journal of Solid-State Circuits*, vol. 34, no. 7, pp. 978–984, 1999.

34. R. Reulke, "Design and Application of High-Resolution Imaging Systems," in *Proceedings of Image and Vision Computing New Zealand*, pp. 169–176, 2001.
35. H. Lin, N. J. Wu, and A. Ignatiev, "A ferroelectric-superconducting photodetector," *Journal of Applied Physics*, vol. 80, 1996.
36. M. Frank, N. Kaiser, W. Buss, R. Eberhardt, U. Fritzsch, B. Kriegel, O. Mollenhauer, R. Roeder, and G. Woldt, "High-speed industrial color and position sensors," in *Proceedings of The International Society for Optical Engineering*, vol. 3649, pp. 50–57, 1999.
37. W. Neeser, M. Boecker, P. Buchholz, P. Fischer, P. Holl, J. Kemmer, P. Klein, H. Koch, M. Loecker, G. Lutz, H. Matthaey, L. Strueder, M. Trimpl, J. Ulrici, and N. Wermes, "DEPFET pixel bioscope," *IEEE Transactions on Nuclear Science*, vol. 47, no. 3III, pp. 1246–1250, 2000.
38. M. Ambrosio, C. Aramo, F. Bracci, P. Facal, R. Fonte, G. Gallo, E. Kemp, G. Mattiae, D. Nicotra, P. Privitera, G. Raia, E. Tusi, and G. Vitali, "The camera of the Auger fluorescence detector," *IEEE Transactions on Nuclear Science*, vol. 48, no. 3I, pp. 400–405, 2001.
39. S. Lauxtermann, G. Israel, P. Seitz, H. Bloss, J. Ernst, H. Firla, and S. Gick, "A mega-pixel high speed CMOS imager with sustainable Gigapixel/sec readout rate," in *Proc of IEEE Workshop on Charge-Coupled Devices and Advanced Image Sensors*, 2001.
40. B. Mahesh and W. Pearlman, "Image coding on a hexagonal pyramid with noise spectrum shaping," *Proceedings of the International Society for Optical Engineering*, vol. 1199, pp. 764–774, 1989.
41. R. M. Mersereau, "The processing of Hexagonally Sampled Two-Dimensional Signals," *Proceedings of the IEEE*, vol. 67, pp. 930–949, June 1979.
42. R. L. Stevenson and G. R. Arce, "Binary display of hexagonally sampled continuous-tone images," *Journal of the Optical Society of America A*, vol. 2, pp. 1009–1013, July 1985.
43. S. C. M. Bell, F. C. Holroyd, and D. C. Mason, "A digital geometry for hexagonal pixels," *Image and Vision Computing*, vol. 7, pp. 194–204, 1989.
44. E. Miller, "Alternative Tilings for Improved Surface Area Estimates by Local Counting Algorithms," *Computer Vision and Image Understanding*, vol. 74, no. 3, pp. 193–211, 1999.
45. A. Nel, "Hexagonal Image Processing," in *COMSIG*, IEEE, 1989.
46. A. Rosenfeld and J. L. Pfaltz, "Distance Functions on Digital Pictures," *Pattern Recognition*, vol. 1, pp. 33–61, 1968.
47. A. Rosenfeld, "Connectivity in Digital Pictures," *Journal of the Association for Computing Machinery*, vol. 17, no. 1, pp. 146–160, 1970.
48. J. Serra, "Introduction to Mathematical Morphology," *Computer Vision, Graphics, and Image Processing*, vol. 35, pp. 283–305, 1986.
49. R. Staunton, "Hexagonal Sampling in Image Processing," *Advances in Imaging and Electro Physics*, vol. 107, pp. 231–307, 1999.
50. C. A. Wüthrich and P. Stucki, "An Algorithmic Comparison between Square- and Hexagonal-Based Grids," *CVGIP : Graphical Models and Image Processing*, vol. 53, pp. 324–339, July 1991.
51. I. Her, "Geometric Transforms on the Hexagonal Grid," *IEEE Transactions on Image Processing*, vol. 4, pp. 1213–1222, September 1995.
52. P. J. Burt, "Tree and Pyramid Structures for Coding Hexagonally Sampled Binary Images," *Computer Graphics and Image Processing*, vol. 14, pp. 271–280, 1980.

53. L. Gibson and D. Lucas, "Spatial data processing using generalized balanced ternary," in *Proceedings of PRIP 82. IEEE Computer Society Conference on Pattern Recognition and Image Processing.*, no. 566-571, 1982.
54. L. Gibson and D. Lucas, "Pyramid algorithms for automated target recognition," in *Proceedings of the IEEE 1986 National Aerospace and Electronics Conference*, vol. 1, pp. 215–219, 1986.
55. D. Lucas and L. Gibson, "Template Decomposition and Inversion over Hexagonally Sampled Images," *Image Algebra and Morphological Image Processing II*, vol. 1568, pp. 257–263, 1991.
56. D. Lucas, "A Multiplication in N-Space," *IEEE Transactions on Image ProcessingProceedings of the American Mathematical Society*, vol. 74, pp. 1–8, April 1979.
57. D. E. Knuth, *The Art of Computer Programming : Seminumerical Algorithms*, vol. 2. Addison Wesley, 1969.
58. P. Sheridan, T. Hintz, and D. Alexander, "Pseudo-invariant image transforms on a hexagonal lattice," *Image and Vision Computing*, vol. 18, pp. 907–917, 2000.
59. D. E. Dudgeon and R. M. Mersereau, *Multidimensional Digital Signal Processing.* Prentice-Hall International, Inc., 1984.
60. G. Rivard, "Direct Fast Fourier Transform of Bivariate Functions," *IEEE Transactions on Accoustics, Speech, and Signal Processing*, vol. ASSP-25, no. 3, pp. 250–252, 1977.
61. R. Hodgson, R. Chaplin, and W. Page, "Biologically Inspired Image Processing," in *Image Processing and It's Applications*, no. 410, IEE, 1995.
62. M. J. E. Golay, "Hexagonal Parallel Pattern Transforms," *IEEE Transactions on Computers*, vol. C-18, pp. 733–740, August 1969.
63. K. Preston, "Feature Extraction by Golay Hexagonal Pattern Transforms," *IEEE Transactions on Computers*, vol. C-20, no. 9, pp. 1007–1014, 1971.
64. K. Preston, M. Duff, S. Levialdi, P. Norgren, and J. Toriwaki, "Basics of Cellular Logic with some Applications in Medical Image Processing," *Proceedings of the IEEE*, vol. 67, no. 5, pp. 826–856, 1979.
65. J. Serra, *Image analysis and mathematical morphology.* Academic Press (London), 1982.
66. J. Serra, *Image analysis and mathematical morphology: Theoretical Advances.* Academic Press (London), 1986.
67. R. Staunton, "An Analysis of Hexagonal Thinning Algorithms and Skeletal Shape Representation," *Pattern Recognition*, vol. 29, no. 7, pp. 1131–1146, 1996.
68. R. Staunton, "A One Pass Parallel Thinning Algorithm," in *Image Processing and Its Applications*, pp. 841–845, 1999.
69. R. C. Staunton, "Hexagonal image sampling : a practical proposition," *Proc. SPIE*, vol. 1008, pp. 23–27, 1989.
70. R. C. Staunton, "The design of hexagonal sampling structures for image digitisation and their use with local operators," *Image and Vision Computing*, vol. 7, no. 3, pp. 162–166, 1989.
71. G. Borgefors, "Distance Transformations on Hexagonal Grids," *Pattern Recognition Letters*, vol. 9, pp. 97–105, 1989.
72. A. J. H. Mehnert and P. T. Jackway, "On computing the exact Euclidean distance transform on rectangular and hexagonal grids," *Journal of Mathematical Imaging and Vision*, vol. 11, no. 3, pp. 223–230, 1999.

73. G. Borgefors and G. Sanniti di Baja, "Skeletonizing the Distance Transform on the Hexagonal Grid," in *Proc. 9th International Conference on Pattern Recognition*, pp. 504–507, 1988.
74. I. Sintorn and G. Borgefors, "Weighted distance transforms for volume images digitzed in elongated voxel grids," *Pattern Recognition Letters*, vol. 25, no. 5, pp. 571–580, 2004.
75. G. Borgefors and G. Sanniti di Baja, "Parallel Analysis of Non Convex Shapes Digitized on the Hexagonal Grid," in *Proc. IAPR Workshop on Machine Vision Applications (MVA '92)*, pp. 557–560, 1992.
76. E. S. Deutsch, "Thinning Algorithms on Rectangular, Hexagonal, and Triangular Arrays," *Communications of the ACM*, vol. 15, no. 9, 1972.
77. H. S. Wu, "Hexagonal discrete cosine transform for image coding," *Electronics Letters*, vol. 27, no. 9, pp. 781–783, 1991.
78. J. C. Ehrhardt, "Hexagonal Fast Fourier transform with rectangular output," *IEEE Transactions on Signal Processing*, vol. 41, no. 3, pp. 1469–1472, 1993.
79. A. M. Grigoryan, "Efficient algorithms for computing the 2-D hexagonal Fourier transforms," *IEEE Transactions on Signal Processing 2*, vol. 50, no. 6, pp. 1438–1448, 2002.
80. L. Zapata and G. X. Ritter, "Fast Fourier Transform for Hexagonal Aggregates," *Journal of Mathematical Imaging and Vision*, vol. 12, pp. 183–197, 2000.
81. J. Kovacevic, M. Vetterli, and G. Karlsson, "Design of multidimensional filter banks for non-separable sampling," in *Proceedings of the IEEE International Symposium on Circuits and Systems*, pp. 2004–2008, 1990.
82. E. P. Simoncelli and E. H. Adelson, "Non-separable Extensions of Quadrature Mirror Filters to Multiple Dimensions," in *Proceedings of the IEEE (Special Issue on Multi-dimensional Signal Processing)*, vol. 78, pp. 652–664, 1990.
83. A. Cohen and J. M. Schlenker, "Compactly supported bidimensional wavelet bases with hexagonal symmetry," *Constructive Approximation*, vol. 9, no. 2-3, pp. 209–236, 1993.
84. J. Kovacevic and M. Vetterli, "Nonseparable multidimensional perfect reconstruction filter banks and wavelet bases for r^n," *IEEE Transactions on Information Theory*, vol. 38, no. 2, pp. 533–555, 1992.
85. S. Schuler and A. Laine, *Time-Frequency and wavelet transforms in Biomedical engineering*, ch. Hexagonal QMF banks and wavelets. IEEE Press (New York), 1997.
86. F. Smeraldi, "Ranklets: orientation selective non-parametric features applied to face detection," in *Proceedings of the 16th International Conference on Pattern Recognition (ICIP '02)*, vol. 3, pp. 379–382, 2002.
87. Y. Kimuro and T. Nagata, "Image processing on an omni-directional view using a spherical hexagonal pyramid: vanishing points extraction and hexagonal chain coding," in *Proceedings 1995 IEEE/RSJ International Conference on Intelligent Robots and Systems*, vol. 3, pp. 356–361, 1995.
88. L. Middleton, "The co-occurrence matrix in square and hexagonal lattices," in *Proceedings of the 6th International Conference on Control, Automation, Robotics and Vision*, 2002.
89. L. Middleton, "Markov Random Fields for Square and Hexagonal Textures," in *Proceedings of the 6th International Conference on Control, Automation, Robotics and Vision*, 2002.

90. A. Almansa, "Image resolution measure with applications to restoration and zoom," in *Proc. of IEEE International Geoscience and Remote Sensing Symposium*, vol. 6, pp. 3830–3832, 2003.
91. O. Hadar and G. D. Boreman, "Oversampling requirements for pixelated-imager systems," *Optical Engineering*, pp. 782–785, 1999.
92. B. Kamgarparsi and B. Kamgarparsi, "Quantization-error in hexagonal sensory configurations," *IEEE Transactions on Pattern Analysis and Machine Intelligence*, vol. 14, no. 6, pp. 655–671, 1992.
93. B. Kamgarparsi, B. Kamgarparsi, and W. A. Sander III, "Quantization error in spatial sampling: comparison between square and hexagonal grids," in *Proc. International Conference on Vision and Pattern Recognition (CVPR '89)*, pp. 604–611, 1989.
94. A. Almansa, *Echantillonnage, Interpolation et D'étection. Applications en Imagerie Satellitaire.* PhD thesis, Ecole Normale Supérieure de Cachan, 2002.
95. R. M. Gray, P. C. Cosman, and K. L. Oehler, *Digital Images and Human Vision*, pp. 35–52. MIT Press, 1993.
96. L. Middleton and J. Sivaswamy, "A framework for practical hexagonal-image processing," *Journal of Electronic Imaging*, vol. 11, no. 1, pp. 104–114, 2002.
97. L. Middleton and J. Sivaswamy, "Edge Detection in a Hexagonal-image Processing Framework," *Image and Vision Computing*, vol. 19, no. 14, pp. 1071–1081, 2001.
98. L. Middleton and J. Sivaswamy, "The FFT in a Hexagonal-image Processing Framework," in *Proceedings of Image and Vision Computing New Zealand*, pp. 231–236, 2001.
99. L. Middleton, J. Sivaswamy, and G. Coghill, "Saccadic Exploration using a Hexagonal Retina," in *Proceedings of ISA 2000, International ICSC Congress on Intelligent Systems and Applications*, 2000.
100. L. Middleton, J. Sivaswamy, and G. Coghill, "Shape Extraction in a Hexagonal-Image Processing Framework," in *Proceedings of the 6th International Conference on Control, Automation, Robotics and Vision, ICARV*, 2000.
101. L. Middleton, J. Sivaswamy, and G. Coghill, "Logo Shape Discrimination using the HIP Framework," in *Proceedings of the 5th Biannual Conference on Artificial Neural Networks and Expert Systems (ANNES 2001)* (N. K. Kasabov and B. J. Woodford, eds.), pp. 59–64, 2001.
102. B. Grünbaum and G. Shephard, *Tilings and Patterns.* W. H. Freeman and Company (New York), 1987.
103. J. V. Field, *Kepler's Geometrical Cosmology.* Athlone (London), 1988.
104. M. Jeger, *Transformation Geometry.* Allen and Unwin (London), English ed., 1966.
105. G. E. Martin, *Transformation Geometry: An introduction to symmetry.* Springer-Verlag (New York), 1982.
106. P. B. Yale, *Geometry and Symmetry.* Holden-Day (San Francisco), 1968.
107. J. G. Proakis and D. G. Manolakis, *Digital Signal Processing.* Macmillan Publishing Co. (New York), 2nd ed., 1992.
108. C. E. Shannon, "Communication in the Presence of Noise," *Proceedings of the IRE*, vol. 37, pp. 10–21, 1949.
109. D. Whitehouse and M. Phillips, "Sampling in a two-dimensional plane," *Journal of Physics A : Mathematical and General*, vol. 18, pp. 2465–2477, 1985.
110. A. Rosenfeld and A. Kak, *Digital Picture Processing.* Academic Press, 2nd ed., 1982.

111. R. Mersereau and T. Speake, "A Unified Treatment of Cooley-Tukey Algorithms for the Evaluation of the Multidimensional DFT," *IEEE Transactions on Acoustics, Speech, and Signal Processing*, vol. ASSP-29, no. 5, pp. 1011–1018, 1981.
112. C. Von der Malsburg, "Self-organisation of orientation specific cells in the striate cortex," *Kybernetik*, vol. 14, pp. 85–100, 1973.
113. J. R. Parker, *Algorithms for Image Processing and Computer Vision.* John Wiley & Sons Inc. (Canada), 1996.
114. D. Marr, *Vision.* W.H Freeman & Co. (San Francisco), 1982.
115. J. Canny, "A Computational Approach to Edge Detection," *IEEE Transactions on Pattern Analysis and Machine Intelligence*, vol. 8, no. 6, 1986.
116. H. Blum, "A transformation for extracting new descriptors of shape," in *Proceedings of the Symposium on Models for the Perception of Speech and Visual Form*, MIT Press, 1964.
117. L. Lam, S.-W. Lee, and C. Suen, "Thinning Methodologies - a comprehensive survey," *IEEE Transactions on Pattern Analysis and Machine Intelligence*, vol. 14, no. 9, pp. 869–885, 1992.
118. R. C. Gonzalez and R. E. Woods, *Digital Image Processing.* Prentice Hall (New Jersey), 2nd ed., 2001.
119. J. Cooley, P. Lewis, and P. Welch, "The Fast Fourier Transform and its Applications," *IEEE Transactions on Education*, vol. 12, no. 1, pp. 27–34, 1969.
120. J. Cooley, P. Lewis, and P. Welch, "Historical Notes on the Fast Fourier Transform," *IEEE Transactions on Audio and Electroacousitcs*, vol. AU-15, no. 2, pp. 76–79, 1967.
121. W. H. Calvin, *The cerebral code: thinking a thought in the mosaics of the mind.* MIT Press (Cambridge, Mass.), 1997.
122. D. Felleman and D. Van Essen, "Distributed Hierarchical Processing in the Primate Cerebral Cortex," *Cerebral Cortex*, vol. 1, pp. 1–47, Jan/Feb 1991.
123. M. Oram and D. Perrett, "Modelling visual recognition from neurobiological constraints," *Neural Networks*, vol. 7, no. 6/7, pp. 945–972, 1994.
124. S. M. Kosslyn and O. Koenig, *Wet Mind: The New Cognitive Neuroscience.* The Free Press (New York), 1995.
125. A. L. Yarbus, *Eye Movements and Vision.* Plenum Press (New York), 1967.
126. L. Itti, C. Gold, and C. Koch, "Visual attention and target detection in cluttered natural scenes," *Optical Engineering*, vol. 40, no. 9, pp. 1784–1793, 2001.
127. N. Bruce, "Computational visual attention," Master's thesis, University of Waterloo, Canada, 2003.
128. E. Erwin, K. Obermayer, and K. Schulten, "Models of Orientation and Ocular Dominance Columns in the Visual Cortex : A Critical Comparison," *Neural Computation*, vol. 7, pp. 425–468, 1995.
129. J. P. Eakins and M. E. Graham, "Similarity Retrieval of Trademark Images," *IEEE Multimedia*, vol. 52, pp. 53–63, April-June 1998.
130. V. Gudivada and V. V. Raghavan, "Content-based image retrieval systems," *IEEE Computer*, vol. 28, no. 9, pp. 18–22, 1995.
131. A. K. Jain and A. Vallaya, "Image retrieval using colour and shapes," *Pattern Recognition*, vol. 29, no. 8, pp. 1233–1244, 1996.
132. Y. S. Kim and W. Y. Kim, "Content-based trademark retrieval system using a visually salient feature," *Image and Vision Computing*, vol. 16, pp. 931–939, 1998.

133. H. S. Hoffman, *Vision and the Art of Drawing.* Prentice Hall (Englewood Cliffs, N.J.), 1989.
134. I. Biederman, "Recognition by component : a theory of human image understanding," *Artificial Intelligence*, vol. 94, pp. 115–145, 1987.
135. D. Marr and H. Nishihara, "Representation and recognition of the spatial organisation of 3-dimensional shapes," *Proceedings of the Royal Society of London, B*, vol. 200, pp. 269–294, 1979.
136. J. Eakins, J. Boardman, and K. Shields, "Retrieval of Trademark Images by Shape Feature - The Artisan Project," in *IEE Colloquium on Intelligent Image Databases*, pp. 9/1–9/6, 1996.
137. M. Morrone and D. Burr, "Feature Detection in Human Vision: A phase dependent energy model," *Proceedings of the Royal Society*, vol. B235, pp. 221–245, 1988.
138. W. Chan, G. Coghill, and J. Sivaswamy, "A simple mechanism for curvature detection," *Pattern Recognition Letters*, vol. 22, no. 6-7, pp. 731–739, 2001.
139. G. Wolberg, *Digital Image Warping.* IEEE Computer Society Press, 1990.
140. T. Kohonen, *Self-organizing maps.* Springer (Berlin), 2nd ed., 1997.
141. T. Kohonen, J. Hynninen, J. Kangas, J. Laaksonen, and K. Torkkola, "LVQ_PAK: The Learning Vector Quantization Program Package," Tech. Rep. A30, Helsinki University of Technology, Laboratory of Computer and Information Science, FIN-02150 Espoo, Finland, 1996.
142. D. Doermann, "UMD Logo Database." `http://documents.cfar.umd.edu/resources/database/UMDlogo.html`.
143. R. Keys, "Cubic convolution interpolation for digital image processing," *IEEE Transactions of Acoustics, Speech, and Signal Processing*, vol. ASSP-29, pp. 1153–1160, 1981.
144. J. H. Conway and N. J. A. Sloane, *Sphere Packings, Lattices, and Groups*, p. 8. Springer-Verlag, 1988.
145. J. E. Bresenham, "Algorithm for Computer Control of a Digital Plotter," *IBM Systems Journal*, vol. 4, no. 1, pp. 25–30, 1965.
146. J. D. Foley and A. van Dam, *Fundamentals of interactive computer graphics.* Addison-Wesley Pub. Co., 1982.
147. J. E. Bresenham, "A Linear Algorithm for Incremental Digital Display of Circular Arcs," *Communications of the ACM*, vol. 20, no. 2, pp. 100–106, 1977.
148. B. Jang and R. Chin, "Analysis of Thinning Algorithms using Mathematical Morphology," *IEEE Transactions on Pattern Analysis and Machine Intelligence*, vol. 12, no. 6, pp. 541–551, 1990.
149. B. Jang and R. Chin, "One-Pass Parallel Thinning : Analysis, Properties, and Quantitative Evaluation," *IEEE Transactions on Pattern Analysis and Machine Intelligence*, vol. 14, no. 11, pp. 1129–1140, 1992.
150. M. T. Heideman, D. H. Johnson, and C. S. Burrus, "Gauss and the history of the fast Fourier transform," *IEEE ASSP Magazine*, vol. 1, no. 4, pp. 14–21, 1984.
151. E. Brigham and R. Morrow, "The fast Fourier transform," *IEEE Spectrum*, vol. December, pp. 63–70, 1967.
152. A. K. Jain, *Fundamentals of Digital Image Processing.* Prentice-Hall Inc., 1989.
153. R. Strand, "Surface skeletons in grids with non-cubic voxels," in *Proceedings of 17th International Conference on Pattern Recognition (ICPR)*, vol. 1, pp. 548–551, 2004.

Index

Addressing
  hexagonal samples, 15, *see* HIP addressing
Aggregate, 35
  examples, 36
  HIP, 37
    lambda-level, 39
    locus of centres, 202
    locus of centres, 40
Aliasing, 14, 15, 96, 189
Axes
  skewed, 140

basis images, 183
Basis vector, 31, 136, 152
  hexagonal lattice, 36
  quincunx lattice, 137
  square vs. hexagonal lattice, 128
Boundary pixel, 55, 170
Brick wall, 11

Canny edge detector, 178
Canny edge detector, 75
  algorithm, 76
Chain code, 126
Circle drawing
  hexagonal vs. square, 167
Circle drawing algorithm
  for HIP image, 215
  for square grid, 161
Cones, 6
  arrangement, 8
Convolution, 10, 74–76, 122
  definition
    for HIP, 61
    for square grid, 172
  hexagonal vs. square, 170
Coordinate
  Burt, 17
  Cartesian, 17, 27, 147
  Her, 16, 20
  skewed axes, 16, 27
Critical point, 108
  extraction, 111
Curve representation, 161
Cutoff frequency, 92, 95, 118

Decimation in space, 85
  for HDFT, 88
Distance function, 19, 155
  hexagonal vs. square, 155
Distance measures, 56
  on skewed axes, 57
  with Her coordinates, 57
Distance transform, 19
Down-sampling, 97, 166, 193

Edge
  comparison of detectors, 76
  compass operators, 73
  concept, 71
  derivative operators, 72
  detection, 19, 71, 108
    hexagonal vs. square, 175
  detection on noisy images, 178
  gradient, 72
  gradient magnitude, 73
  mask, 72, 75

noise, 72, 73
second derivative, 74
zero crossings, 74
Energy signature, 119
External pixel, 55, 172

Fast Fourier transform, *see* HFFT
Feature, 107, 113
curved, 199
curved vs. linear, 119
edge vs. line, 121
vector, 123
Filter
averaging, 99
bank, 113, 119
Fourier basis images
HIP, 183
on square grid, 184
Fovea, 6, 106
Foveation/fixation, 106
Frequency domain lattice, *see* Reciprocal lattice

Gabor filter, 121
Gaussian smoothing, 75
GBT, 18, 40
Geometric transformation, 20, 116

Hexagonal image acquisition, 11, *see* Resampling
Hexagonal image display, 21, *see* HIP image display, 141
brickwall, 142
Hexagonal sampling theorem, 18
HFFT, 83, 122, 182
decomposition, 90
example, 88
twiddle factor, 85
weight function, 86
Hierarchical aggregation, 18, 36
Highpass filter, 92, 188, 190
HIP addressing, 40
definition, 41, 49
example, 41
fractional, 48
frequency domain, 65
example, 65
notation, 41
polar equivalent, 213
vector equivalent, 211
HIP address conversion
to Cartesian coordinates, 54
to Her's coordinates, 54, 206
to skewed coordinates, 54
HIP aggregate
cardinality, 201
HIP arithmetic
addition, 43, 211
example, 44
table, 45
code, 219
division, 48
multiplication, 46, 213
table, 47
scalar multiplication, 47
subtraction, 45
HIP closed arithmetic, 49
addition, 50
addition example, 50
multiplication, 51
multiplication example, 51
subtraction, 50
HIP image
code to display, 236
display, 169
display using hyperpixel example, 147
storage, 43
Hyperpixel
definition, 142

Image coding, 20, 193
Image compression, 11, 96, 198
Image retrieval
shape-based, 117
Image transform, *see* Transform
Interpolation, 130
bi-cubic, 13, 133
bi-linear, 133, 169
linear, 12
nearest neighbour, 131, 140
Isometry, 29
examples, 28
Isotropic kernel, 74

Laplacian of Gaussian
mask, 74
Laplacian of Gaussian, 74, 177

algorithm, 75
Lattice
definition, 29
hexagonal
advantages, 2
spatial vs. frequency domain, 63
quincunx, 137
Linear phase, 209
Linear filtering, 186
definition, 91
results, 95
Line drawing
hexagonal vs. square, 167
Line drawing algorithm
for hexagonal grid, 157
for HIP image, 215
for square grid, 157
Line drawing comparison, 159
Line representation, 21, 156
Local energy, 119
Lowpass filter, 92, 118, 188
Lowpass fiter, 92
LVQ classifier, 123
codebook, 124

Mask, 61
Masking operation, 73
Morphological
processing, 19, 100
Morphological operator
closing, 103
dilation, 101
erosion, 101
opening, 102
Multiresolution, 19, 20, 96

Neighbourhood, 59, 172, 211
N_n, 59
N_n^h, 60
N_n^r, 61
Neighbourhood operations, 172
Nyquist rate, 34

p-norm, 56
Periodicity matrix, 63, 84
Prewitt operator, 72, 113, 176
algorithm, 73
mask, 73
Pyramid, 11, 17, 20, 21, 96, 108, 113, 192
Pyramid decomposition
by averaging, 98
by subsampling, 97

Quantisation error, 21

Ranklets, 20
Reciprocal lattice, 62
examples, 33
Reciprocal lattice, 32, 62
Representation
object-centred, 111, 115
shape, 106, 114
Resampling, 11, 128, 168
hexagonal to square, 138
square to hexagonal, 128
square to HIP image, 230
Ringing, 95, 189
Rods, 6
Rotation matrix, 206

Saccades, 106
Sampling, 11, 27
example as a tiling, 34
hexagonal vs. square, 152
matrix, 62, 152
quincunx, 12
Sampling density, 198
hexagonal vs. square, 154
Sampling lattices
hexagonal vs. square, 153
Sensor
array, 1, 6
CCD, 198
CMOS, 14
fovea type, 25
photoreceptors, 1
Set operations, 100
Shape extraction, 111
Shape analysis, 19
Shape discrimination, 117
Skeleton, 181
Skeletonisation, 79, 108, 113, 180
algorithm, 81
example, 82
Skewed axes, 15, 17, 52
Spatial averaging, 166

Spiral, 40, 205
Spline, 13, 22, 134
  hexagonal, 21
Structural element, 103
Subband, 198
Subband coding, 20
Symmetry, 29
  hexagonal aggregate, 36
  reflectional
    examples, 30

Tessellation, 29
Texture, 19
  co-occurrence matrix, 21
Thinning, 19, *see* Skeletonisation
Thresholding
  in edge detection, 72, 74, 75
Tiling, 28
  dihedral, 28
    example, 30
  monohedral, 28
  periodic, 29
  prototile, 28, 34, 152
  regular, 28
  symmetric, 29
Transform
  cortex, 20
  DCT, 19
  DFT, 66, 82, 174
    expression, 62
  Fourier, 31
    example, 118
    of sampled signal, 32
  HDFT, 19, 174
    expression, 68
    fast algorithm, 83
    fast algorithm, 18
    matrix formulation, 83
    properties, 68, 208
    separability, 68
  Walsh, 19
Translation matrix, 39

Viewpoint invariance, 115
Visual perception, 8, 24
Voronoi cell, 152

Wavelet, 20, 198

Printed in the United States
By Bookmasters